ENDOCRINOLOGY RESEARCH AND CLINICAL DEVELOPMENTS

GLUCOSE UPTAKE

REGULATION, SIGNALING PATHWAYS AND HEALTH IMPLICATIONS

ENDOCRINOLOGY RESEARCH AND CLINICAL DEVELOPMENTS

Additional books in this series can be found on Nova's website under the Series tab.

Additional e-books in this series can be found on Nova's website under the e-book tab.

ENDOCRINOLOGY RESEARCH AND CLINICAL DEVELOPMENTS

GLUCOSE UPTAKE

REGULATION, SIGNALING PATHWAYS AND HEALTH IMPLICATIONS

CARTER C. JOHNSON
AND
DAVIS B. WILLIAMS
EDITORS

New York

For permission to use material from this book please contact us:
Telephone 631-231-7269; Fax 631-231-8175
Web Site: http://www.novapublishers.com

Library of Congress Cataloging-in-Publication Data

ISBN: 978-1-62618-670-5

Library of Congress Control Number: 2013935917

Published by Nova Science Publishers, Inc. † New York

Contents

Preface

Insulin stimulates the uptake of glucose in several target tissues, including skeletal muscle and adipose tissue, thereby contributing to the decline in the blood glucose level. In this book, the authors present current research in the study of the regulation, signaling pathways and health implications of glucose uptake. Topics include glucose uptake and transport regulation; the mechanisms that contribute to glucose homeostasis; expression and regulation of neuronal glucose transporters in health and disease; role for the microvasculature in glucose uptake in skeletal muscle; implications of diabetes on sperm glucose uptake and metabolism; the beneficial role of estrogen signaling in glucose homeostasis and energy expenditure; glucose uptake and androgen responsiveness of prostate cancer cells; regulation of intestinal glucose uptake by leptin; and the Ras superfamily of small GTP-binding proteins in glucose transporter type 4-mediated glucose uptake in insulin-responsive tissues.

Chapter 1 – Glucose is an essential nutrient of great importance for the organism. It is the major energy source for many cell types, which are dependent on a steady glucose supply. With the diabetes epidemic affecting the world population, there is an increasing demand for means to regulate glucose metabolism.

This chapter is intended to give an overview about glucose uptake and transport mechanisms, highlighting some important aspects of their regulation focusing, lastly, on the role of bone cells in the regulation of energy homeostasis.

Uptake and storage of glucose via glycogenesis, its release via glycogenolysis and gluconeogenesis and its catabolism via glycolysis are complex mechanisms strictly regulated. Knowledge of how signaling and trafficking pathways of these processes are adjusted and regulated will be essential to better understand a wide range of metabolic disorders. Cellular uptake of glucose and transport across cell membranes is mediated by two families of specialized carrier proteins: the facilitative glucose transporters (GLUTs) and the sodium-coupled glucose cotransporters (SGLTs). The various transporters exhibit different substrate specificities, kinetic properties, and tissue expression profiles. The whole family of human GLUT proteins consists of at least 14 members, indicating that glucose delivery into cells is a procedure of considerable complexity.

There is substantial amount of data confirming that the skeleton participates in the regulation of energy metabolism, mainly by controlling glucose homeostasis, energy expenditure, and fat mass. Although other bone derived molecules and/or hormones cannot be excluded, osteocalcin is a main osteoblast-derived hormone through which bone signals to pancreas and adipose tissue to exert this function. Since bone is a metabolically highly active tissue which requires a large amount of energy for remodeling; it becomes an important player in coordinating global energy utilization. In view of an increasing number of investigations, presented are further in this chapter studies addressing glucose uptake and transport regulation in bone cells by hormones and growth factors. This information can be essential to understanding the pathophysiologic derangements that may result from glucose uptake or transport disorders. Conditions such as diabetes are characterized by an imbalance in glucose regulation, causing chronic hyperglycemia and ultimately leading to multiorgan damage.

Chapter 2 – The cellular energy from glucose is obtained through glycogen breakdown and also from skeletal muscle through intermediary metabolites. This waste of whole-body energy is replenished by digested and absorbed dietary glucose on intestine and distributed among muscle, adipose tissue, liver, kidney, brain and others tissues involved in glycemia by means of glucose uptake, metabolism, storage and excretion. Glucose is unable to readily diffuse through the plasma membranes due to its hydrophilic nature. So, at least, two main types of assistance transport proteins are needed: facilitative transporters (glucose carriers (GLUTs) from the SLC2 gene family and active transport proteins (glucose carriers (SGLTs) from the SCL5 gene family. Among twelve known GLUT molecules, GLUT4 is the major glucose transporter for skeletal muscle, adipose tissue (insulin dependent for these both tissues) and cardiac muscle. Other ones, such as GLUTs 1, 2, 3, and 8, facilitate the glucose entry into the brain and liver, for example. Significant amounts of glucose ($\geq$ 80%) are transported into the skeletal muscle where it may be used for energy supply or for glycogen stores. On the other hand, the glucose uptake on adipose tissue is around 5% although it plays an important influence on insulin sensitivity in muscle and liver. Furthermore, several transporters are involved on glucose flux from cells to the blood and from organelles to the cytosol. The glucose homeostasis is maintained by a complex network among pancreatic cells, intestine, kidney, skeletal muscle, adipose tissue, liver and lately also the central glucose regulation by brain has been carefully concerned. The reabsorption of glucose by kidney occurs through the sodium glucose cotransporter SGLT2 (the majority) and the remaining ones by SGLT1 (around 10%). SGLT1 is also extensively expressed in the small intestine. The SGLT active glucose transport system takes advantage of sodium electrochemical potential gradient across the brush border membrane and also utilize sodium:potassium adenosine triphosphatase pumps at the basolateral membrane to the suitable glucose uptake. This review provides a discussion about the benefits and what is going on with glucose uptake and health implications, the mechanisms that contribute to glucose homeostasis and the potential impact of glucose transporters as pharmacological targets able to correct glucose imbalance.

Chapter 3 – The brain is an energy demanding organ and relies on glucose as its main energy source. Glucose is taken up by brain cells through glucose transporters, which are expressed in every cell of the central nervous system. Maintaining a constant influx of energetic substrates is necessary for the survival and function of every cell in the body. However, glucose uptake can be modified under different physiological and pathological conditions.

In this chapter the authors will discuss neuronal glucose transporter expression and function, with descriptions of neuronal glucose transport regulation during periods of both rest and activity. They will give some insight into the signalling pathways involved in the regulation of glucose transporter function and availability. And finally, by focusing on neurodegenerative disorders, with special emphasis on Alzheimer's disease, the authors will describe glucose transporter function, regulation and signalling under disease conditions.

Chapter 4 – Skeletal muscle is an important site for insulin action and is responsible for ~80% of insulin-mediated glucose disposal following a meal. Insulin stimulates muscle glucose uptake in two ways: firstly insulin improves the delivery of glucose to muscle cells by increasing blood flow to the microvasculature and transport across the endothelium in contact with the myocyte, and secondly by increasing glucose uptake across the muscle cell membrane via activation of the insulin signalling pathway leading to the recruitment of a specific glucose transporter (GLUT4) to the cell membrane. During insulin resistance and type 2 diabetes, both of these responses to insulin can become impaired. Here the authors review the considerable body of literature showing that insulin-stimulated microvascular recruitment improves the delivery of insulin and glucose to the myocyte to augment insulin's metabolic action, and that vascular actions of insulin are lost during insulin resistance and type 2 diabetes. Furthermore the authors discuss more recent evidence to suggest that microvascular insulin responses are lost prior to and independently of the development of myocyte insulin resistance.

Chapter 5 – Diabetes mellitus (DM) is described as a metabolic disorder characterized by hyperglycaemia resulting from defective insulin secretion, resistance to insulin, or both, and represents one of the greatest threats to modern global health as its incidence is rapidly rising worldwide. Type 1 DM results from an absolute deficiency of insulin due to an autoimmune destruction of the pancreatic beta cells while type 2 DM is characterized by impaired insulin secretion and increased insulin resistance.

It is well known that glucose regulation is crucial for normal spermatogenesis and fertility. In this process, insulin plays a crucial role since its dysfunction is connected with decreased cellular glucose transport. Both clinical and experimental reports suggest that fertility is highly decreased in patients or animals with DM. Both DM types impair male fertility and numerous studies in male diabetic individuals have demonstrated a marked reduction in fecundity, as well as impairment of sperm quality and higher percentage of spermatozoa with nuclear DNA damage. All these effects are known to be related with metabolic signaling pathways in testis that results in defective sperm metabolism. The different regions and structures of the sperm flagellum are of great importance because the metabolic pathways are compartmentalized in them. Spermatozoa metabolize several substrates as energy sources, such as hexoses (glucose, mannose, and fructose) or other metabolites (lactate and citrate). Any alteration in the ability of the spermatozoa to utilize substrates involved in ATP production is expected to compromise motility and subsequently fertility. For glycolysis to occur in sperm, these cells need specific carriers to transport energy sources through the cellular membrane, namely glucose transporters (GLUTs). DM is known to modulate spermatozoa substrate consumption and/or production due to altered glycolysis. The transport of hexoses via GLUTs is also known to be highly dysregulated in diabetic male individuals.

Throughout this chapter the authors will discuss the effects of DM in sperm glucose uptake and metabolism. Understanding the functioning and regulation of these processes is crucial to identify the key mechanisms associated with male (in)fertility in order to develop possible therapeutics.

Chapter 6 – Glucose is the most important fuel of mammalian cells and its uptake is crucial for cellular metabolism, differentiation, survival and proliferation. The insulin signaling pathway regulates the uptake of glucose and whole-body metabolic homeostasis by transducing extracellular signals through the insulin receptor (IR) to downstream intracellular targets. Dysregulation of insulin secretion or alterations of IR signal transduction pathways is associated with self-generating, progressive insulin resistance in thorough interplay with an imbalance of sexual steroid production. Insulin resistance and the associated metabolic and hormonal alterations predispose patients to a variety of life threatening diseases; such as metabolic syndrome, type-2 diabetes, cardiovascular lesions and malignancies at different sites. Estrogens have beneficial effects on energy metabolism and glucose homeostasis by several pathways. In the central nervous system, hypothalamic nuclei are the pivotal regulators of food intake and energy expenditure by means of their estrogen receptors (ERs). In the pancreatic islet cells, ERs advantageously regulate the biosynthesis and secretion of insulin and maintain the equilibrium of glycogen synthesis and glycogenolysis in the liver by the balanced activation of glycogen synthase and glycolytic enzymes. In the peripheral tissues, ERs advantageously modulate the insulin stimulated glucose uptake through regulation of the phosphorylation of IR protein and increase the expression and translocation of intracellular glucose transporters (GLUTs). Considering the regulatory effects of ERs on food intake, insulin secretion, glucose uptake and metabolic processes, estrogen administration seems to be a therapeutic avenue to repair insulin sensitivity in patients with dysmetabolism and diabetes mellitus.

Chapter 7 – Prostate cancer (PCa) onset and development requires a sequence of genetic changes and additional metabolic adaptations to ensure survival of the modified cells and to finally achieve a malignant phenotype. This concept may be associated with the "Warburg phenomenon", which consists of an increased glycolytic capacity and reduced mitochondrial oxidative phosphorylation, maintained even in the presence of high oxygen concentrations. It has been proposed that this alteration may be a metabolic mechanism of tumor cells to survive and grow under hypoxic conditions through the modulation of key enzymes and transporters involved in glycolytic metabolic pathways. PCa progression is characterized by a transition from an androgen-dependent to an androgen-independent phenotype, which corresponds to early and advanced stages of PCa respectively, and an increase of glucose uptake and consumption is required for rapid proliferation of androgen-dependent cells. In addition, androgen-responsive and androgen-nonresponsive PCa cells are known to have distinct glycolytic characteristics and present different levels of oxidative stress, which have been associated with a deregulated expression of transporters and enzymes related with glucose uptake and utilization. Identifying the metabolic alterations that occur in the transition of androgen-responsive to androgen-nonresponsive stage is a fundamental step for the development of new diagnosis and therapeutic approaches of PCa.

Chapter 8 – The endocrine and paracrine regulation of the intestinal absorption of nutrients is being subject of intense research leading to the identification of new actors. In this chapter, the authors present the findings regarding the regulation of intestinal glucose uptake by leptin, a hormone discovered in 1994 that is implicated in the control of food intake and energy balance.

Shortly after leptin identification, the authors demonstrated that leptin inhibits sugar absorption in rat intestine *in vitro* by short-term regulation of the Na^+/glucose cotransporter SGLT-1. In the same year, Bado et al. (1998) reported that the stomach chief cells secrete leptin into the gastric lumen in response to a meal or endocrine and neurocrine secretagogues. In line with those results, the authors described that leptin receptors are expressed in both the apical and basolateral membrane of human and murine enterocytes. This distribution of leptin receptors led us to characterize the effects of leptin on SGLT-1 from luminal and serosal side of the intestinal mucosa, using the polarized Ussing chamber system. The authors demonstrated a rapid leptin receptor-mediated action of leptin from the luminal side and a slower effect from the serosal side that required leptin-induced CCK release. This study was followed by the finding that the gastric chief cells also secrete the leptin soluble receptor, which binds to leptin in the secretory granules before its release into the gastric lumen, and explains how leptin can remain stable in the gastric juice protected from this adverse proteolytic environment. Further studies also showed the control by luminal leptin of glucose transporters from the GLUT family.

All these data supported the authors' initial discovery about the role of luminal leptin as regulator of the intestinal sugar absorption and lead to the demonstration of the implication of PKC, PKA and AMPK on this leptin action, and leptin modulation of sugar transport in health and disease that will be presented and discussed in this chapter.

Chapter 9 – Insulin stimulates the uptake of glucose in several target tissues, including skeletal muscle and adipose tissue, thereby contributing to the decline in the blood glucose level. Insulin signaling is mediated by a highly integrated network composed of a variety of signal transducing proteins. Induction of glucose uptake in skeletal muscle and adipose tissue primarily depends on the redistribution of the glucose transporter type 4 (GLUT4) from intracellular storage sites to the plasma membrane. This process consists of several discernible steps, such as vesicle transport, docking, and fusion, each of which is regulated rigorously by the signal transduction machinery. Despite intensive research for decades, mechanisms underlying the regulation of GLUT4-mediated glucose uptake remain largely elusive. The Ras superfamily of small GTP-binding proteins serves as a molecular switch in many facets of signal transduction processes in various types of mammalian cells. Emerging evidence supports the notion that insulin also employs small GTP-binding proteins in the intracellular signaling network to exert its actions in skeletal muscle and adipose cells. Two types of proteins, guanine nucleotide exchange proteins (GEFs) and GTPase-activating proteins (GAPs), regulate on/off switching of small GTP-binding proteins, respectively. In fact, diverse GEFs and GAPs have been characterized as key components in insulin signaling.

In: Glucose Uptake
Editors: Carter C. Johnson and Davis B. Williams
ISBN: 978-1-62618-670-5

Chapter 1

Glucose Uptake and Transport Regulation

Evangelos Zoidis *
Department of Nutritional Physiology and Feeding,
Faculty of Animal Science and Aquaculture,
Agricultural University of Athens,
Athens, Greece

Abstract

Glucose is an essential nutrient of great importance for the organism. It is the major energy source for many cell types, which are dependent on a steady glucose supply. With the diabetes epidemic affecting the world population, there is an increasing demand for means to regulate glucose metabolism.

This chapter is intended to give an overview about glucose uptake and transport mechanisms, highlighting some important aspects of their regulation focusing, lastly, on the role of bone cells in the regulation of energy homeostasis.

Uptake and storage of glucose via glycogenesis, its release via glycogenolysis and gluconeogenesis and its catabolism via glycolysis are complex mechanisms strictly regulated. Knowledge of how signaling and trafficking pathways of these processes are adjusted and regulated will be essential to better understand a wide range of metabolic disorders. Cellular uptake of glucose and transport across cell membranes is mediated by two families of specialized carrier proteins: the facilitative glucose transporters (GLUTs) and the sodium-coupled glucose cotransporters (SGLTs). The various transporters exhibit different substrate specificities, kinetic properties, and tissue expression profiles. The whole family of human GLUT proteins consists of at least 14 members, indicating that glucose delivery into cells is a procedure of considerable complexity.

There is substantial amount of data confirming that the skeleton participates in the regulation of energy metabolism, mainly by controlling glucose homeostasis, energy expenditure, and fat mass. Although other bone derived molecules and/or hormones cannot be excluded, osteocalcin is a main osteoblast-derived hormone through which

* E-mail: ezoidis@aua.gr, Tel: +30 210 529 44 15, Fax: +30 210 529 44 13.

bone signals to pancreas and adipose tissue to exert this function. Since bone is a metabolically highly active tissue which requires a large amount of energy for remodeling; it becomes an important player in coordinating global energy utilization. In view of an increasing number of investigations, presented are further in this chapter studies addressing glucose uptake and transport regulation in bone cells by hormones and growth factors. This information can be essential to understanding the pathophysiologic derangements that may result from glucose uptake or transport disorders. Conditions such as diabetes are characterized by an imbalance in glucose regulation, causing chronic hyperglycemia and ultimately leading to multiorgan damage.

Introduction

Glucose is the most abundant molecule on earth as it constitutes the main product of carbon fixation by the photosynthetic organisms. Glucose can be found as monomer or as polymer, including cellulose, starch, glycogen and others. It is a major metabolic substrate of great importance since it acts as an intermediate for the synthesis of many other molecules. Of pivotal significance is that glucose comprises the first fuel used by most organisms serving as quick substrate for ATP production (also under anaerobic conditions) through glycolysis and the subsequent metabolic pathways (Martínez-Quintana and Yepiz-Plascencia, 2012).

Breakdown of endogenous liver glycogen stores delivers glucose which is also indirectly released in the muscles through intermediary metabolites. These whole-body energy stores are replenished from dietary glucose (Triplitt, 2012). In higher organisms, glucose results directly from the diet when disaccharides and polysaccharides are hydrolyzed. After that, glucose travels long ways to reach its target cells to be stored or metabolized. Transfer from the lumen of the small intestine to epithelium cells is the first step, followed by transport to each organ, tissue and cell by the circulation of blood. Due to glucose's hydrophilic nature, these processes involve the transfer of glucose across the phospholipid bilayer cell membranes. This transport is mediated by specialized integral transporter proteins classified - based on functional and phylogenetic data- in two families: the sodium-coupled glucose cotransporters (SGLTs) and the facilitative glucose transporters (GLUTs) (Martínez-Quintana and Yepiz-Plascencia, 2012; Wilson-O'Brien et al., 2010).

Glucose homeostasis aims to keep blood glucose concentrations constant within a narrow range for each organism. This is maintained by a dynamic equilibrium between glucose utilization and endogenous glucose production, which involves several metabolic events. Furthermore, glucose homeostasis is mainly controlled by the coordinated regulation of three processes: first, glucose absorption via the small intestine; second, glucose production in the liver; and third, consumption of glucose by nearly all tissues (Scheepers et al., 2004). These processes are regulated by hormones, metabolites such as glucose, free fatty acids (FFA) and carbon-3 compounds, such as lactate, alanine and glycerol (Rodan and Bernroiden, 2003). Insulin and glucagon are the most relevant hormones, among many others, involved in the regulation of glucose handling (Triplitt, 2012). Most notably, insulin reduces blood glucose levels by increasing transport of glucose into insulin-sensitive cells and facilitating its conversion to storage compounds via glycogenesis and lipogenesis (Guyton and Hall, 2006). On the other hand, glucagon is produced in response to low normal glucose levels or hypoglycemia and aims to increase glucose levels by promoting glycogenolysis and gluconeogenesis (Triplitt, 2012). However, following a glucose rich meal, glucagon secretion is inhibited by hyperinsulinemia, which contributes to suppression of hepatic glucose production and maintenance of normal postprandial glucose tolerance (Guyton and Hall, 2006). The bulk of glucose uptake in peripheral tissues occurs in skeletal muscles, where glucose can be used directly for energy production or stored as glycogen (DeFronzo, 2009). Skeletal muscles are insulin-dependent, and therefore require insulin for the activation of glycogen synthase (GS) which is the crucial enzyme regulating glycogen formation. While adipose tissue is responsible for a limited amount of peripheral glucose uptake, it has a big contribution in total glucose homeostasis by regulating the release of FFA (which increase gluconeogenesis) from stored triglycerides, influencing insulin sensitivity in liver and muscles (DeFronzo, 2004). The kidneys play also an important role in glucose homeostasis via release of glucose into the circulation (gluconeogenesis), uptake of glucose from the circulation to cover renal energy demands, and reabsorption of glucose at the proximal tubule (Wright et al., 2007). The kidneys also support in elimination of excess glucose by facilitating its excretion in the urine (Guyton and Hall, 2006). While the liver does not require insulin to facilitate glucose uptake, it does need insulin to regulate glucose output (DeFronzo, 2004). Accordingly, when insulin concentrations are low, hepatic glucose output increases. Furthermore, insulin aids the storage of the most of the absorbed glucose in the form of glycogen in the liver (Guyton and Hall, 2006).

Hepatic Glucose Uptake

The liver is mainly responsible for maintaining normal blood glucose concentrations by storing glucose as glycogen and to produce and deliver glucose from glycogenolysis or gluconeogenic precursors. In the postprandial state, blood glucose is taken up by the hepatocyte via the glucose transporter type 2 (GLUT2) – a membrane bound transporter with high capacity and low affinity for glucose (Bechmann et al., 2012). In contrast to GLUT4, which is expressed mainly by muscle and adipose tissue, the expression and activity of GLUT2 is independent of insulin signaling. In pancreatic islet cells, GLUT2 is thus also referred to as "glucose sensor" (Leturque et al., 2009). Once taken up by the hepatocyte, glucose is phosphorylated to glucose-6-phosphate (G6P) by liver glucokinase (L-GCK), the rate limiting enzyme for hepatic glucose utilization (Agius, 2008). In contrast to other hexokinases, GCK is not feed-back inhibited, which allows for postprandial glycogen storage within the hepatocyte. In the fasting state, L-GCK is inactive and bound to glucokinase regulatory protein (GCKR) within the hepatocyte nucleus. Postprandial glucose abundance and insulin-action synergistically cause rapid dissociation of L-GCK from GCKR and translocation to the cytoplasm (Chu et al., 2004). L-GCK is transcriptionally regulated by sterol-response-binding-protein-1c (SREBP-1c), hepatic nuclear factor-4-alpha (HNF4a), hepatic nuclear factor 6 (HNF6), forkhead box proteinO1 (FoxO1), and upstream stimulatory factor 1 (USF1). Indeed, mutations in the GCK gene have been associated with insulin resistance (IR) and the pathogenesis of maturity-onset diabetes of the young (MODY) in several studies (Bechmann et al., 2012).

Glycolysis and Glycogenesis

G6P is either further processed in glycolysis or utilized for glycogenesis, depending on the systemic metabolic state. The process of glycolysis metabolizes glucose to pyruvate with a net gain of two ATP and two NADH molecules per glucose molecule. Glycolysis is regulated, as stated above, by L-GCK, which provides G6P, phosphofructokinase, which is inhibited by its product fructose-1,6-bisphosphate, AMP and pyruvate kinase (PK), the final enzyme in this ten-step metabolic pathway. PK is activated by its substrate and inhibited by ATP abundance. Insulin, epinephrine, and glucagon also regulate PK via the phosphatidylinositol 3′-kinase (PI3K) pathway and carbohydrate responsive element binding protein (CREBP) induces PK transcription in the presence of glucose (Dentin et al., 2005; Bechmann et al., 2012). Pyruvate is further decarboxylized to acetyl-CoA and then processed in the tricarboxylic acid (TCA) cycle or utilized for *de novo* lipogenesis (DNL). The pentose phosphate pathway is an alternative way for degradation of G6P in hepatocytes, which provides the cell with reducing equivalents, in the form of NADPH, an important antioxidant and co-substrate for DNL, cholesterol-, nucleotide- and aromatic amino acid synthesis. GS catalyzes the synthesis of glycogen after conversion of G6P to UDP-glucose (Roach, 2002). GS is regulated by the allosteric activator G6P and is inactive in the phosphorylated state. Glycogen synthase kinase 3 (GSK3) phosphorylates GS and is a downstream target of protein kinase Akt/PI3K and thus insulin signaling. GSK3 is a multifunctional kinase, involved in cell senescence, apoptosis and lipid metabolism via phosphorylation of SREBP-1c (Kim et al., 2010; Bechmann et al., 2012). Other protein kinases that phosphorylate GS are AMP-activated protein kinase (AMPK) and protein kinase A (PKA). Insulin activates glycogenesis via repression of PKA. Finally, the growing glycogen polymer is further branched by a specific glycosyltransferase (branching enzyme).

Glycogenolysis and Gluconeogenesis

In the fasting state, the liver supplies the body with glucose by glycogenolysis, and following prolonged fasting by gluconeogenesis (Raddatz and Ramadori, 2007). Glycogenolysis is catalyzed by glycogen phosphorylase, which cleaves glucose from the glycogen polymer and produces glucose-1-phosphate, which is then converted to G6P by phosphoglucomutase. A debranching enzyme with glucosyltransferase and glucosidase activity cleaves the last four glucose monomers. Glycogen phosphorylase is regulated through allosteric activation by AMP and via phosphorylation by PKA, which is inhibited by insulin. Hepatic gluconeogenesis occurs during prolonged fasting episodes and begins in the mitochondria by the induction of pyruvate carboxylase in abundance of acetyl-CoA. Interestingly, in mice, inhibition of mitochondrial fatty acid oxidation significantly represses hepatic gluconeogenesis via inhibition of carnitine palmitoyltransferase-1 (CPT-1) (Raddatz and Ramadori, 2007; Conti et al., 2011). Gluconeogenesis is further regulated via allosteric and transcriptional activation of phosphoenolpyruvate carboxykinase (PEPCK), fructose-1,6-bisphosphatase and glucose-6-phosphatase (G6Pase). Overexpression of PEPCK promotes IR in mice (Bechmann et al., 2012). Insulin's action represses PEPCK expression via Akt-mediated FoxO1 phosphorylation. This pathway is suppressed by activation of peroxisome proliferator-activated receptor α (PPARα) signaling-mediated induction of the pseudokinase tribbles-homologue 3 (TRB3), and TRB3 expression is associated with IR in patients (Stayrook et al., 2005; Oberkofler et al., 2010). Other nuclear receptors [PPARγ co-activator-1α (PGC1α), farnesoid X receptor (FXR)], glucagon and glucocorticoids also effect the expression of PEPCK and G6Pase (Cao et al., 2010; Raddatz and Ramadori, 2007; Bechmann et al., 2012). FoxO1, a transcriptional activator of G6Pase and PEPCK, is both directly and indirectly activated by PGC1a, HNF4a, CREBP and PPARα (Puigserver et al., 2003). Accordingly, PEPCK knock-out mice exhibit not only reduced gluconeogenesis, but more significantly a decreased removal of TCA anions, which causes accumulation of hepatic triglycerides and steatosis (fatty degeneration) (Hakimi et al., 2005).

Insulin and Normal Insulin Signaling

Insulin is a key hormone with a fundamental role in the growth and development of tissues and the control of glucose balance (Rains and Jain, 2011; Pirola et al., 2004). Insulin's actions at the cellular level include carbohydrate, lipid and amino acid metabolism and mRNA transcription and translation (Wilcox, 2005). Insulin is secreted by the β-cells of pancreas as preproinsulin, an inactive single-chain precursor, with a signal sequence that directs its passage into secretory vesicles. Proteolytic cleavage of this signal sequence results in the formation of proinsulin. Increase in blood glucose or amino acid concentration results in secretion of proinsulin and its conversion into the active insulin by special proteases. The active insulin molecule is a small protein (molecular weight of 5808 Da) composed of 51 amino acids. It is a dimer of an A- and a B-chain, which are linked together by two disulfide bonds (Melloul et al., 2002). The primary role of insulin is to control glucose homeostasis by stimulating glucose transport into muscle and adipose cells, while reducing hepatic glucose production via gluconeogenesis and glycogenolysis. Insulin regulates lipid metabolism by increasing lipid synthesis in liver and fat cells while inhibiting lipolysis. Insulin is also essential for amino acid uptake and protein synthesis (Sesti, 2006). The pleiotropic actions of insulin are all crucial for maintenance of normal cell homeostasis and allow cellular proliferation and differentiation.

Normal insulin signaling occurs through activation of a specific insulin receptor, which belongs to a subfamily of receptor tyrosine kinases (Ogawa et al., 1998). Insulin binds to the α subunit of the receptor, releasing the inhibition of tyrosine autophosphorylation by the β subunit (Bloch-Damti and Bashan, 2005; Saltiel and Kahn, 2001). The receptor is autophosphorylated at distinct tyrosine residues. The activated insulin receptor subsequently directly phosphorylates messenger proteins, namely insulin receptor substrates (IRSs) on multiple tyrosine residues (Van Cromphaut, 2009). The subcellular distribution of these proteins between the cytoplasm and low-density membrane compartments of the cell has been shown to play a vital role in transmitting the proper insulin response (Kriauciunas et al., 2000; Anai et al., 1998). Tyrosine-phosphorylated IRS proteins then act as a binding site for signaling molecules containing SH-2 (Src-homology-2) domains such as PI3K, growth factor receptor-bound protein 2 / guanine nucleotide exchange factor SOS (GRB-2/mSOS), and protein tyrosine phosphatase SHP-2. These molecules bind the phosphorylated tyrosine residues of IRS proteins, composing a signaling complex to mediate downstream signaling.

PI3K is the main signal mediator of the metabolic and mitogenic actions of insulin. It is composed of a p85 regulatory subunit, which binds to IRS proteins, and a p110 catalytic subunit. After the association of p85 with IRS-1/2, the p110 subunit displays elevated catalytic activity. This allows phosphorylation of its substrate, phosphatidylinositol-4,5-bisphosphate (PtdIns(4,5)P_2), on the 3′ position of the inositol ring to generate phosphatidylinositol-(3,4,5)-trisphosphate (PtdIns(3,4,5)P_3) (Bloch-Damti and Bashan, 2005). The second messenger, PtdIns(3,4,5)P_3, recruits the serine kinases phosphoinositide-dependent kinase 1 (PDK-1), PKB/Akt, and PKC to the plasma membrane via their PH domains. The activation of these kinases results in several of insulin's responses, such as GLUT4 translocation to the membrane, glycogen synthesis by phosphorylation of GSK-3, and lipogenesis by up-regulating synthesis of the fatty acid synthase gene (Pessin and Saltiel, 2000). In addition to insulin signaling via PI3K, insulin can activate the mitogen-activated protein kinase (MAPK) ERK, which leads to the gene expression of various cellular proliferation or differentiation components. After phosphorylation of IRS-1/2, the adaptor proteins GRB-2 and SOS are recruited and operate together with a stimulated tyrosine phosphatase, SHP-2, to activate membrane-bound Ras. Activated Ras leads to a kinase cascade, allowing ERK to translocate to the nucleus for gene expression (Saltiel and Kahn, 2001). Insulin's main action of glucose uptake also requires activation of another signaling pathway involving tyrosine phosphorylation of the Cbl proto-oncogene. Cbl is associated with the adaptor protein c-Cbl-associated protein (CAP), which contains three SH3 domains and a sorbin homology (SoHo) domain. The SoHo domain of the phosphorylated Cbl-CAP complex allows translocation to lipid rafts and association with the protein flotillin. A signaling complex is composed at the site of the lipid raft, resulting in the activation of a small G protein, TC10. While under basal conditions, GLUT4 is concentrated in intracellular vesicles, with only about 5% appearing in the plasma membrane (Van Cromphaut, 2009), activated TC10 is thought to act as a second signal in the large recruitment of GLUT4 protein to the membrane as well as in elevated GLUT4 activity (Pirola et al., 2004; Saltiel and Kahn, 2001).

Insulin Signaling and Mitochondrial Dysfunction

It has been known for years that mitochondrial dysfunction in genetic diseases leads to IR, glucose homeostasis impairment and is an underlying cause in the development of diabetes (Rains and Jain, 2011). Mitochondria are the main power supply for the cells and play a central role in cell life and death (Duchen, 2004; Pieczenik and Neustadt, 2007). They provide energy for almost all cellular processes, from muscle contraction to maintenance of ionic gradients for vesicle fusion, and the cycling necessary for secretion of hormones and neurotransmitters. The mitochondria also play an eminent role in how β-cells release insulin in response to glucose levels and the sensing of oxygen tension in the carotid body and pulmonary vasculature (Duchen, 2004; Lowell and Shulman, 2005). Increasing amounts of evidence have demonstrated that mitochondrial dysfunction is associated with disease states such as IR and type-2 diabetes (T2D) (Duchen, 2004; Pieczenik and Neustadt, 2007; Lowell and Shulman, 2005; Kim et al., 2008; Lim et al., 2006; Sirikul et al., 2006; Turner and Heilbronn, 2008). Genetic factors, oxidative stress, mitochondrial biogenesis, and aging are all factors that have been shown to affect mitochondrial function and lead to IR with severe effects on glucose handling (Kim et al., 2008). Furthermore, aging is also a factor in decreased mitochondrial biogenesis and ATP production. Respiration is decreased in isolated mitochondria from elderly subjects who have reduced mitochondrial number and function (Mogensen et al., 2007). The decreased number of mitochondria, density, and mitochondrial gene expression may all play important roles in contributing to the development of IR and diabetes. On the other hand, calorie restriction may be a positive mechanism by which to increase mitochondrial biogenesis and insulin sensitivity. Two approaches have been proposed to attenuate the effects of excessive nutrients leading to IR, one being weight loss to enhance insulin sensitivity and the other being alterations in the macronutrient content of diets to avoid stimulating compensatory insulin mechanisms (Reaven, 2005). These forms of therapeutic intervention may be applied to improve insulin sensitivity and delay or suspend the onset of IR (Rains and Jain, 2011).

Regulation of Blood Glucose

Hepatic glucose disposal and peripheral euglycemia is preserved by numerous factors, including glycemia, hormonal, neural and hepatic auto-regulatory mechanisms and the routes of glucose delivery (Mizock, 2001). The effects of oxidative stress, exercise, the circulating levels of various nutrient substrates, and the presence of first-phase insulin secretion further modify glucose uptake. These regulatory steps present potential therapeutic targets for improvement of glucose disposal in individuals with diabetes or impaired glucose tolerance (Moore et al., 2003). As already stated, blood glucose level is the net result of the difference between rates of glucose entry into and removal from the circulation. Furthermore, when the renal reabsorptive capacity for glucose is exceeded (renal threshold), urinary loss of glucose becomes an additional factor influencing the maintenance of the blood glucose concentration (Polakof et al. 2011).

The levels of plasma insulin vary rapidly in response to fluctuations in glucose level. Insulin reduces blood glucose levels through glucose uptake and glycogen synthesis, and suppression of glycogenolysis and gluconeogenesis. Conversely, glucagon, catecholamines, cortisol and growth hormone raise blood glucose by enhanced glycogenolysis and gluconeogenesis while inhibiting peripheral insulin-mediated glucose uptake. Importantly, the secretion of these counter-regulatory hormones requires the presence of hypoglycemia or vital stimuli.

The enzymes glycogen phosphorylase and GS, promoting glycogen breakdown and synthesis, respectively, are submitted to hormonal modulation. Insulin stimulates glycogen synthase and inactivates glycogen phosphorylase (Schulman et al., 1995; Van Cromphaut, 2009). Glucagon and epinephrine have the opposite effect on glycogen metabolism. Insulin inhibits PEPCK gene transcription, whereas glucagon induces the synthesis of this gluconeogenetic enzyme. Glucagon is the primary hormonal stimulator of hepatic gluconeogenesis, without affecting renal gluconeogenesis. Then again, epinephrine represents the primary stimulus for renal gluconeogenesis (Gerich, 1993; Stumvoll et al., 1997).

Glucosensing and Glycemia

The organism continually adjusts its metabolism to keep blood glucose levels near a constant value. Glucose homeostasis is maintained in most mammals by a feedback mechanism. Central figure of this homeostatic control is the existence of glucosensors dispersed in different parts of the body that continuously monitor glycemia. They respond to alterations in blood glucose levels by triggering hormonal secretion or activation of the autonomic nervous system to control glucose utilization, production, uptake and/or transport, and also to control energy expenditure and food intake (Polakof et al., 2011).

The molecular basis of glucodetection is relatively well understood for insulin secretion by pancreatic β-cells. Additionally, glucosensing cells have been found in peripheral locations such as the L-cell of the intestine (Reimann et al., 2008), glucose-inhibited α-cells (Rorsman et al., 2008), hepatoportal vein (Donovan, 2002), liver (Magnuson and Matschinsky, 2004) and carotid body (Pardal and López-Barneo, 2002), as well as in central locations within the brain where glucosensing neurons have been located to areas such as the septum, amygdala, striatum, motor cortex, hindbrain and hypothalamus (Levin et al., 2004; Moran, 2010).

More specifically, central and peripheral glucosensors monitor glucose availability and guide neural glucoregulation (Niijima, 1989). Hyperglycemia releases the inhibitory sympathetic tone on the pancreas, triggering insulin secretion through increased activity in the ventromedial hypothalamic nucleus. On the other hand, hypoglycemia provokes firing of the nucleus tractus solitarius and lateral hypothalamic glucosensors, increasing splanchnic sympathetic and decreasing parasympathetic activity, which results in glycogenolysis and reduced insulin release (Niijima, 1989).

Decreased firing of peripheral glucose sensors in the portal vein, the liver and small intestine in response to a rise in glucose concentration is communicated to the medullar nucleus solitaries and to the hypothalamus. This signal inhibits adrenal nerve activity and activates vagal pancreatic efferents suppressing catecholamine release and increasing insulin secretion, respectively (Niijima, 1989; Van Cromphaut, 2009). Accordingly, the CNS receives input regarding elevated portal vein glycemia and responds by promoting hepatic glucose uptake. Hepatic autoregulation allows the liver to respond directly and autonomously to an increase in circulating glucose by decreasing hepatic glucose production. As stated previously, the enzymes glucokinase and G6Pase regulate the interconversion of glucose and G6P (derived from gluconeogenesis, glycogenolysis or plasma glucose). This glucose cycle incorporates a critical adaptation mechanism for hepatic glucose output and glucokinase controls the final common pathway for glucose release into the circulation, underscoring its central role in hepatic autoregulation of glucose production (Van Cromphaut, 2009). Inhibition of glycogenolysis explains the decline in hepatic glucose output brought about by hyperglycemia. Apparently, high intra-hepatic G6P levels from enhanced flux of plasma glucose through glucokinase stimulate GS, inhibit glycogen phosphorylase and more than double glucose-G6P cycling rate, whereas gluconeogenesis remains unaffected (Rossetti et al., 1993). Second, hepatic autoregulation prevents increased hepatic glucose output secondary to an increase in gluconeogenic precursors, such as lactate or glycerol (Tappy et al., 1995). In fact, hyperglycemia with substantial hepatic glycogen stores (overnight fast) merely inhibits hepatic glucose output by decreased glycogenolysis or enhanced glucose cycling.

By contrast, hyperglycemia with depleted glycogen stores (prolonged fast) autoregulates hepatic glucose production through suppression of gluconeogenesis or decreased gluconeogenic substrate uptake (Shulman et al., 1980; Van Cromphaut, 2009). The ability to accurately detect circulating glucose levels and its fluctuations is vital to guarantee the correct adjustment of the organism to environmental challenges and changes in glucose availability and/or demand.

Dietary Antioxidants and Glucose Metabolism

Over the last decades, numerous *in vitro* and *in vivo* studies have demonstrated that a broad set of dietary antioxidants, either as extracts or as part of the food, have advantageous effects on glucose metabolism, including the prevention of T2D. The most extensively studied category is the group of polyphenols which have shown interesting results (Avignon et al., 2012).

Only to mention the most recent studies, quercetin, one of the most common flavonoids in the human diet, has shown beneficial effects on glucose metabolism by attenuating TNF-α-mediated inflammation and IR in primary human adipocytes (Chuang et al., 2010); cyanidin-3-O-β-glucoside, a thoroughly studied anthocyanin, has shown insulin-like activities by PPARγ activation in human omental adipocytes (Scazzocchio et al., 2011); the citrus flavonoids naringenin and hesperetin have shown a direct inhibition of tumor necrosis factor (TNF)-α-stimulated FFA secretion in cultured rodent adipocytes (Yoshida et al., 2010), elevated uptake of glucose by AMPK in cultured skeletal muscle cells (Zygmunt et al., 2010), and ameliorated insulin sensitivity by increasing tyrosine phosphorylation in fructose-fed rats (Kannappan et al., 2010). Moreover, the flavonoid luteolin, has been demonstrated to increase insulin sensitivity by an activation of PPARγ transcriptional activity in 3T3-L1 adipocytes (Ding et al., 2010) and caffeic acid phenethyl-ester effectively has been shown to up-regulate glucose uptake in cultured skeletal muscle cells through the AMPK pathway (Eid et al., 2010) and in insulin resistant mouse hepatocytes, similarly to cinnamic acid (Huag et al., 2009). Undoubtedly, one of the phenolic compounds whose effects have been extensively investigated on glucose handling is resveratrol, a phenolic compound of the stilbene family that, although exerting antioxidant properties, can directly interact with many metabolic pathways (Szkudelski and Szkudelska, 2011). There have been studies showing how resveratrol ameliorates energy homeostasis, increases mitochondrial activity, and protects against diet induced obesity and IR showing how this compound may play a critical role in the prolepsis of metabolic diseases (Lagouge et al., 2006; Baur et al., 2006). Resveratrol´s various effects on glucose metabolism seem not restricted to its antioxidant capacity but rather are mediated through sirtuin 1 and thus connected with gene silencing (Chuang et al., 2010; de Kreutzenberg et al., 2010; Mercader et al., 2011; Vigilanza et al., 2011; Brasnyó et al., 2011).

Hence, the ideal antioxidant supplement for IR, the metabolic syndrome, and diabetes prevention may be one that is capable to reproduce the natural combination of antioxidants found in dietary plants even if it cannot be excluded that the beneficial effects of the latter might be synergistic with other compounds (Puchau et al., 2010).

Glucose Cotransporters

Since the eukaryotic phospholipid bilayer of the plasma membrane is impermeable for hydrophilic molecules, glucose is transported across the plasma membrane by membrane associated carrier proteins, the glucose transporters. They are divided into two families: the Na^+/glucose cotransporters (SGLTs) and the facilitative diffusion glucose transporters (GLUTs) (Bell et al., 1990). Both types of transporters belong to families of the solute carrier gene series (SLC), which comprise 43 families (SLC1-SLC43) (Sabino-Silva et al., 2010).

Glucose transporters play an important role in the maintenance of euglycemia, not solely by determining glucose uptake in all cell types, but also by liberating glucose from the liver when circulating glucose levels decline. In addition, they are responsible for the intestinal glucose absorption from the diet and for the glucose reabsorption from the glomerular filtrate in kidneys (Hediger and Rhoads, 1994). The coordinate action of SGLTs are responsible for transepithelial glucose transport in cells from the small intestine, the renal proximal tubules and salivary gland ducts allowing the influx of glucose through the luminal membrane, and the efflux of glucose through the basolateral membrane (Hediger and Rhoads, 1994).

SGLTs

SGLTs comprise a large family of membrane associated proteins involved in the transport of glucose, amino acids, osmolytes, vitamins, and some ions across the brush border membrane of the small intestine epithelium and the renal proximal tubules. Although 6 isoforms of Na^+/glucose cotransporters have been described, the SGLT1 and SGLT2 proteins, encoded by the solute carrier genes SLC5A1 and SLC5A2, respectively, seem to be the most important ones and have been extensively studied focusing on glucose fluxes under both physiological and pathological conditions. Pioneer studies postulated that active glucose transport across the intestinal epithelium is driven by the sodium gradient across the membrane, by means of a Na^+/glucose cotransport (Crane et al., 1960). Consecutive studies demonstrated that the sodium electrochemical potential gradient across the brush border membrane promotes sodium influx, providing the energy, by chemiosmosis, which drives glucose into the enterocytes. Sodium that enters the cell along with glucose is then transported into the interstitium by Na^+/K^+ ATPase at the basolateral membrane, maintaining the electrochemical gradient for luminal sodium influx. Thus, the energy for the whole process comes from the ATP consumed by Na^+/K^+ ATPase (Sabino-Silva et al., 2010). Glucose accumulated within the enterocyte is then transported across the basolateral membrane by GLUTs (Wright et al., 2004). This process ensures electroneutrality, since one mole glucose and two moles sodium are transported across the enterocytes from the lumen into the interstitium followed by water and two moles anions (Wright et al., 2004).

SGLT1

SGLT1 is a 75-kDa membrane protein with a Na^+/glucose stoichiometry of 2:1. It contains 14 transmembrane α-helices, a single glycosylation site between transmembrane helices 5 and 6, and two phosphorylation sites between transmembrane helices 6 and 7 and between transmembrane helices 8 and 9 (Wright et al., 2004; Turk and Wright, 1997). The

NH_2- and COOH termini are extracellular and intramembrane, respectively, and the glucose binding and translocation domains involve the five transmembrane segments, which are next to the COOH terminus (Wright et al., 2004). The SGLT1 protein, encoded by the SLC5A1 gene, is located mainly in the intestine, but has also been detected in the kidney, parotid and submandibular salivary glands as well as in the heart (Wright et al., 2004; Zhou et al., 2003; Sabino-Silva et al., 2009). It has high affinity for glucose (and galactose), but low capacity of transporting it, and is specifically inhibited by phlorizin (Diez-Sampedro et al., 2000). The net result of the transport through the SGLT1 is two sodium ions for each molecule of glucose. This cotransport is a secondary Na^+/K^+ ATPase-dependent process and is able to transport glucose into the cell against its concentration gradient (Wood and Trayhurn, 2003). In the steady state, Na^+ ions entering the cell across the luminal membrane are pumped out by Na^+/K^+ ATPase in the basolateral membrane. Glucose accumulates within the cell, a process that allows it to move downhill to the interstitium through the GLUTs (Bell et al., 1990). It seems that Na^+ and glucose transport by the SGLT1 occurs by a sequential mechanism in which two Na^+ ions on the extracellular side bind to the transporter right before glucose, inducing a conformational change in the glucose binding site and increasing the transporter affinity for this substrate. Then, another conformational change determines the shift of the Na^+ and glucose binding sites to the inner surface of the membrane, with the glucose molecule and immediately after the two Na^+ ions being released within the cell (Wright et al., 2004; Diez-Sampedro et al., 2000). This mechanism allows the recycling of SGLT1 protein about one thousand times per second at 37°C (Wright et al., 2004). Moreover, it has been demonstrated that, besides the two Na^+ ions plus the one glucose molecule, the SGLT1 protein is also able to transport about 264 water molecules. This coupled water transport occurs due to conformational changes in the protein structure induced by sodium and glucose during the transport cycle (Wright et al., 2004; Zeuthen et al., 2001). Therefore, it seems clear that SGLT1 plays an important role in water absorption in the brush border membrane of enterocytes (Wright et al., 2004).

SGLT2

The SGLT2 protein was originally described as an amino acid cotransporter (Kong et al., 1993), but was recognized later as a glucose transporter (Mackenzie et al., 1994). It consists of 672 amino acids and both NH_2 and COOH termini are extracellular (Turk and Wright, 1997). Differently from SGLT1, SGLT2 is a low-affinity and high-capacity glucose transporter, which transports one Na^+ ion for each glucose molecule. The SGLT2 encoding gene (SLC5A2) is predominantly expressed in kidneys, but low mRNA expression levels have also been noted in mammary glands, liver, lungs, intestine, skeletal muscle, and spleen (Zhao et al., 2007). SGLT2 is highly expressed in the luminal membrane of the S1 segment of the renal proximal tubule. This isoform is mainly responsible for the reabsorption of glucose from the glomerular filtrate. Actually, more than 90% of the filtered glucose is reabsorbed in the initial segments of the proximal tubule via SGLT2 (Wright et al., 2004; Wells et al., 1992).

Transepithelial Transport of Glucose

Both SGLTs and GLUTs are involved in transepithelial transport of glucose. Renal glucose reabsorption occurs mainly in the S1 segment of the proximal tubule by the coordinated action of the SGLT2 and GLUT2 located in the luminal and basolateral membranes, respectively (Sabino-Silva et al., 2010). Only a residual amount of glucose is reabsorbed in the S3 segment, where SGLT1 is present in the luminal membrane, co-expressed with GLUT1 in the basolateral membrane (Wright et al., 2004). Intestinal glucose absorption occurs mostly in the duodenum and in the initial portion of the jejunum, and involves the co-expression of SGLT1 and GLUT2 (Wright et al., 2004). In all these processes, SGLTs present in the luminal membrane transport glucose from the lumen into the intracellular medium, where glucose concentrates generating a gradient that favors its transport through the GLUTs in the basolateral membrane, from the cytoplasm to the interstitium (Wright et al., 2004).

SGLTs and Protein Kinases

The activity of membrane transporter proteins seems to be regulated either directly or indirectly by protein kinases (PK) (Hirsch et al., 1996). Direct effects involve phosphorylation of the transporter, thus changing its kinetic parameters. In rat small intestines, it has been shown that the activation of β-adrenoreceptors induces phosphorylation of SGLT1 via PKA, enhancing cotransporter's function, and thus increasing glucose transport (Ishikawa et al., 1997). Indirect regulation occurs via changes in the rate of the insertion into or retrieval from the plasma membrane. Membrane proteins are cotranslationally inserted into the endoplasmic reticulum and processed by the Golgi apparatus, where vesicles are formed to deliver the proteins to the plasma membrane (Ishikawa et al., 1997). In *Xenopus laevis* oocytes, it has been demonstrated that PK regulate the SGLT1-induced glucose transport by controlling the distribution of transporters between intracellular compartments and the plasma membrane. Activation of PKA and PKC, respectively, increases and decreases the maximum rate of Na^+/glucose transport, accompanied by an increase and decrease, respectively, in the content of SGLT1 protein in the plasma membrane (Hirsch et al., 1996). However, SGLT1 translocation seems to be independent of the phosphorylation of potential phosphorylation sites present in the transporter (Hirsch et al., 1996). Although some studies propose that these effects might be extended to other isoforms of SGLTs (Wright et al., 2004; Hirsch et al., 1996), these mechanisms seemed clearly demonstrated only for the SGLT1 protein. As a physiological consequence of these rapid regulations of SGLT1, it is hypothesized that the anticipative and the postprandial increases in β-adrenergic activity could contribute to enhancing intestinal glucose absorption by translocating SGLT1 (Sabino-Silva et al., 2010).

Transcriptional Regulation of SGLTs

There is a number of transcription factors reported to participate in the regulation of SGLT expression. However, hepatocyte nuclear factors 1α and 1β (HNF-1α and HNF-1β) may be the most essential regulators of SGLT1 expression (Balakrishnan et al., 2008). HNF-1α and HNF-1β are developmentally expressed, and their transcriptional effects have been primarily related directly to the level of protein expression. HNF-1α and HNF-1β contain an NH_2-terminal dimerization domain, and they bind to DNA as homo- or heterodimers. Isoforms α and β have been shown to possess distinct transcriptional activity, and thus the pattern of dimerization plays a central role in the final effect on the transcription of target genes (Rhoads et al., 1998; Hayashi et al., 1999). In the SLC5A1 gene promoter, there are two cis-regulatory elements for HNF-1α/β, which were demonstrated to be functional by transient promoter assays (Balakrishnan et al., 2008; Martin et al., 2000; Vayro et al., 2001). The ratio of HNF-1α and HNF-1β dimerization is particularly important in modulating the transcription of SGLT1. Homodimers of HNF-1α are associated with increased SGLT1 transcription, whereas heterodimers of HNF-1α/β decrease SGLT1 transcription (Rhoads et al., 1998). Notably, intestinal SGLT1 expression was reported to show a circadian periodicity in *ad libitum*-fed rats: the higher transcriptional rate correlated with the appearance of homodimers in the morning and the lower rate with heterodimers in the afternoon (Rhoads et al., 1998). However, the mechanism of these dimerization changes was not determined. The interaction of HNF-1 with other proteins is an additional mechanism that regulates transcriptional activity. Key nuclear proteins have been identified, such as HNF-3, HNF-4, HNF-6, CCAAT/enhancer binding protein (C/EBP), and D binding protein (DBP), which can cooperatively enhance the transcriptional activity of HNF-1α and/or HNF-1β (Hayashi et al., 1999; Gautier-Stein et al., 2006). HNF1-α interacts also with caudal-type homeobox protein 2 (CDX2) and GATA-binding protein (GATA) family members to regulate the expression of some intestinal genes (Balakrishnan et al., 2008). In addition, proteins from the Sp1 family of transcription factors, which bind in GC boxes, in the presence of HNF-1 also act synergistically by up-regulating the transactivation of the promoter of target genes (Martin et al., 2000). Finally, a substantial role was attributed to the enhancer of rudimentary homolog (erh) protein, which, by interacting with the dimerization cofactor of HNF-1 (coactivator DCoH), represses the transcriptional activity of HNF-1 (Wan et al., 2005). The majority of these synergistic mechanisms have been reported to regulate the expression of the SGLT1 gene (Balakrishnan et al., 2008; Martin et al., 2000). The above described interactions exhibit how extensive and complex the HNF-1 transcriptional regulation is. Nevertheless, little is known about the mechanisms that control these interactions. In this regard, the inhibition of protein phosphatases 1/2A in Caco-2 cells has been demonstrated to enhance HNF-1α/β phosphorylation, decreasing their DNA-binding capacity, and thus repressing the expression of the gene encoding sucrase-isomaltase (Carriere et al., 2001). Likewise, in HepG2 cells, mutation of the HNF-1 binding site decreases the cAMP/PKA-induced HNF-1 promoter activity, displaying that there is an involvement of this pathway in the transcriptional activity (Gautier-Stein et al., 2006). Fasting- and refeeding- as well as insulin - and glucose -induced regulation of HNF-1 transcriptional activity have been reported mainly in the liver and intestine. Most of these studies involved genes related to glucose metabolism, including the SLC5A1 gene that codes for the SGLT1 protein (Balakrishnan et al., 2008; Rhoads et al., 1998; Vayro et al., 2001; Gautier-Stein et al., 2006; Carriere et al., 2001). However, the role of dietary lipids or proteins, as well as of insulin and its countervailing hormones, on HNF-1 transcriptional activity is largely unknown. Little is known about HNF-1 regulation of the SLC5A2 gene, which codes for the SGLT2 protein. Impaired renal tubular glucose reabsorption was observed in HNF-1α- deficient mice, suggesting that this transcription factor also regulates renal SGLT2 expression (Pontoglio et al., 1996). Further studies, in renal tubular cells, confirmed decreased SGLT2 expression of HNF-1α-deficient mice (Friedlander et al., 1999). Subsequently, it was shown, in mice and humans, that HNF-1α directly regulates SGLT2 expression (Pontoglio et al., 2000). Lately, it was demonstrated that increased HNF-1α expression and binding activity in the SLC5A2 promoter is involved in the diabetes-induced overexpression of SGLT2 (Freitas et al., 2008). Moreover, it was shown, in rats, that diabetes-induced changes in HNF-1α activity and SGLT2 expression are reversed by reducing glycemia, independently of insulinemia (Freitas et al., 2008), signifying that glucose is a major regulator of this transcriptional activity, as described for SGLT1.

Dietary Regulation of SGLTs

It has been reported that increased dietary sugar enhances the intestinal sodium-dependent glucose transport, suggesting that SGLT1 expression may be increased (Solberg and Diamond, 1987). Actually, changes in the mRNA and protein expression of SGLT1 as a result of dietary carbohydrate content were shown in ruminants (Dyer et al., 2007; Lescale-Matys et al., 1993). Besides induction by dietary carbohydrates, intestinal expression of SGLT1 exhibits circadian periodicity in its activity. Furthermore, since the daily variation in SGLT1 activity persists with starvation, this variation could be described as an early response and is transcriptionally regulated (Rhoads et al., 1998). Significant changes in the intestinal and renal glucose cotransporters can be also induced by sodium intake. Studies in rats have shown a significant decrease in SGLT1 mRNA in the upper small intestine along with high salt intake. Additionally, SGLT1 mRNA increased in the middle segments with low salt intake (Sabino-Silva et al., 2010). These results correspond with a decrease in mRNA half-life in the upper segment (on high salt) and an increase in the middle segment with low salt intake. Nevertheless, this regulation was not associated with alterations in intestinal glucose absorption (Sabino-Silva et al., 2010). Moreover, it was also demonstrated, in rats, that a high sodium diet increased the expression of SGLT2, GLUT2 and GLUT1 in renal proximal tubules (Vestri et al., 2001). In the renal cortex from rats fed a high sodium diet, the very high sodium filtration rate improved the cellular glucose influx in the early S1 segment. In this segment, most of the filtered glucose rapidly entered into the epithelial cells and the high-capacity/low-affinity GLUT2 transporter could not proportionally increase the cellular efflux of glucose. For that reason, the S1 epithelial cells may be transiently submitted to a high glucose concentration, which is adequate to determine the increase of SGLT2 and GLUT2 expression (Sabino-Silva et al., 2010). Recently, a diet-induced mechanism of SGLT1 expression regulation has been described, involving the intestinal luminal glucose content and a sugar sensor located at the luminal membrane of the enterocytes. It has been shown that the sweet taste receptor subunit T1R3 and the taste G protein gustducin, expressed in enteroendocrine cells, underlie cellular sugar sensing (Margolskee et al., 2007), triggering a signal linked to the cAMP-PKA pathway, which eventually causes an increase in the expression of SGLT1 mRNA and protein (Dyer et al., 2007; Sabino-Silva et al., 2010).

GLUTs

Facilitated diffusion of glucose and related hexoses across biological membranes is managed by glucose transporters or GLUTs which are members of the SLC2 family, (Uldry and Thorens, 2004). GLUTs act as simple carriers and the movement of hexoses across the plasma membrane proceeds in the direction enforced by its electrochemical gradient. All GLUTs share a common structure feature of 12 transmembrane domains (TM) with both the NH_2- and COOH terminals present on the cytosolic side and a unique N-linked oligosaccharide side-chain present either in the first or the fourth extracellular loop. Glucose transporters are ubiquitously expressed, as might be expected from the fundamental role of glucose in providing metabolic energy and building materials for biosynthesis. The specific physiological role of the isoforms expressed in tissues involved in the control of glucose homeostasis, i.e. liver, muscle, adipose tissue, pancreatic β-cells and brain, has been investigated in detail. In fact, GLUTs play key roles in the control of glucose uptake and utilization, glucose production and glucose sensing and their aberrant expression in these tissues may underlie pathogenetic mechanisms leading to development of diseases such as diabetes mellitus, but also other specific monogenic diseases (Uldry and Thorens, 2004). There have been 14 members of the human SLC2A gene family identified, some of them displaying discreet tissue expression patterns and function (Augustin, 2012; Martínez-Quintana and Yepiz-Plascencia, 2012). All GLUT protein members have been named by consensus GLUT1-14 (genes SLC2A1-14) and GLUT13 is named HMIT (H^+-coupled *myo*-inositol transporter).

Class I Family Members

Isoforms of class I GLUTs belong to the well characterized transporters with GLUT1 being the first isoform cloned and described in 1985 (Mueckler et al., 1985).

GLUT1 (SLC2A1)

GLUT1, also known as the HepG2 or erythrocyte sugar transporter, is highly expressed in erythrocyte membranes making up ~ 5% of their total proteins. It was the first transporter to be characterized by molecular cloning, and its cDNA was isolated from an expression library using antibodies against the humans' erythrocyte glucose transporter (Mueckler et al., 1985). Although cloned from a hepatoma cDNA library, GLUT1 is not expressed in normal hepatocytes. It is, however, induced during oncogenic transformation of most cell types and its expression correlates with the increase in glucose metabolism observed in tumor cells (Flier et al., 1987). GLUT1 is found in almost every tissue with different levels of expression in different cell types. The expression level usually correlates with the rate of cellular glucose metabolism. It is also highly expressed in blood-tissue barriers, in particular in the endothelial cells forming the blood-brain barrier and in endothelial and epithelial-like barriers of the eye, peripheral nerve and placenta (Maher et al., 1994; Uldry and Thorens, 2004; Augustin, 2012). GLUT1 transports glucose with a Km of ~3 mM. Other transported molecules are galactose, mannose and glucosamine (Uldry et al., 2002). Glucose transport by GLUT1 is sensitive to several inhibitors that also block transport by other isoforms. Many of them are competitive inhibitors of sugar binding, either to the extracellular or the cytosolic sugar binding sites.

A wide spectrum of heterozygous mutations, including nonsense, missense, insertion, deletion and splice site mutations and hemizygosity of the GLUT1 gene, have been identified (Augustin, 2012). In humans, mutations in the GLUT1 gene are responsible for an autosomal-dominant disorder characterized by infantile seizures, developmental delay, acquired microcephaly, ataxia and which is due to a decreased rate of glucose transport from the blood into cerebrospinal fluid (De Vito et al., 1991; Seidner et al., 1998; Pascual et al., 2004). As ketone bodies enter the brain via a monocarboxylic acid transporter (MCT1), bypassing the blood-brain barrier, they supply an alternative energy source for the brain under conditions of GLUT1 deficiency.

Accordingly, a ketogenic diet is effective in controlling the seizures and other symptoms of the GLUT1 deficiency syndrome (Klepper, 2008). However, this treatment is less effective regarding neurobehavioral symptoms. A few years ago, GLUT1 homozygous and heterozygous knockout mice were characterized (Wang et al., 2006). While homozygous loss of GLUT1 was associated with embryonic lethality, heterozygous knockout mice did not exhibit any differences in development and growth. However, brain weights were decreased but no histological abnormalities were noted. Moreover, in the heterozygous animals, plasma glucose levels were normal while they were decreased in the cerebrospinal fluid. Positron emission tomography (PET) scan analysis revealed that glucose uptake/metabolism was reduced in brains of the heterozygous GLUT1 knockout animals. These animals also displayed deficits in motor activity, balance and coordination as well as spontaneous cortical seizures. Overall, heterozygosity for GLUT1 in mice resembles features of humans with the GLUT1 deficiency syndrome (Augustin, 2012).

GLUT2 (SLC2A2)

Initially, GLUT2 has been detected mainly in the liver, kidney, and intestine (Fukumoto et al., 1988), but later its presence was demonstrated specifically in pancreatic β-cells where it catalyzes the first step in glucose-stimulated insulin secretion, a glucose metabolism-dependent signaling process where phosphorylation of glucose by glucokinase represents the rate controlling step (Orci et al., 1989; Johnson et al., 1990). In hepatocytes, GLUT2 is present in the sinuosidal membrane where it is involved in both glucose uptake and glucose release into the blood (Guillam et al., 2000; Matschinsky, 1996). Functional studies indicated the requirement for GLUT2 in other glucose-sensing units, in particular those in the hepatoportal vein, the hypothalamus and brain stem. These glucose-sensing units are involved in the control of counter-regulation, food intake and stimulation of glucose uptake by peripheral tissues, as described before (Burcelin et al., 2000; Burcelin and Thorens, 2001).

GLUT2 is a low-affinity, high capacity transporter, with a Km in the range of 17 mM, which is the highest Km for glucose among the known members of the GLUT family. It also transports galactose, D-mannose, and D-fructose (Johnson et al., 1990); while recently, GLUT2 was shown to transport glucosamine with high affinity (Uldry et al., 2002).

Homozygous or compound heterozygous mutations of the GLUT2 gene are responsible for a type of glycogen storage diseases. The first patient was described by Fanconi and Bickel's; therefore, the GLUT2 deficiency is referred to as Fanconi-Bickel Syndrome (FBS) (Manz et al., 1987). The clinical manifestation of the FBS include hepatomegaly secondary to glycogen accumulation, glucose and galactose intolerance, fasting hypoglycemia, a tubular nephropathy, and severely stunted growth (Santer et al., 1997). Opposite to the metabolism of glucose and galactose, utilization of administered fructose is normal in FBS patients (Manz et al., 1987). Glucose homeostasis in FBS patients is heavily disturbed and postprandial hyperglycemia is frequently observed. There is no specific treatment for patients with FBS, while symptomatic treatment is directed toward a stabilization of glucose homeostasis and compensation for renal losses of various solutes. The improvement of the consequences of tubular nephropathy includes the replacement of water and electrolytes. To control renal glucose loss and hepatic glycogen accumulation, FBS patients receive a diet with adequate caloric intake with slowly absorbed carbohydrates (Augustin, 2012).

Whole body GLUT2 knockout in mice results in offspring that seems normal at birth, but soon neonates develop symptoms similar to T2D and do not survive beyond the third week of age (Manolescu et al., 2007a). Homozygous GLUT2 deficiency in mice results in hyperglycemia and elevated plasma levels of FFA and β-hydroxybutyrate. *In vivo*, glucose tolerance is abnormal, and *in vitro* β-cells exhibit a gradual loss of control of insulin gene expression by glucose. GLUT2 knockout mice show marked hyperglucagonemia, which is associated with alterations in the postnatal development of pancreatic islets, evidenced by a gradual inversion of the α- to β-cell ratio. Recently, a direct link has been demonstrated between diet-induced IR and β-cell dysfunction via disturbed GLUT2 localization in the plasma membrane. High-fat feeded mice demonstrated intracellular retention of the transporter due to improper glycosylation, thereby, leading to impaired glucose stimulated insulin secretion (Ohtsubo et al., 2005).

In humans, GLUT2 deficiency is associated with a marked hypoglycemia in the fasting state owing to a diminished hepatic glucose output (Santer et al., 2002) and a failure of glucagon to increase plasma glucose (Hosokawa et al., 2002). However, FBS patients do not develop diabetes and do not show a complete loss of their β-cell function, indicating that functionality of human pancreatic β-cells is not entirely dependent on GLUT2 (De Vos et al., 1995). Currently, the observation of normal glucose output in GLUT2 deficient murine hepatocytes remains elusive. However, the normal response of FBS patients to intravenously administered fructose might indicate that besides GLUT2 another GLUT exists that contributes to hepatic fructose handling in men (Santer et al., 2002) and potentially mice.

GLUT3 (SLC2A3)

GLUT3 is regarded as a neuron-specific glucose transporter because of its dominant expression in the brain in various species (Shepherd et al., 1992; Haber et al., 1993; McCall et al., 1994). However, GLUT3 is also expressed in tissues with high demand for glucose such as testis, placenta (Illsley, 2000), or certain cancer cells and cancer tissues (Kayano et al., 1988; Ciampi et al., 2008). Its tissue distribution and function correlates with its high affinity and transport capacity for glucose. Other substrates of GLUT3 are galactose, mannose, maltose, xylose, and dehydroascorbic acid but not fructose (Colville et al., 1993).

Heterozygous GLUT3 knockout mice displayed significantly enhanced cerebrocortical activity and were more sensitive to an acoustic startle stimulus, but their behavior regarding coordination, reflexes, motor abilities, anxiety, learning, and memory were normal (Schmidt et al., 2008a). Examination of counterregulatory responses in expression of sugar transport facilitators or monocarboxylate transporters (MCT) revealed no alterations on mRNA levels for GLUT1, GLUT8, and MCT1 or MCT2 expression (Schmidt et al., 2008a), whereas other studies noted elevated protein amounts for GLUT1, GLUT8, and MCT2 in heterozygous GLUT3 knockout mice (Zhao et al., 2010). Additionally, it was reported that heterozygosity for GLUT3 does not cause any alterations in body weight, blood glucose, and insulin levels. In addition, food intake was unchanged by intracerebroventricular administration of 2-DG or D-glucose, and food choice for carbohydrates/fat was not affected (Schmidt et al., 2008b). However, it was shown that male but not female heterozygous GLUT3 mice at an age of nine to eleven month display an increased total body fat mass associated with IR and increased concentrations for insulin, triglycerides, and total cholesterol (Ganguly and Devaskar, 2008). The different phenotypical manifestations of the heterozygous GLUT3 knockout mice reported might be ascribed to the knockout strategy and/or purity of the genetic background (Augustin, 2012).

In humans, a whole-genome association analysis in 200 dyslexic children, focusing on mismatch negativity (MMN) measurements, identified two single-nucleotide polymorphisms (SNPs) that both showed a significant association with mRNA-expression levels of SLC2A3 on chromosome 12. It was suggested that a possible transregulation effect on SLC2A3 might lead to glucose deficits in dyslexic children that might cause their attenuated MMN in passive listening tasks (Roeske et al., 2011).

GLUT4 (SLC2A4)

GLUT4 has an eminent physiological role namely, the rate-limiting step in insulin-stimulated glucose uptake of skeletal and cardiac muscle, brown and white adipose tissue (Huang and Czech, 2007). Thereby, impaired GLUT4 translocation is causally linked to IR and therefore to noninsulin-dependent diabetes mellitus. GLUT4 was cloned from human (Fukumoto et al., 1989), rat (Birnbaum, 1989; Charron et al., 1989), and mouse tissues (Kaestner et al., 1989) and shows a similar affinity for glucose as GLUT1 (Km of 5 mM) whilst it also transports dehydroascorbic acid and glucosamine.

GLUT4 contains unique sorting motifs at its NH_2- (FQQI) and COOH-end (dileucine) that are responsible for its ability to traffic between specific intracellular compartments and to translocate to the plasma membrane in response to different stimuli, such as insulin (Huang and Czech, 2007). GLUT4 translocation to the plasma membrane can also be stimulated by exercise.

Observations made by total interference microscopy (TIRF) supported a view where insulin not only induces recruitment of GLUT4 vesicles from an intracellular pool but also controls the integration of those vesicles into the plasma membrane in a stepwise, fine adjusted mode of tethering, docking, and fusion (Watson and Pessin, 2007).

Insulin-stimulated glucose uptake depends on a complex signaling cascade (Saltiel and Pessin, 2002) that is still not fully understood. A defect in the ability of insulin to regulate this metabolic event is one of the key physiological dysfunctions of type-2 diabetes. A decreased expression of GLUT4 may also cause IR. However, in T2D, decreased expression of GLUT4 is observed only in adipose tissue and not in muscle, although the latter is responsible for ~90% of glucose utilization in the postprandial state and is thus a major site of IR (Uldry and Thorens, 2004).

Furthermore, although the significance of GLUT4 for insulin-stimulated glucose uptake in muscle and adipose tissue is well established, thus far, no polymorphisms within the GLUT4 gene have been characterized that would be certainly associated with compromised glucose homeostasis in humans under circumstances such as T2D, increased fasting blood glucose levels, or obesity (Friedel et al., 2002; Kusari et al., 1991).

GLUT14 (SLC2A14)

Searching the human genome for additional genes coding for glucose transporters, a putative gene was identified and cloned showing 95% identity on nucleotide level to GLUT3 (Wu and Freeze, 2002). The SLC2A14 gene maps to chromosome 12p13.3 and seems to be a consequence of a gene duplication of GLUT3. GLUT14 was shown to be specifically expressed in testis (Wu and Freeze, 2002) and two alternatively spliced forms of GLUT14 were identified. Interestingly, GLUT14 has no orthologue in mice; a finding that has also been made for GLUT11 (Scheepers et al., 2005).

Class II Family Members

A common feature of class II transporters is their ability to transport fructose, with GLUT5 representing the bona fide fructose transporter. Interestingly, class II isoforms were demonstrated not to transport 2-DG and galactose (Manolescu et al., 2007b). Unique for some members (GLUT9 and GLUT11) is the alternative splicing / promoter usage, where either two or three different mRNAs are transcribed, respectively (Keembiyehetty et al., 2006; Augustin et al., 2004; Scheepers et al., 2005).

GLUT5 (SLC2A5)

GLUT5 was cloned from human intestinal epithelial cells and from rat and rabbit jejunum cDNA libraries (Kayano et al., 1990; Rand et al., 1993; Miyamoto et al., 1994). When expressed in oocytes, human GLUT5 does not display glucose transport activity (Burant et al., 1992) but is a fructose transporter (Km ~6 mM) and is not inhibited by cytochalasin B, phloretin or phlorizin (Mate et al., 2001). Besides fructose, the rat GLUT5 transports glucose, an activity that can be blocked by cytochalasin B (Rand et al., 1993). In human, rat, and mouse, GLUT5 is primarily expressed in the jejunal region of the small intestine. Lower protein levels are expressed in kidney, skeletal muscle, brain, and adipose tissue (Burant et al., 1992; Rand et al., 1993; Corpe et al., 2002). GLUT5 mediates fructose absorption in the jejunum at the apical, potentially also basolateral, membrane of the epithelial cells into the portal vein (Blakemore et al., 1995).

GLUT5 deficiency (Barone et al., 2009) is associated with decreased fructose absorption in animals challenged by a high-fructose diet. Wild-type mice upon high fructose feeding exhibit an enhanced salt absorption in their jejuna and develop systemic hypertension, whereas GLUT5 knockout mice do not display fructose stimulated salt absorption. Instead, the animals show impaired nutrient absorption together with hypotension. Absence of GLUT5 leads to a massive dilatation of the cecum and colon, consistent with severe malabsorption. On a normal chow diet, GLUT5 deficient mice have normal blood pressure and exhibit physiological weight gain. The phenotype of GLUT5 deficient mice shows that this isoform is vital for fructose absorption by the intestine and, thereby, fundamentally involved in fructose-induced hypertension (Augustin, 2012).

GLUT7 (SLC2A7)

The human GLUT7 has been cloned from an intestinal cDNA library using a PCR-based strategy (Li et al., 2004). GLUT7 is primarily expressed in the small intestine and colon, although mRNA has been detected in the testis and prostate as well. The protein has been localized to the apical membrane of the small intestine and colon (Li et al., 2004). GLUT7 shows a rather high affinity for glucose (Km 0.3 mM) and fructose, whereas galactose, 2-DG, and xylose are not transported. Sugar transport by GLUT7 is not inhibited by cytochalasin B or phloretin (Li et al., 2004, Manolescu et al., 2005).

GLUT9 (SLC2A9)

Human GLUT9 cDNA was isolated by PCR amplification from a human kidney cDNA library (Phay et al., 2000). GLUT9 mRNA is almost exclusively detected in the kidney and liver and at lower levels in the small intestine, lung, placenta and leucocytes. It is localized to the insulin-secreting β-cells of human and mouse islets, where down-regulation of the protein by siRNA in rat and mouse insulinoma cells leads to a reduced glucose-stimulated insulin secretion (Evans et al., 2009).

Initial characterization of GLUT9 assigned a rather low affinity for 2-DG (Keembiyehetty et al., 2006; Augustin et al., 2004), but a high affinity for glucose and fructose with Km values of 0.6 and 0.4 mM, respectively (Manolescu et al., 2007b). More recently, GLUT9 has been defined as a high-affinity uric acid transporter with a Km of 0.9 and 0.6 mM for the human (Vitart et al., 2008) and mouse (Bibert et al., 2009) protein, respectively. Although transport of glucose and fructose has not been observed by some investigators (Bibert et al., 2009), GLUT9 presumably exchanges glucose or fructose for urate.

The classical inhibitor cytochalasin B does not block GLUT9 function (Augustin et al., 2004; Bibert et al., 2009). GLUT9 has been determined to represent a major regulator of urate homeostasis in mice (Preitner et al., 2009) and men (Anzai et al., 2008). Whether GLUT9 can play a role as a (glucose)/fructose transporter that possibly links fructose uptake by the liver and urate metabolism remains to be established (Augustin, 2012).

GLUT11 (SLC2A11)

Human GLUT11 cDNA was isolated by PCR amplification based on sequence information from expressed sequence tags (ESTs) and a genomic sequence (Doege et al., 2001). Three variants, GLUT11-A, -B, and -C, differing only in their N-terminal sequences have been identified (Sasaki et al., 2001; Wu et al., 2002). The three alternative spliced isoforms are expressed in a tissue specific manner: GLUT11-A is present in heart, skeletal muscle, and kidney; GLUT11-B in kidney, adipose tissue, and placenta; and GLUT11-C in adipose tissue, heart, skeletal muscle, and pancreas (Scheepers et al., 2005; Sasaki et al., 2001; Wu et al., 2002). When GLUT11, transiently expressed in HEK293 cells, was reconstituted in liposomes, glucose transport activity was detected and GLUT11 seemed to have low affinity for this substrate (Doege et al., 2001). This property has been verified for the splice variants GLUT11-A and -B. When measured in *Xenopus laevis* oocytes, GLUT11 showed affinity for glucose with a Km of 0.16 mM (Manolescu et al., 2007b). GLUT11 also transports fructose but not galactose (Manolescu et al., 2007b) and exhibits a rather low affinity for cytochalasin B (Doege et al., 2001). Human GLUT11 is expressed exclusively in slow-twitch muscle fibers and is unaffected by physiological and pathophysiological conditions except in primary myopathy (Gaster et al., 2004). Because of its specific expression in muscle, the protein might play a role in the regulation of the glucose homeostasis (Uldry and Thorens, 2004). Surprisingly, the human SLC2A11 gene has not orthologue in the rat and mouse genomes (Scheepers et al., 2005).

Class III Family Members

GLUT isoforms of class III family share unique characteristics. Structurally all isoforms carry their N-glycosylation side at the fifth extracellular loop and bear an internalization signal (dileucine or YSRI in case of GLUT10) that retains these transporters, under steady state conditions, to an intracellular localization. Yet, a translocation stimulus of class III members leading to plasma membrane localization has only been proposed for GLUT13 (Uldry et al., 2004), which remains to be confirmed (Di Daniel et al., 2009).

GLUT6 (SLC2A6)

GLUT6 (formerly designated GLUT9) was cloned from human leucoyte cDNA by (RACE)-PCR from murine ESTs and a human genomic sequence. Human GLUT6 mRNA is expressed predominantly in the brain, spleen and peripheral leucocytes. When reconstituted in liposomes, GLUT6 transport activity was found only in the presence of 5 mM but not 1 mM glucose and displays a low cytochalasin B binding affinity. However, when GLUT6, mutated for its N-terminal dileucine internalization motif, was expressed in *Xenopus* oocytes, it was expressed at the cell surface but no transport activity for glucose was detected over a wide range of concentrations, nor for fructose, galactose and mannose (Uldry and Thorens, 2004). As stated before, GLUT6 contains an N-terminal dileucine motif, which is responsible for intracellular retention of the protein when overexpressed in primary rat adipocytes. GLUT6 is only detected at the plasma membrane when the dileucine residues are mutated to alanine or when clathrin dependent endocytosis is blocked by overexpression of a dominant-negative dynamin (a GTPase responsible for endocytosis) mutant. Nevertheless, no cell-surface translocation of GLUT6 was observed in response to stimuli such as insulin, phorbol ester, or hyperosmolarity (Lisinski et al., 2001).

GLUT8 (SLC2A8)

GLUT8 (formerly GLUTX1) was the first member of the SLC2A family to be isolated by database mining (Doege et al., 2000; Ibberson et al., 2000). It is mainly expressed in the testis, but lower levels are found also in the brain (cerebellum), liver, spleen, lung, adrenal gland and brown adipose tissue (Ibberson et al., 2000). Functional characterization of GLUT8 in *Xenopus laevis* oocytes revealed the intracellular retention of the transporter due to the presence of a dileucine based motif in its N-terminal end. Mutations of the two leucine residues leaded to plasma membrane localization of GLUT8 (Ibberson et al., 2000), which allowed the determination of transport kinetics for the protein. GLUT8 shows a high affinity for glucose (Km of 2 mM) and the transport could be competed with fructose, galactose and cytochalasin B (Ibberson et al., 2000). The intracellular localization of GLUT8 raised the question whether it would be insulin responsive and, thereby, compensate for GLUT4 in the respective knockout mice that show near normal glucose tolerance (Katz et al., 1995). Indeed, in mice blastocysts, GLUT8 has been found to account for insulin-stimulated glucose uptake (Carayannopoulos et al., 2000). GLUT8 may play a crucial role in glucose metabolism of blastocysts since its suppression by antisense oligonucleotides leads to an increased rate of apoptosis (Pinto et al., 2002). However, extensive studies performed failed to identify a stimulus that leads to a plasma membrane translocation of GLUT8 (Widmer et al., 2005; Lisinski et al., 2001; Augustin et al., 2005; Shin et al., 2004). The intracellular retention signal of GLUT8 contains the consensus sequence [D/E]XXXL[L/I] that represents a late-endosomal/lysosomal sorting motif (Augustin et al., 2005). The DEXXXLL sorting signal of GLUT8 interacts with the adaptor proteins AP-1 and AP-2 and controls trafficking of the protein (Diril et al., 2009; Schmidt et al., 2006). Interestingly, both GLUT8 and GLUT12 contain a [D/E]XXXL[L/I] sorting signal. However, it seems that the exact composition of the XXX residues is responsible to adjust the routing and trafficking of the two proteins (Flessner and Moley, 2009).

GLUT8 deficiency is associated with behavioral alterations indicating a significant physiological role for this isoform in the central nervous system (Membrez et al., 2006).

Further, GLUT8 knockout mice exhibit normal growth, body weight development and glycemic control, demonstrating that GLUT8 may not play a critical role for maintenance of whole body glucose homeostasis (Gawlik et al., 2008). However, its cellular function and physiological importance remain thus far unknown (Schmidt et al., 2009).

GLUT10 (SLC2A10)

GLUT10 has been cloned from human liver cDNA based on an EST identified via homology searches with known GLUT protein sequences (Mc Vie-Wylie et al., 2001). GLUT10 mRNA was detected in human heart, lung, brain, liver, skeletal muscle, kidney, pancreas, and placenta (Mc Vie-Wylie et al., 2001; Dawson et al., 2001). Expression of GLUT10 was also present in human and mouse white adipose tissue as well as adipocyte cell lines (Wood et al., 2003). One notable structural characteristic was the absence of the PESPR motif just after helix 6 that is conserved for all the GLUT isoforms. Heterologous expression of GLUT10 in *Xenopus laevis* oocytes revealed high affinity 2-DG transport (Km ~0.3 mM). 2-DG uptake could be competed with glucose and galactose and inhibited with phloretin (Dawson et al., 2001). Immunocytochemical studies indicated a rather intracellular localization for GLUT10 under steady-state conditions. The presence of the potential internalization motif YSRI at the C-terminal end of the transporter supported these findings (Coucke et al., 2006). Interestingly, the chromosome localization (20q12-13.1) of GLUT10 is in a loci associated with T2D in Caucasian Americans and is thus a possible candidate for a susceptibility gene involved in this disease (Dawson et al., 2001).

GLUT12 (SLC2A12)

GLUT12 cDNA was cloned from a human embryonic cDNA library and detected in breast cancer cells. It is expressed in skeletal muscle, heart, small intestine, and prostate (Rogers et al., 2002). GLUT12 showed glucose transport when functionally characterized in *Xenopus laevis* oocytes that can be competed with fructose, galactose, 2-DG, and cytochalasin B. However, thus far, the affinity of GLUT12 for glucose is unknown. GLUT12, as class III member, contains a dilecuine motif at the N- and C-terminal end of the protein. Endogenous and overexpressed GLUT12 protein localizes to intracellular compartments and to the plasma membrane in various cell lines (Flessner and Moley, 2009; Rogers et al., 2002; Chandler et al., 2003). As described above, the N-terminal dileucine signal of GLUT12 is similar to that of GLUT8 and represents a late endosomal/ lysosomal sorting signal. However, GLUT12 does not co-localize with GLUT8, but rather resides in the Golgi network and at the plasma membrane (Flessner and Moley, 2009). Plasma membrane-associated GLUT12 is not endocytosed, suggesting the absence of a continuous cycling mechanism for GLUT12. In human skeletal muscle, insulin induces an increase in plasma membrane GLUT12 which is comparable to the insulin-stimulated GLUT4 translocation (Stuart et al., 2009). Although GLUT 12 expression in skeletal muscle is unaffected under pathophysiological conditions such as of obesity and T2D (Gaster et al., 2004), the data indicate that an additional transporter is expressed in human muscle that is insulin-responsive in a PI3-kinase dependent manner (Stuart et al., 2009).

GLUT13; HMIT (SLC2A13)

Human and rat HMIT (SLC2A13) cDNAs were cloned by screening spleen and frontal cortex cDNA libraries from both species (Uldry et al., 2001). Its deduced amino acid sequence presents all motifs important for glucose transport activity. HMIT is predominantly expressed in the brain, while lower expression levels have been detected in the white, brown and epididymal adipose tissues and in the kidney. As for other class III isoforms, HMIT is restricted to an intracellular location. HMIT protein functional characterization in *Xenopus* oocytes and in mammalian cells has been possible by introducing various mutations that yielded significant plasma membrane expression. Surprisingly, no sugar transport activity has been found for HMIT. Instead, it has been identified as a H^+-coupled *myo*-inositol symporter (Km ~0.1 mM). In addition, HMIT has been shown to transport inositol-3-phosphate (IP3) (Di Daniel et al., 2009) and is inhibited by the common GLUT inhibitors phloretin, phlorizin, and cytochalasin B, although at high concentrations (Uldry et al., 2001).

Translocation of HMIT to the plasma membrane has been shown both in glial cells and primary neurons upon depolarization or PKC activation, resulting in functional HMIT at the plasma membrane and evidenced by increased *myo*-inositol uptake in those cells (Uldry et al., 2004). However, no stimulus that induces plasma membrane translocation of the transporter could be observed (Di Daniel et al., 2009). Because of its predominant expression in the brain compared to two other *myo*-inositol transporters that are sodium-coupled (SMIT1 and SMIT2), interest has been raised as to whether HMIT might play a role in the regulation of *myo*-inositol/ phosphatidylinositol physiology in neurons. Phosphatidylinositol and inositol polyphosphates are key regulators of various signaling pathways and processes, in particular, vesicle endo- and exocytosis, polymerization of microfilaments, regulation of ion channels and signal transduction by Gq-coupled receptors (Cremona and De Camilli, 2001).

Glucose Metabolism and Bone

The skeleton is a highly metabolic tissue acknowledged as an important player in the coordination of global energy use through its hormonal interactions with other tissues (Fukumoto and Martin, 2009; Lee and Karsenty, 2008). The first hint in the search for mechanisms that could mediate a co-regulation of bone and energy balance originated from two familiar clinical observations namely, that bone remodeling is dysregulated upon depletion of sex-steroids, and that obesity protects against the bone loss accompanied this dysregulation (Wei and Ducy, 2010). Considering these observations together they suggest, in molecular terms, that bone mass, body weight and gonadal function could be regulated by one or several common molecules (Fulzele and Clemens, 2012).

Osteocalcin

The discovery that the skeleton participated in global energy homeostasis occurred when leptin (a 16 kDa hormone secreted by adipocytes) showed to alter bone mass through a hypothalamic-osteoblast endocrine loop (Ducy et al., 2000). This conceptual breakthrough implied that bone cells might secrete additional hormones that might control energy metabolism (Fulzele and Clemens, 2012). Searching for such chemical signals, mouse genome profiling (Lee et al., 2007) identified two genes whose expression was mainly restricted to osteoblasts. The first gene encoded the well-known osteoblast protein osteocalcin, already on the list of candidates; because osteocalcin knock-out mice were fatter than their wild-type littermates (Ducy et al., 1996). Like other hormones, osteocalcin is initially synthesized as pre-pro-osteocalcin and then processed into pro-osteocalcin in the endoplasmic reticulum. Before its secretion by osteoblasts, osteocalcin undergoes vitamin-K-dependent carboxylation that confers greater affinity for bone matrix (Hauschka et al., 1989). Nevertheless, a minor portion of osteocalcin remains undercarboxylated and is secreted into the circulation (Delmas et al., 1983).

Importantly, cell culture experiments demonstrated that osteocalcin acts directly on β-cells and adipocytes to regulate insulin secretion and/ or gene expression (Lee et al., 2007; Ferron et al., 2008). These studies also demonstrated that it is the uncarboxylated form of osteocalcin (ucOC) that appears to act as a hormone and, accordingly, injecting mice with warfarin, an inhibitor of Vitamin K-dependent c-carboxylation, caused a decline in blood glucose concentration (Lee et al., 2007; Hinoi et al., 2008). In this regard, osteocalcin seems to represent the first bone-derived energy regulating hormone sufficiently important that it requires insulin to regulate its production in a feed-forward loop (Motyl et al., 2010).

Although there is clear evidence that bone-derived ucOC functions as a hormone to regulate energy metabolism, the osteocalcin receptor remained unknown until very recently, when a study revealed that, in addition to regulating energy metabolism, osteocalcin binds to a specific receptor present on Leydig cells of the testes to favor testosterone synthesis (Oury et al., 2011). In addition to studies in rodents, a growing body of evidence in the clinical literature has confirmed a favorable link between osteocalcin and energy balance, implying that the metabolic functions of the skeleton are retained in humans (Silva and Kousteni, 2012).

Diseases of dysregulated glucose handling, such as T2D, have been for years connected with defects in bone strength or mass. Osteocalcin acts as a hormone that communicates the signal of the osteoblasts to the peripheral tissues that control energy metabolism: pancreas, liver, muscle, and white adipose tissue, while it favors insulin secretion and sensitivity, proliferation of β-cells, and energy expenditure (Silva and Kousteni, 2012). Further, insulin signaling in osteoblasts is a positive regulator of osteocalcin production and activity, affirming the existence of a pancreas-bone feedback loop.

At the molecular level, the widely expressed FoxO1- and the osteoblast-enriched ATF4 transcription factor have been shown to control energy homeostasis by regulating the activity of osteocalcin through their expression in osteoblasts (Rached et al., 2010; Yoshizawa et al., 2009). FoxO1 (one of the four FoxO isoforms of the Forkhead family of transcription factors) is highly expressed in insulin responsive tissues, including pancreas, liver, skeletal muscle, white and brown adipose tissue, and hypothalamus. In these tissues, it is the main transcriptional modulator of insulin actions, harmonizing the transcriptional cascades that regulate glucose metabolism. Insulin down-regulates FoxO1 activity by inducing its phosphorylation, which results in nuclear exclusion of FoxO1 and suppression of its transcriptional activity (Kousteni, 2011; Kousteni, 2012). Besides its role in regulating glucose metabolism in the classical insulin-target organs, FoxO1 was shown to influence the energy handling through its expression in osteoblasts (Rached et al., 2010; Silva and Kousteni, 2012).

Regulation of Energy Metabolism by the Skeleton in Humans

Although studies in rodents identified the ucOC as the regulator of glucose metabolism and fat mass, the majority of clinical studies evaluated the correlation of total osteocalcin levels with metabolic markers in human subjects. Only recently, a few studies evaluated the levels of circulating ucOC and, in fact, confirmed that not only total but also ucOC is associated with energy metabolism in humans (Kanazawa et al., 2011; Pollock et al., 2011).

Evidence that the skeleton, in mice, acts as an endocrine organ to control glucose balance has stimulated the search of this link in humans.

Affirmation of this matter was provided by a study respecting Esp gene and insulin signaling (Ferron et al., 2010). The Esp gene encodes the osteotesticular protein tyrosine phosphatase (OST-PTP) whose expression is stimulated during osteoblast differentiation *in vitro*, and downregulated in mineralizing osteoblasts (Mauro et al., 1994; Wheeler et al., 2002).

Since the ESP is a pseudogene in humans (Cousin et al., 2004), the hypothesis tested was that PTP1B, a tyrosine phosphatase expressed in human osteoblasts, could be the functional homologue of Esp in this species. Indeed, PTP1B acts in the same way as the OST-PTP (the product of the Esp gene) in mice, and induces dephosphorylation of the insulin receptor. Additionally, examination of subjects with osteopetrosis demonstrated that the deficient bone resorption observed was associated with decreased ucOC and serum insulin levels after feeding (Ferron et al., 2010). Moreover, a growing number of clinical studies have shown a link between osteocalcin and energy metabolism in adults and children, healthy or affected by conditions characterized by increased IR (Silva and Kousteni, 2012).

Adiponectin and Undercarboxylated Osteocalcin

Studies have shown that adiponectin may be responsible for the effects of ucOC on insulin sensitivity (Motyl et al., 2010; Lee et al., 2007). Adiponectin is an adipocytokine that is elevated in subjects with better insulin sensitivity and reduced with diabetes, opposite to the inflammatory adipocytokines, such as TNF-α, IL-6 and plasminogen activator inhibitor (PAI)-1, that are increased with IR (Ronti et al., 2006). Adiponectin may also protect against development of diabetes (Mather et al., 2008; Nakashima et al., 2006).

Furthermore, positive correlations between serum adiponectin and osteocalcin concentrations have been noted for non-diabetic women and men (Richards et al., 2007; Fernandez-Real et al., 2008), while in type-2 diabetic patients; adiponectin was correlated with ucOC/OC in men but not with OC or ucOC (Kanazawa et al., 2011). In women, adiponectin was correlated with OC, but not ucOC or ucOC/OC. During improved glycemic control in diabetic patients, increases in OC but not in ucOC were associated with decreases in adiponectin (Kanazawa et al., 2009).

Other Hormones Affecting Glucose Handling in Bone Cells

In bone cells, insulin stimulates the transport of 2DG (Ituarte et al., 1989), but glucose transport could not be attributed to a specific GLUT, although more recent findings indicated that both GLUT1 and GLUT3 (mRNA and protein) were expressed in rat osteogenic sarcoma cells (Thomas et al., 1996). GLUT3 was also detected in bone, and it was suggested that insulin regulates GLUT3-mediated glucose uptake in bone cells (Zoidis et al., 2011). Furthermore, IGF I has been shown to imitate the stimulatory effect on glycogen synthesis from [^{14}C] glucose in rat osteoblast-like cells and was more potent than insulin in this respect (Schmid and Ernst, 1992; Schmid et al., 1982). Moreover, it has been shown that insulin stimulated only at relatively high doses glucose uptake and metabolism in non-osteosarcoma osteoblastic cells (Zoidis et al., 2011; Schmid et al., 1983; Froesch et al., 1985), although insulin receptors in osteoblasts are considered to play an important role in energy homeostasis (apparently, via an osteocalcin-dependent mechanism) (Ferron et al., 2010; Fulzele et al., 2010).

The major portion of glucose taken up by bone cells is metabolized to lactate and CO_2. Parathyroid hormone (PTH) was found to increase lactate production and to decrease citrate decarboxylation by bone and osteoblastic cells (Wong et al., 1977; Borle et al., 1960a, 1960b; Luben et al., 1976; Felix et al., 1978). Moreover, it stimulated 2DG transport in rat osteogenic sarcoma cells after 1 h, and inhibited 2DG transport after 16 h (Thomas et al., 1995). Further, PTH increased [^{14}C] glucose incorporation into glycogen in cultured rat calvaria cells but not in fibroblasts (Schmid et al., 1982).

Current studies showed that IGF I stimulated glucose transport in rat bone-derived osteoblastic (PyMS) cells and that it was effective at much lower concentrations than insulin, not only regarding glycogen and DNA synthesis but also with regard to enhancing glucose uptake: 10–20 times higher concentrations of insulin than of IGF I were required for significant stimulation of DNA synthesis and 2DG uptake; even 50–100 times higher concentrations of insulin than of IGF I were needed for half maximal stimulation of DNA synthesis and 2DG uptake (Zoidis et al., 2011). Noteworthy, these studies revealed that GLUT4 was absent in osteoblastic cells, and insulin seemed unlikely to play a physiological role in stimulating glucose transport in bone tissue.

Sine bone is a metabolically highly active tissue which requires a large amount of energy for remodeling; it becomes an important player in coordinating global energy utilization. In view of an increasing number of investigations and the ubiquitous expression of enzymes activating and inactivating thyroid hormones and of thyroid hormone receptors, there have been several proponents of the relative importance of systemic and local generation, activity and inactivation of thyroid hormones, in the brain, liver, bone, and elsewhere. Some studies addressed glucose uptake regulation in bone cells by triiodothyronine (T3) (Zoidis et al., 2012). T3 may directly affect most tissues, although a recent study has suggested that the well-known effects of thyroid hormones to increase energy expenditure may be predominantly explained via the brain (López et al., 2010). Nevertheless, T3 increased 2DG uptake in PyMS cells and at the same time, T3 at 1 nM increased the abundance of GLUT1 and GLUT3 in membranes prepared from osteoblastic PyMS cells. Therefore, increased glucose uptake induced by T3 in osteoblasts may be mediated by the high-affinity glucose transporters GLUT1 and GLUT3.

Considering that thyroid hormones increase metabolism (Lanni et al., 2003) and bone turnover, it was speculated that the cells meet their increased energy demand by increasing glucose uptake. *In vivo*, T3 has been shown to stimulate metabolic rate with a concomitant increase in fuel (lipids as well as carbohydrates) oxidation. T3 also stimulates numerous genes and proteins involved in substrate/energy metabolism, including proteins involved in glycolysis and fatty acid oxidation, and uncoupling proteins (Lebon et al., 2001; Clement et al., 2002; Visser et al., 2009). Considering overall (whole body) metabolism, liver and skeletal muscles are particularly important target tissues of T3. In skeletal muscle, T3 increases the phosphorylation of both AMPK (an enzyme central for cellular energy status, maintaining the balance between ATP production and consumption, as stated before) and of Akt/PKB and, by activating these signal pathways, stimulates both fatty acid and glucose metabolism. *In vivo*, T3 also increases serum-free fatty acids (De Lange et al., 2008). It was further speculated that T3 may target mitochondria and act on uncoupling proteins (UCP) (Lanni et al., 2003; Lebon et al., 2001), while increased uptake of glucose would be required to meet the elevated energy demand. Proteins uncoupling oxygen consumption from ATP generation were first discovered and their role (catecholamine and thyroid hormone dependency) characterized in more detail in brown adipose tissue (UCP-1, dissipating energy as heat), but T3 has been found to regulate UCP-2 expression in a variety of tissues including adipose tissue, heart and skeletal muscle, liver, and lung (Xiao et al., 2004). Therefore, T3 may also increase UCP expression in bone although this issue has not been addressed in these studies (Zoidis et al., 2012).

Conclusion

Glucose is the essential energy source for many cells and tissues. Its homeostasis, uptake and transport are processes tightly regulated via a complex interplay between pancreatic β- and a-cells, associated organs (e.g., liver, skeletal muscle, intestine, adipose tissue, and bone), transcription factors, and hormones (i.e., insulin, glucagon, T3, osteocalcin and others). Beyond these primary sites of glucose regulation, a multitude of transport proteins (SGLTs and GLUTs) are responsible for transepithelial glucose transport through otherwise impermeable cellular membranes. The principal tissues involved in glucose utilization include the brain, muscle, fat, and the skeleton, with muscle comprising the most demanding site of peripheral glucose uptake. Knowledge of the fundamentals of normal glucose homeostasis is substantial to understanding the pathophysiologic derangements that may result from glucose imbalance disorders. Conditions such as insulin resistance, T2D mellitus and the metabolic syndrome are characterized, among others, by an imbalance in glucose handling. Various factors are involved in the development of metabolic disorders, including insulin resistance and insulin signaling, impaired hepatocyte glucose metabolism, mitochondrial dysfunction and others. A continuously better comprehending of these processes should help both researchers and clinicians to better manage these disorders and ameliorate clinical outcomes.

References

Agius L. (2008). Glucokinase and molecular aspects of liver glycogen metabolism. *Biochemical Journal*, 414, 1–18.

Anai M., Ono H., Funaki M., Fukushima Y., Inukai K., Ogihara T., Sakoda H., Onishi Y., Yazaki Y., Kikuchi M., Oka Y., Asano T. (1998). Different subcellular distribution and regulation of expression of insulin receptor substrate (IRS)-3 from those of IRS-1 and IRS-2. *Journal of Biological Chemistry,* 273, 29686–29692.

Anzai N., Ichida K., Jutabha P., Kimura T., Babu E., Jin C.J., Srivastava S., Kitamura K., Hisatome I., Endou H., Sakurai H. (2008). Plasma urate level is directly regulated by a voltage-driven urate efflux transporter URATv1 (SLC2A9) in humans. *Journal of Biological Chemistry*, 283, 26834–26838.

Augustin R. (2012). The protein family of glucose transport facilitators: It's not only about glucose after all. *IUBMB Life*, 62, 315–333.

Augustin R., Carayannopoulos M.O., Dowd L.O., Phay J.E., Moley J.F., Moley K.H. (2004). Identification and characterization of human glucose transporter-like protein-9 (GLUT9): alternative splicing alters trafficking. *Journal of Biological Chemistry*, 279,16229–16236.

Augustin R., Riley J., Moley K.H. (2005). GLUT8 contains a [DE]XXXL[LI] sorting motif and localizes to a late endosomal/lysosomal compartment. *Traffic*, 6, 1196–1212.

Avignon A., Hokayem M., Bisbal C., Lambert K. (2012). Dietary antioxidants: Do they have a role to play in the ongoing fight against abnormal glucose metabolism? *Nutrition*, 28, 715–721.

Balakrishnan A., Stearns A.T., Rhoads D.B., Ashley S.W., Tavakkolizadeh A. (2008). Defining the transcriptional regulation of the intestinal sodium-glucose cotransporter using RNA-interference mediated gene silencing. *Surgery,* 144, 168-173.

Barone S., Fussell S.L., Singh A.K., Lucas F., Xu J., Kim C., Wu X., Yu Y., Amlal H., Seidler U., Zuo J., Soleimani M. (2009). Slc2a5 (Glut5) is essential for the absorption of fructose in the intestine and generation of fructose-induced hypertension. *Journal of Biological Chemistry*, 284, 5056–5066.

Baur J.A., Pearson K.J., Price N.L., Jamieson H.A., Lerin C., Kalra A. (2006). Resveratrol improves health and survival of mice on a high-calorie diet. *Nature,* 444, 337–342.

Bechmann L.P., Hannivoort R.A., GerkenG., Hotamisligil G.S., Trauner M., Canbay A. (2012). The interaction of hepatic lipid and glucose metabolism in liver diseases. *Journal of Hepatology*, 56, 952–964.

Bell G.I., Kayano T., Buse J.B., Burant C.F., Takeda J., Lin D. (1990). Molecular biology of mammalian glucose transporters. *Diabetes Care*, 13, 198-208.

Bibert S., Hess S.K., Firsov D., Thorens B., Geering K., Horisberger J.D., Bonny O. (2009). Mouse GLUT9: evidences for a urate uniporter. *American Journal of Physiology - Renal Physiology*, 297, 612–619.

Birnbaum M.J. (1989). Identification of a novel gene encoding an insulin-responsive glucose transporter protein. *Cell,* 57, 305–315.

Blakemore S.J., Aledo J.C., James J., Campbell F.C., Lucocq J.M., Hundal H.S. (1995). The GLUT5 hexose transporter is also localized to the basolateral membrane of the human jejunum. *Biochemical Journal*, 309, 7–12.

Bloch-Damti A., Bashan N. (2005). Proposed mechanisms for the induction of insulin resistance by oxidative stress. *Antioxidants and Redox Signaling*, 7, 1553–1567.

Borle A.G., Nichols N., Nichols G. Jr. (1960a). Metabolic studies of bone *in vitro* I. Normal bone. *Journal of Biological Chemistry*, 235, 1206–1210.

Borle A.G., Nichols N., Nichols G. Jr. (1960b). Metabolic studies of bone *in vitro* II. The metabolic pattern of accretion and resorption. *Journal of Biological Chemistry,* 235, 1211–1214.

Brasnyó P., Molnár G.A., Mohás M., Markó L., Laczy B., Cseh J. (2011). Resveratrol improves insulin sensitivity, reduces oxidative stress and activates the Akt pathway in type 2 diabetic patients. *British Journal of Nutrition,* 106, 383–389.

Burant C.F., Takeda J., Brot-Laroche E., Bell G.I., Davidson N.O. (1992). Fructose transporter in human spermatozoa and small intestine is GLUT5. *Journal of Biological Chemistry*, 267, 14523–14526.

Burcelin R., Dolci W., Thorens B. (2000). Glucose sensing by the hepatoportal sensor is GLUT2-dependent: *in vivo* analysis in GLUT2-null mice. *Diabetes,* 49, 1643–1648.

Burcelin R., Thorens B. (2001). Evidence that extrapancreatic GLUT2-dependent glucose sensors control glucagon secretion. *Diabetes,* 50, 1282–1289.

Cao R., Cronk Z.X., Zha W., Sun L., Wang X., Fang Y. (2010). Bile acids regulate hepatic gluconeogenic genes and farnesoid X receptor via G(alpha)i-protein-coupled receptors and the AKT pathway. *Journal of Lipid Research,* 51, 2234–2244.

Carayannopoulos M.O., Chi M.M., Cui Y., Pingsterhaus J.M., McKnight R.A., Mueckler, M., Devaskar S.U., Moley K.H. (2000). GLUT8 is a glucose transporter responsible for insulin-stimulated glucose uptake in the blastocyst. *Proceedings of the National Academy of Sciences of the USA,* 97, 7313–7318.

Carriere V., Lacasa M., Rousset M. (2001). Activity of hepatocyte nuclear factor 1alpha and hepatocyte nuclear factor 1beta isoforms is differently affected by the inhibition of protein phosphatases 1/2A. *Biochemical Journal,* 354, 301-308.

Chandler J.D., Williams E.D., Slavin J.L., Best J.D., Rogers S. (2003). Expression and localization of GLUT1 and GLUT12 in prostate carcinoma. *Cancer,* 97, 2035–2042.

Charron M.J., Brosius F.C. III, Alper S.L., Lodish H.F. (1989). A glucose transport protein expressed predominately in insulinresponsive tissues. *Proceedings of the National Academy of Sciences USA*, 86, 2535–2539.

Chu C.A., Fujimoto Y., Igawa K., Grimsby J., Grippo J.F., Magnuson M.A. (2004). Rapid translocation of hepatic glucokinase in response to intraduodenal glucose infusion and changes in plasma glucose and insulin in conscious rats. *American Journal of Physiology - Gastrointestinal and Liver Physiology,* 286, 627–634.

Chuang C.C., Martinez K., Xie G., Kennedy A., Bumrungpert A., Overman A. (2010). Quercetin is equally or more effective than resveratrol in attenuating tumor necrosis factor-a–mediated inflammation and insulin resistance in primary human adipocytes. *American Journal of Clinical Nutrition,* 92, 1511–1521.

Ciampi R., Vivaldi A., Romei C., Del Guerra A., Salvadori P., Cosci B., Pinchera A., Elisei R. (2008). Expression analysis of facilitative glucose transporters (GLUTs) in human thyroid carcinoma cell lines and primary tumors. *Molecular and Cellular Endocrinology,* 291, 57–62.

Clement K., Viguerie N., Diehn M., Alizadeh A., Barbe P., Thalamas C., Storey J.D., Brown P.O., Barsh G.S., Langin D. (2002). *In vivo* regulation of human skeletal muscle gene expression by thyroid hormone. *Genome Research,* 12, 281–291.

Colville C.A., Seatter M.J., Jess T.J., Gould G.W., Thomas H.M. (1993). Kinetic analysis of the liver-type (GLUT2) and braintype (GLUT3) glucose transporters in *Xenopus* oocytes: substrate specificities and effects of transport inhibitors. *Biochemical Journal*, 290, 701–706.

Conti R., Mannucci E., Pessotto P., Tassoni E., Carminati P., Giannessi F. (2011). Selective reversible inhibition of liver carnitine palmitoyl-transferase 1 by teglicar reduces gluconeogenesis and improves glucose homeostasis. *Diabetes,* 60, 644–651.

Corpe C.P., Bovelander F.J., Munoz C.M., Hoekstra J.H., Simpson I.A., Kwon O., Levine M., Burant, C.F. (2002). Cloning and functional characterization of the mouse fructose transporter, GLUT5. *Biochimica et Biophysica Acta,* 1576, 191–197.

Coucke P.J., Willaert A., Wessels M.W., Callewaert B., Zoppi N., De Backer J., Fox J.E., Mancini G.M., Kambouris M., Gardella R., Facchetti F., Willems P.J., Forsyth R., Dietz H.C., Barlati S., Colombi M., Loeys B., De Paepe A. (2006). Mutations in the facilitative glucose transporter GLUT10 alter angiogenesis and cause arterial tortuosity syndrome. *Nature Genetics,* 38, 452–457.

Cousin W., Courseaux A., Ladoux A, Dani C., Peraldi P. (2004). Cloning of hOST-PTP: the only example of a protein-tyrosine-phosphatase the function of which has been lost between rodent and human. *Biochemical and Biophysical Research Communications*, 321, 259–265.

Crane R.K., Miller D., Bihler I. (1960). The restrictions on possible mechanism of intestinal active transport of sugars. *Proceedings of the Membrane Transport and Metabolism Symposium.* Prague, August 22-27, 1960. Czech Academy of Sciences, 439-449.

Cremona O., De Camilli P. (2001). Phosphoinositides in membrane traffic at the synapse. *Journal of Cell Science*, 114, 1041–1052.

Dawson P.A., Mychaleckyj J.C., Fossey S.C., Mihic S.J., Craddock A.L., Bowden D.W. (2001). Sequence and functional analysis of GLUT10: a glucose transporter in the Type 2 diabetes-linked region of chromosome 20q12–13.1. *Molecular Genetics and Metabolism*, 74, 186–199.

de Kreutzenberg S.V., Ceolotto G., Papparella I., Bortoluzzi A., Semplicini A., Dalla Man C. (2010). Downregulation of the longevity-associated protein sirtuin 1 in insulin resistance and metabolic syndrome: potential biochemical mechanisms. *Diabetes*, 59, 1006–1015.

De Lange P., Senese R., Cioffi F., Moreno M., Lombardi A., Silvestri E., Goglia F., Lanni A. (2008). Rapid activation by 3,5,3'-L-triiodothyronine of adenosine 5'-monophosphate-activated protein kinase/acetyl-coenzyme A carboxylase and Akt/protein kinase B signaling pathways: relation to changes in fuel metabolism and myosin heavy-chain protein content in rat gastrocnemius muscle *in vivo*. *Endocrinology*, 149, 6462–6470.

De Vivo D.C., Trifiletti R.R., Jacobson R.I., Ronen G.M., Behmand R.A., and Harik S.I. (1991). Defective glucose transport across the blood-brain barrier as a cause of persistent hypoglycorrhachia, seizures, and developmental delay. *New England Journal of Medicine*, 325, 703–709.

De Vos A., Heimberg H., Quartier E., Huypens P., Bouwens L., Pipeleers D., Schuit F. (1995). Human and rat beta cells differ in glucose transporter but not in glucokinase gene expression. *Journal of Clinical Investigation*, 96, 2489–2495.

DeFronzo R.A. (2004). Diabetes: pathogenesis of type-2 diabetes mellitus. *Medical Clinics of North America*, 88, 787-835.

DeFronzo R.A. (2009). Banting Lecture. From the triumvirate to the ominous octet: a new paradigm for the treatment of type-2 diabetes mellitus. *Diabetes*, 58, 773-795.

Delmas P.D., Wilson D.M., Mann K.G., Riggs B.L. (1983). Effect of renal function on plasma levels of bone Gla-protein. *Journal of Clinical Endocrinology and Metabolism*, 57, 1028–1030.

Dentin R., Girard J., Postic C. (2005). Carbohydrate responsive element binding protein (ChREBP) and sterol regulatory element binding protein-1c (SREBP-1c): two key regulators of glucose metabolism and lipid synthesis in liver. *Biochimie*, 87, 81–86.

Di Daniel E., Mok M.H., Mead E., Mutinelli C., Zambello E., Caberlotto L.L., Pell T.J., Langmead C.J., Shah A., Duddy G., Kew J.N., Maycox P.R. (2009). Evaluation of expression and function of the H1/myo-inositol transporter HMIT. *BMC Cell Biology*, 10, 54.

Diez-Sampedro A., Lostao M.P., Wright E.M., Hirayama B.A. (2000). Glycoside binding and translocation in Na^+-dependent glucosecotransporters: comparison of SGLT1 and SGLT3. *Journal of Membrane Biology*, 176, 111-117.

Ding L., Jin D., Chen X. (2010). Luteolin enhances insulin sensitivity via activation of PPARγ transcriptional activity in adipocytes. *The Journal of Nutritional Biochemistry*, 21, 941–947.

Diril M.K., Schmidt S., Krauss M., Gawlik V., Joost H.G., Schurmann A., Haucke V., Augustin R. (2009). Lysosomal localization of GLUT8 in the testis—the EXXXLL motif of GLUT8 is sufficient for its intracellular sorting via AP1- and AP2-mediated interaction. *FEBS Journal*, 276, 3729–3743.

Doege H., Bocianski A., Scheepers A., Axer H., Eckel J., Joost H.G., Schurmann A. (2001). Characterization of human glucose transporter (GLUT) 11 (encoded by SLC2A11), a novel sugar-transport facilitator specifically expressed in heart and skeletal muscle. *Biochemical Journal*, 359, 443–449.

Doege H., Schurmann A., Bahrenberg G., Brauers A., Joost H.G. (2000). GLUT8, a novel member of the sugar transport facilitator family with glucose transport activity. *Journal of Biological Chemistry*, 275, 16275–16280.

Donovan C.M. (2002). Portal vein glucose sensing. *Diabetes, Nutrition and Metabolism*, 15, 308–312.

Duchen M.R. (2004). Roles of mitochondria in health and disease. *Diabetes*, 53, 96–102.

Ducy P., Amling M., Takeda S., Priemel M., Schilling A.F., Beil F.T. (2000). Leptin inhibits bone formation through a hypothalamic relay: a central control of bone mass. *Cell*, 100, 197–207.

Ducy P., Desbois C., Boyce B., Pinero G., Story B., Dunstan C. (1996). Increased bone formation in osteocalcin-deficient mice. *Nature*, 382, 448–452.

Dyer J., Daly K., Salmon K.S., Arora D.K., Kokrashvili Z., Margolskee R.F. (2007). Intestinal glucose sensing and regulation of intestinal glucose absorption. *Biochemical Society Transactions*, 35, 1191-1194.

Eid H.M., Vallerand D., Muhammad A., Durst T., Haddad P.S., Martineau L.C. (2010). Structural constraints and the importance of lipophilicity for the mitochondrial uncoupling activity of naturally occurring caffeic acid esters with potential for the treatment of insulin resistance. *Biochemical Pharmacology*, 79, 444–454.

Evans S.A., Doblado M., Chi M.M., Corbett J.A., Moley K.H. (2009). Facilitative Glucose Transporter 9 Expression Affects Glucose Sensing in Pancreatic β-Cells. *Endocrinology,* 150, 5302–5310.

Felix R., Neuman W.F., Fleisch H. (1978). Aerobic glycolysis in bone: lactic acid production by rat calvaria cells in culture. *American Journal of Physiology,* 234, 51–55.

Fernandez-Real J.M., Izquierdo M., Ortega F., Gorostiaga E., Gomez-Ambrosi J., Moreno-Navarrete J.M., Fruhbeck G., Martinez C., Idoate F., Salvador J., Forga L., Ricart W., Ibanez J. (2008). The relationship of serum osteocalcin concentration to insulin secretion, sensitivity, and disposal with hypocaloric diet and resistance training. *Journal of Clinical Endocrinology and Metabolism*, 94, 237–245.

Ferron M., Hinoi E., Karsenty G., Ducy P. (2008). Osteocalcin differentially regulates beta cell and adipocyte gene expression and affects the development of metabolic diseases in wild-type mice. *Proceedings of the National Academy of Sciences of the USA*, 105, 5266–5270.

Ferron M., Wei J., Yoshizawa T., Del Fattore A., DePinho R.A., Teti A., Ducy P., Karsenty G. (2010). Insulin signaling in osteoblasts integrates bone remodeling and energy metabolism. *Cell,* 142, 296–308.

Flessner L.B., Moley K.H. (2009). Similar [DE]XXXL[LI] motifs differentially target GLUT8 and GLUT12 in Chinese hamster ovary cells. *Traffic* 10, 324–333.

Flier J.S., Mueckler M.M., Usher P., Lodish H.F. (1987). Elevated levels of glucose transport and transporter messenger RNA are induced by ras or src oncogenes. *Science,* 235, 1492–1495.

Freitas H.S., Anhe G.F., Melo K.F., Okamoto M.M., Oliveira-Souza M., Bordin S. (2008). Na^+-glucose transporter-2 messenger ribonucleic acid expression in kidney of diabetic rats correlates with glycemic levels: involvement of hepatocyte nuclear factor-1α expression and activity. *Endocrinology,* 149, 717-724.

Friedel S., Antwerpen B., Hoch A., Vogel C., Grassl W., Geller F., Hebebrand J., Hinney A. (2002). Glucose transporter 4 gene: association studies pertaining to alleles of two polymorphisms in extremely obese children and adolescents and in normal and underweight controls. *Annals of the New York Academy of Sciences,* 967, 554–557.

Friedlander G., Runembert I., Vrtovsnik F., Terzi F. (1999). Renal tubular cells cultured from genetically modified animals. Experimental Nephrology, 7, 407-412.

Froesch E.R., Schmid C., Schwander J., Zapf J. (1985). Actions of insulin-like growth factors. *Annual Review of Physiology*, 47, 443–467.

Fukumoto H., Kayano T., Buse J.B., Edwards Y., Pilch P.F., Bell G.I., Seino S. (1989). Cloning and characterization of the major insulin-responsive glucose transporter expressed in human skeletal muscle and other insulin-responsive tissues. *Journal of Biological Chemistry*, 264, 7776–7779.

Fukumoto H., Seino S., Imura H., Seino Y., Eddy R.L., Fukushima Y., Byers M.G. Shows T.B., Bell G.I. (1988). Sequence, tissue distribution, and chromosomal localization of mRNA encoding a human glucose transporter-like protein. *Proceedings of the National Academy of Sciences USA*, 85, 5434–5438.

Fukumoto S., Martin T.J. (2009). Bone as an endocrine organ. *Trends in Endocrinology and Metabolism*, 20, 230–236.

Fulzele K., Clemens T.L. (2012). Novel functions for insulin in bone. *Bone*, 50, 452-456.

Fulzele K., Riddle R.C., DiGirolamo D.J., Cao X., Wan C., Chen D., Faugere M.-C., Aja S., Hussain M.A., Brüning J.C., Clemens T.L. (2010). Insulin receptor signaling in osteoblasts regulates postnatal bone acquisition and body composition. *Cell,* 142, 309–319.

Ganguly, A., Devaskar, S.U. (2008) Glucose transporter isoform-3-null heterozygous mutation causes sexually dimorphic adiposity with insulin resistance. *American Journal of Physiology - Endocrinology and Metabolism*, 294, 1144–1151.

Gaster M., Handberg A., Schurmann A., Joost H. G., Beck-Nielsen H., Schroder, H.D. (2004). GLUT11, but not GLUT8 or GLUT12, is expressed in human skeletal muscle in a fibre type-specific pattern. *Pflügers Archiv.,* 448, 105–113.

Gautier-Stein A., Zitoun C., Lalli E., Mithieux G., Rajas F. (2006). Transcriptional regulation of the glucose-6-phosphatase gene by cAMP/vasoactive intestinal peptide in the intestine. Role of HNF4alpha, CREM, HNF1alpha, and C/EBPalpha. *Journal of Biological Chemistry,* 281, 31268-31278.

Gawlik V., Schmidt S., Scheepers A., Wennemuth G., Augustin R., Aumuller G., Moser M., Al-Hasani H., Kluge R., Joost H.G., Schurmann A. (2008) Targeted disruption of Slc2a8 (GLUT8) reduces motility and mitochondrial potential of spermatozoa. *Molecular Membrane Biology,* 25, 224–235.

Gerich J.E. (1993). Control of glycemia. *Bailliere's Clinical Endocrinology and Metabolism,* 7, 551–586.

Guillam M.T., Dupraz P., Thorens B. (2000). Glucose uptake, utilization, and signaling in GLUT2-null islets. *Diabetes*, 49, 1485–1491.

Guyton A.C., Hall J.E. *Textbook of Medical Physiology*. 11th ed. Philadelphia, PA: Elsevier Inc; 2006.

Haber R.S., Weinstein S.P., O'Boyle E., Morgello S. (1993). Tissue distribution of the human GLUT3 glucose transporter. *Endocrinology,* 132, 2538–2543.

Hakimi P., Johnson M.T., Yang J., Lepage D.F., Conlon R.A., Kalhan S.C. (2005). Phosphoenolpyruvate carboxykinase and the critical role of cataplerosis in the control of hepatic metabolism. *Nutrition and Metabolism,* 2:33.

Hauschka P.V., Lian J.B., Cole D.E., Gundberg C.M. (1989). Osteocalcin and matrix Gla protein: vitamin K-dependent proteins in bone. *Physiological Reviews,* 69, 990–1047.

Hayashi Y., Wang W., Ninomiya T., Nagano H., Ohta K., Itoh H. (1999). Liver enriched transcription factors and differentiation of hepatocellular carcinoma. *Molecular Pathology*, 52, 19-24.

Hediger M.A., Rhoads D.B. (1994) Molecular physiology of sodium-glucose cotransporters. *Physiological Reviews*, 74, 993-1026.

Hinoi E., Gao N., Jung D.Y., Yadav V., Yoshizawa T., Myers M.G. Jr., Chua S.C. Jr., Kim J.K., Kaestner K.H., Karsenty G. (2008). The sympathetic tone mediates leptin's inhibition of insulin secretion by modulating osteocalcin bioactivity. *Journal of Cellular Biology,* 183, 1235–1242.

Hirsch J.R., Loo D.D., Wright E.M. (1996). Regulation of Na^+/glucose cotransporter expression by protein kinases in *Xenopus laevis* oocytes. *Journal of Biological Chemistry*, 271, 14740-14746.

Hosokawa M., Thorens, B. (2002). Glucose release from GLUT2-null hepatocytes: characterization of a major and a minor pathway. *American Journal of Physiology - Endocrinology and Metabolism*, 282, 794–801.

Huang D-W., Shen S-C., Wu J.S.-B. (2009). Effects of caffeic acid and cinnamic acid on glucose uptake in insulin-resistant mouse hepatocytes. *Journal of Agricultural Food Chemistry*, 57, 7687–7692.

Huang S., Czech M.P. (2007). The GLUT4 glucose transporter. *Cell Metabolism*, 5, 237–252.

Ibberson M., Uldry M., Thorens B. (2000). GLUTX1, a novel mammalian glucose transporter expressed in the central nervous system and insulin-sensitive tissues. *Journal of Biological Chemistry*, 275, 4607–4612.

Illsley N.P. (2000). Glucose transporters in the human placenta. *Placenta*, 21, 14–22.

Ishikawa Y., Eguchi T., Ishida H. (1997). Mechanism of beta-adrenergic agonist-induced transmural transport of glucose in rat small intestine. Regulation of phosphorylation of SGLT1 controls the function. *Biochimica et Biophysica Acta*, 1357, 306-318.

Ituarte E.A., Halstead L.R., Iida-Klein A., Ituarte H.G., Hahn T.J. (1989). Glucose transport system in UMR-106-01 osteoblastic osteosarcoma cells: regulation by insulin. *Calcified Tissue International*, 45, 27–33.

Johnson J.H., Newgard C.B., Milburn J.L., Lodish H.F., Thorens B. (1990). The high Km glucose transporter of islets of Langerhans is functionally similar to the low affinity transporter of liver and has an identical primary sequence. *Journal of Biological Chemistry*, 265, 6548–6551.

Kaestner K.H., Christy R.J., McLenithan J.C., Braiterman L.T., Cornelius P., Pekala P.H., Lane M.D. (1989). Sequence, tissue distribution, and differential expression of mRNA for a putative insulin-responsive glucose transporter in mouse 3T3-L1 adipocytes. *Proceedings of the National Academy of Sciences USA*, 86, 3150–3154.

Kanazawa I., Yamaguchi T., Yamauchi M., Yamamoto M., Kurioka S., Yano S., Sugimoto T. (2011). Serum undercarboxylated osteocalcin was inversely associated with plasma glucose level and fat mass in type 2 diabetes mellitus. *Osteoporosis International*, 22, 187–194.

Kanazawa I., Yamaguchi T., Yamauchi M., Yamamoto M., Kurioka S., Yano S., Sugimoto T. (2009). Adiponectin is associated with changes in bone markers during glycemic control in type 2 diabetes mellitus. *Journal of Clinical Endocrinology and Metabolism*, 94 3031–3037.

Kannappan S., Anuradha C.V. (2010). Naringenin enhances insulin-stimulated tyrosine phosphorylation and improves the cellular actions of insulin in a dietary model of metabolic syndrome. *European Journal of Nutrition*, 49, 101–109.

Katz E.B., Stenbit A.E., Hatton K., DePinho R., Charron M.J. (1995). Cardiac and adipose tissue abnormalities but not diabetes in mice deficient in GLUT4. *Nature*, 377, 151–155.

Kayano T., Burant C.F., Fukumoto H., Gould G.W., Fan Y.S., Eddy R.L., Byers M.G., Shows T.B., Seino S., Bell G.I. (1990). Human facilitative glucose transporters. Isolation, functional characterization, and gene localization of cDNAs encoding an isoform (GLUT5) expressed in small intestine, kidney, muscle, and adipose tissue and an unusual glucose transporter pseudogenelike sequence (GLUT6). *Journal of Biological Chemistry*, 265, 13276–13282.

Kayano T., Fukumoto H., Eddy R.L., Fan Y.S., Byers M.G., Shows T. B., Bell G.I. (1988). Evidence for a family of human glucose transporter-like proteins. Sequence and gene localization of a protein expressed in fetal skeletal muscle and other tissues. *Journal of Biological Chemistry*, 263, 15245–15248.

Keembiyehetty C., Augustin R., Carayannopoulos M.O., Steer S., Manolescu A., Cheeseman C.I., Moley K.H. (2006). Mouse glucose transporter 9 splice variants are expressed in adult liver and kidney and are up-regulated in diabetes. *Molecular Endocrinology*, 20, 686–697.

Kim J., Wei,Y., Sowers J.R. (2008). Role of mitochondrial dysfunction in insulin resistance. *Circulation Research*, 102, 401–414.

Kim Y.M., Seo Y.H., Park C.B., Yoon S.H., Yoon G. (2010). Roles of GSK3 in metabolic shift toward abnormal anabolism in cell senescence. *Annals of the New York Academy of Sciences,* 1201, 65–71.

Klepper J. (2008). Glucose transporter deficiency syndrome (GLUT1DS) and the ketogenic diet. *Epilepsia,* 49 (Suppl 8), 46–49.

Kong C.T., Yet S.F., Lever J.E. (1993). Cloning and expression of a mammalian Na^+/amino acid cotransporter with sequence similarity to Na^+/glucose cotransporters. *Journal of Biological Chemistry*, 268, 1509-1512.

Kousteni S. (2011). FoxO1: a molecule for all seasons. *Journal of Bone and Mineral Research,* 26, 912–917.

Kousteni S. (2012). FoxO1, the transcriptional chief of staff of energy metabolism. *Bone*, 50, 437–443.

Kriauciunas K.M., Myers Jr. M.G., Kahn C.R. (2000). Cellular compartmentalization in insulin action: altered signaling by a lipid-modified IRS-1. *Molecular and Cellular Biology*, 20, 6849–6859.

Kusari J., Verma U.S., Buse J.B., Henry R.R., Olefsky J.M. (1991). Analysis of the gene sequences of the insulin receptor and the insulin-sensitive glucose transporter (GLUT-4) in patients with common-type non-insulin-dependent diabetes mellitus. *Journal of Clinical Investigation*, 88, 1323–1330.

Lagouge M., Argmann C., Gerhart-Hines Z., Meziane H., Lerin C., Daussin F. (2006). Resveratrol improves mitochondrial function and protects against metabolic disease by activating SIRT1 and PGC-1alpha. *Cell,* 127, 1109–1122.

Lanni A., Moreno M., Lombardi A., Goglia F. (2003). Thyroid hormones and uncoupling proteins. *FEBS Letters*, 543, 5–10.

Lebon V., Dufour S., Petersen K.F., Ren J., Jucker B.M., Slezak L.A., Cline G.W., Rothman D.L., Shulman G.I. (2001). Effect of triiodothyronine on mitochondrial energy coupling in human skeletal muscle. *Journal of Clinical Investigation*, 108, 733–737.

Lee N.K., Karsenty G. (2008). Reciprocal regulation of bone and energy metabolism. Trends in Endocrinology and Metabolism, 19, 161–166. Pirola L., Johnston A.M., Obberghen E. (2004). Modulation of insulin action. *Diabetologia,* 47, 170–184.

Lee N.K., Sowa H., Hinoi E., Ferron M., Ahn J.D., Confavreux C., Dacquin R., Mee P.J., McKee M.D., Jung D.Y, Zhang Z., Kim J.K., Mauvais-Jarvis F., Ducy P., Karsenty G. (2007). Endocrine regulation of energy metabolism by the skeleton. *Cell,* 130, 456–469.

Lescale-Matys L., Dyer J., Scott D., Freeman T.C., Wright E.M., Shirazi-Beechey S.P. (1993). Regulation of the ovine intestinal Na^+/ glucose cotransporter (SGLT1) is dissociated from mRNA abundance. *Biochemical Journal*, 291, 435-440.

Leturque A., Brot-Laroche E., Le Gall M. (2009). GLUT2 mutations, translocation, and receptor function in diet sugar managing. *American Journal of Physiology - Endocrinology and Metabolism*, 296, 985–992.

Levin B.E., Routh V.H., Kang L., Sanders N.M. Dunn-Meynell A.A., (2004). Neuronal glucosensing. What do we know after 50 years? *Diabetes,* 53, 2521–2528.

Li Q., Manolescu A., Ritzel M., Yao S., Slugoski M., Young J.D., Chen X.Z., Cheeseman C.I. (2004). Cloning and functional characterization of the human GLUT7 isoform SLC2A7 from the small intestine. *American Journal of Physiology - Gastrointestinal and Liver Physiology,* 287, 236–242.

Lim J., Lee J., Suh Y., Kim W., Song J., Jung M. (2006). Mitochondrial dysfunction induces aberrant insulin signalling and glucose utilisation in murine C2C12 myotube cells. *Diabetologia,* 49, 1924–1936.

Lisinski I., Schurmann A., Joost H.G., Cushman S.W., Al-Hasani H. (2001). Targeting of GLUT6 (formerly GLUT9) and GLUT8 in rat adipose cells. *Biochemical Journal*, 358, 517–522.

López M., Varela L., Vázquez M.J., Rodríguez-Cuenca S., González C.R., Velagapudi V.R., Morgan D.A., Schoenmakers E., Agassandian K., Lage R., Martínez de Morentin P.B., Tovar S., Nogueiras R., Carling D., Lelliott C., Gallego R., Oresic M., Chatterjee K., Saha A.K., Rahmouni K., Diéguez C., Vidal-Puig A. (2010). Hypothalamic AMPK and fatty acid metabolism mediate thyroid regulation of energy balance. *Nature Medicine,* 16, 1001–1008

Lowell B.B., Shulman G.I. (2005). Mitochondrial dysfunction and type 2 diabetes. *Science,* 307, 384–387.

Luben R.A., Wong G.L., Cohn D.V. (1976). Biochemical characterization with parathormone and calcitonin of isolated bone cells: provisional identification of osteoclasts and osteoblasts. *Endocrinology,* 99, 526–534.

Mackenzie B., Panayotova-Heiermann M., Loo D.D., Lever J.E., Wright E.M. (1994). SAAT1 is a low affinity Na^+/glucose cotransporter and not an amino acid transporter. A reinterpretation. *Journal of Biological Chemistry,* 269, 22488-22491.

Magnuson M.A., Matschinsky F.M. (2004). Glucokinase as a glucose sensor: past, present and future. In: Matschinsky F.M., Magnuson M.A. (Eds.), *Glucokinase and Glycemic Disease: From Basics to Novel Therapeutics.* Krager, Basel, pp. 1–17.

Maher F., Vannucci S.J., Simpson I.A. (1994). Glucose transporter proteins in brain. *FASEB Journal,* 8, 1003–1011.

Manolescu A.R., Augustin R., Moley K., Cheeseman C. (2007a). A highly conserved hydrophobic motif in the exofacial vestibule of fructose transporting SLC2A proteins acts as a critical determinant of their substrate selectivity. *Molecular Membrane Biology,* 24, 455–463.

Manolescu A.R., Salas-Burgos A.M., Fischbarg J., Cheeseman, C.I. (2005). Identification of a hydrophobic residue as a key determinant of fructose transport by the facilitative hexose transporter SLC2A7 (GLUT7). *Journal of Biological Chemistry,* 280, 42978–42983.

Manolescu A.R., Witkowska K., Kinnaird A., Cessford T., Cheeseman C. (2007b). Facilitated hexose transporters: new perspectives on form and function. *Physiology,* 22, 234–240.

Manz F., Bickel H., Brodehl J., Feist D., Gellissen K., Gescholl-Bauer B., Gilli G., Harms E., Helwig H., Nutzenadel W., Waldherr R. (1987). Fanconi-Bickel syndrome. *Pediatric Nephrology*, 1, 509–518.

Margolskee R.F., Dyer J., Kokrashvili Z., Salmon K.S., Ilegems E., Daly K. (2007). T1R3 and gustducin in gut sense sugars to regulate expression of Na^+-glucose cotransporter 1. *Proceedings of the National Academy of Sciences of the USA*, 104, 15075-15080.

Martin M.G., Wang J., Solorzano-Vargas R.S., Lam J.T., Turk E., Wright E.M. (2000). Regulation of the human Na^+-glucose cotransporter gene, SGLT1, by HNF-1 and Sp1. *American Journal of Physiology - Gastrointestinal and Liver Physiology,* 278, 591-603.

Martínez-Quintana J.A., Yepiz-Plascencia G. (2012). Glucose and other hexoses transporters in marine invertebrates: A mini review. *Electronic Journal of Biotechnology* ISSN: 0717-3458 doi: 10.2225/vol15-issue5-fulltext-12.

Mate A., de la Hermosa M.A., Barfull A., Planas J.M., Vazquez C.M. (2001). Characterization of D-fructose transport by rat kidney brush-border membrane vesicles: changes in hypertensive rats. *Cellular and Molecular Life Sciences,* 58, 1961–1967.

Mather K.J., Funahashi T., Matsuzawa Y., Edelstein S., Bray G.A., Kahn S.E., Crandall J., Marcovina S., Goldstein B., Goldberg R. (2008). Adiponectin, change in adiponectin, and progression to diabetes in the Diabetes Prevention Program. *Diabetes,* 57, 980–986.

Matschinsky F.M. (1996). Banting Lecture 1995. A lesson in metabolic regulation inspired by the glucokinase glucose sensor paradigm. *Diabetes,* 45, 223–241.

Mauro L.J., Olmsted E.A., Skrobacz B.M., Mourey R.J., Davis A.R., Dixon J.E. (1994). Identification of a hormonally regulated protein tyrosine phosphatase associated with bone and testicular differentiation. *Journal of Biological Chemistry,* 269, 30659–30667.

McCall A.L., Van Bueren A.M., Moholt-Siebert M., Cherry N.J., Woodward W.R. (1994). Immunohistochemical localization of the neuron-specific glucose transporter (GLUT3) to neuropil in adult rat brain. *Brain Research,* 659, 292–297.

McVie-Wylie A.J., Lamson D.R., Chen Y.T. (2001). Molecular cloning of a novel member of the GLUT family of transporters, SLC2a10 (GLUT10), localized on chromosome 20q13.1: a candidate gene for NIDDM susceptibility. *Genomics,* 72, 113–117.

Melloul D., Marshak S., Cerasi E. (2002). Regulation of insulin gene transcription. *Diabetologia,* 45, 309–326.

Membrez M., Hummler E., Beermann F., Haefliger J.A., Savioz R., Pedrazzini T., Thorens B. (2006). GLUT8 is dispensable for embryonic development but influences hippocampal neurogenesis and heart function. *Molecular and Cellular Biology,* 26, 4268–4276.

Mercader J., Palou A, Bonet M.L. (2011). Resveratrol enhances fatty acid oxidation capacity and reduces resistin and retinol-binding protein 4 expression in white adipocytes. *Journal of Nutritional Biochemistry*, 22, 828–834.

Miyamoto K., Tatsumi S., Morimoto A., Minami H., Yamamoto H., Sone K., Taketani Y., Nakabou Y., Oka T., Takeda E. (1994). Characterization of the rabbit intestinal fructose transporter (GLUT5). *Biochemical Journal,* 303, 877–883.

Mizock B.A. (2001). Alterations in fuel metabolism in critical illness: hyperglycemia. *Best Practice and Research Clinical Endocrinology and Metabolism,* 15, 533–551.

Mogensen M., Sahlin K., Fernstrom M., Glintborg D., Vind B.F., Beck-Nielsen H., Hojlund K. (2007). Mitochondrial respiration is decreased in skeletal muscle of patients with type 2 diabetes. *Diabetes,* 56, 1592–1599.

Moore M.C., Cherrington A.D., Charles H., Best C.H., David H. Wasserman D.H. (2003). Regulation of hepatic and peripheral glucose disposal. *Best Practice and Research Clinical Endocrinology and Metabolism*, 17, 343–364.

Moran T.H. (2010). Hypothalamic nutrient sensing and energy balance. *Forum of Nutrition,* 63, 94–101.

Motyl K.J., McCabe L.R., SchwartzA.V. (2010). Bone and glucose metabolism: A two-way street. *Archives of Biochemistry and Biophysics,* 503, 2–10.

Mueckler M., Caruso C., Baldwin S.A., Panico M., Blench I., Morris H.R., Allard W.J., Lienhard G.E., Lodish H.F. (1985). Sequence and structure of a human glucose transporter. *Science,* 229, 941–945.

Nakashima R., Kamei N., Yamane K., Nakanishi S., Nakashima A., Kohno N. (2006). Decreased total and high molecular weight adiponectin are independent risk factors for the development of type 2 diabetes in Japanese-Americans. *Journal of Clinical Endocrinology and Metabolism,* 91, 3873–3877.

Niijima A. (1989). Neural mechanisms in the control of blood glucose concentration. *The Journal of Nutrition*, 119, 833–840.

Oberkofler H., Pfeifenberger A., Soyal S., Felder T., Hahne P., Miller K. (2010). Aberrant hepatic TRIB3 gene expression in insulin-resistant obese humans. *Diabetologia*, 53, 1971–1975.

Ogawa W., Matozaki T., Kasuga M. (1998). Role of binding proteins to IRS-1 in insulin signalling. *Molecular and Cellular Biochemistry*, 182, 13–22.

Ohtsubo K., Takamatsu S., Minowa M.T., Yoshida A., Takeuchi M., Marth, J.D. (2005). Dietary and genetic control of glucose transporter 2 glycosylation promotes insulin secretion in suppressing diabetes. *Cell,* 123, 1307–1321.

Orci L., Thorens B., Ravazzola M., Lodish H.F. (1989). Localization of the pancreatic beta cell glucose transporter to specific plasma membrane domains. *Science,* 245, 295–297.

Oury F., Sumara G., Sumara O., Ferron M., Chang H., Smith C.E, Hermo L., Suarez S., Roth B.L., Ducy P., Karsenty G. (2011). Endocrine regulation of male fertility by the skeleton. *Cell,* 144, 796–809.

Pardal R., López-Barneo J. (2002). Low glucose-sensing cells in the carotid body. *Nature Neuroscience*, 5, 197–198.

Pascual J.M., Wang D., Lecumberri B., Yang H., Mao X., Yang R., De Vivo D.C. (2004). GLUT1 deficiency and other glucose transporter diseases. *European Journal of Endocrinology*, 150, 627–633.

Pessin J.E., Saltiel A.R. (2000). Signaling pathways in insulin action: molecular targets of insulin resistance. *Journal of Clinical Investigation,* 106, 165–169.

Phay J.E., Hussain H.B., Moley, J.F. (2000). Cloning and expression analysis of a novel member of the facilitative glucose transporter family, SLC2A9 (GLUT9). *Genomics,* 66, 217–220.

Pieczenik S.R., Neustadt J. (2007). Mitochondrial dysfunction and molecular pathways of disease. *Experimental and Molecular Pathology*, 83, 84–92.

Pinto A.B., Carayannopoulos M.O., Hoehn A., Dowd L., Moley K.H. (2002). Glucose transporter 8 expression and translocation are critical for murine blastocyst survival. *Biology of Reproduction,* 66, 1729–1733.

Pirola L., Johnston A.M., Obberghen E. (2004). Modulation of insulin action. *Diabetologia,* 47, 170–184.

Polakof S., Mommsen T.P., Soengas J.L. (2011). Glucosensing and glucose homeostasis: From fish to mammals. *Comparative Biochemistry and Physiology, Part B* 160, 123–149.

Pollock N.K., Bernard P.J., Gower B.A., Gundberg C.M., Wenger K., Misra S., Bassali R.W., Davis C.L. (2011). Lower uncarboxylated osteocalcin concentrations in children with prediabetes is associated with beta-cell function. *The Journal of Clinical Endocrinology and Metabolism*, 96, 1092–1099.

Pontoglio M., Barra J., Hadchouel M., Doyen A., Kress C., Bach J.P. (1996). Hepatocyte nuclear factor 1 inactivation results in hepatic dysfunction, phenylketonuria, and renal Fanconi syndrome. *Cell,* 84, 575-585.

Pontoglio M., Prie D., Cheret C., Doyen A., Leroy C., Froguel P. (2000). HNF1alpha controls renal glucose reabsorption in mouse and man. *EMBO Reports,* 1, 359-365.

Preitner F., Bonny O., Laverriere A., Rotman S., Firsov D., Da Costa A., Metref S., Thorens B. (2009). Glut9 is a major regulator of urate homeostasis and its genetic inactivation induces hyperuricosuria and urate nephropathy. *Proceedings of the National Academy of Sciences USA,* 106, 15501–15506.

Puchau B., Zulet M.A., de Echavárri A.G., Hermsdorff H.H.M., Martínez J.A. (2010). Dietary total antioxidant capacity is negatively associated with some metabolic syndrome features in healthy young adults. *Nutrition*, 26, 534–541.

Puigserver P., Rhee J., Donovan J., Walkey C.J., Yoon J.C., Oriente F. (2003). Insulin regulated hepatic gluconeogenesis through FOXO1-PGC-1alpha interaction. *Nature,* 423, 550–555.

Rached M.T., Kode A., Silva B.C., Jung D.Y., Gray S., Ong H., Paik J.-H., DePinho R.A., Kim J.K., Karsenty G., Kousteni S. (2010). FoxO1 expression in osteoblasts regulates glucose homeostasis through regulation of osteocalcin in mice. *Journal of Clinical Investigation*, 120, 357–368.

Raddatz D., Ramadori G. (2007). Carbohydrate metabolism and the liver: actual aspects from physiology and disease. *Zeitschrift für Gastroenterologie,* 45, 51–62.

Rains J.L., Jain S.K. (2011). Oxidative stress, insulin signaling, and diabetes. *Free Radical Biology and Medicine*, 50, 567–575.

Rand E.B., Depaoli A.M., Davidson N.O., Bell G.I., Burant C.F. (1993). Sequence, tissue distribution, and functional characterization of the rat fructose transporter GLUT5. *Americal Journal of Physiology*, 264, 1169–1176.

Reaven G.M. (2005). The insulin resistance syndrome: definition and dietary approaches to treatment. *Annual Review of Nutrition*, 25, 391–406.

Reimann F., Habib A.M., Tolhurst G., Parker H.E., Rogers G.J., Gribble F.M. (2008). Glucose sensing in L cells: a primary cell study. *Cell Metabolism,* 8, 532–539.

Rhoads D.B., Rosenbaum D.H., Unsal H., Isselbacher K.J., Levitsky L.L. (1998). Circadian periodicity of intestinal Na^+/glucose cotransporter 1 mRNA levels is transcriptionally regulated. *Journal of Biological Chemistry,* 273, 9510-9516.

Richards J.B., Valdes A.M., Burling K., Perks U.C., Spector T.D. (2007). Serum adiponectin and bone mineral density in women. *Journal of Clinical Endocrinology and Metabolism,* 92, 1517–1523.

Roach P.J. (2002). Glycogen and its metabolism. *Current Molecular Medicine,* 2, 101–120.

Rodan M., Bernroiden E. (2003). Hepatic glucose metabolism in humans—its role in health and disease. *Best Practice and Research Clinical Endocrinology and Metabolism,* 17, 365–383.

Roeske D., Ludwig K.U., Neuhoff N., Becker J., Bartling J., Bruder J., Brockschmidt F.F., Warnke A., Remschmidt H., Hoffmann P., Muller-Myhsok B., Nothen M.M., Schulte-Korne G. (2011). First genome-wide association scan on neurophysiological endophenotypes points to trans-regulation effects on SLC2A3 in dyslexic children, *Molecular Psychiatry,* 16, 97-107.

Rogers S., Macheda M.L., Docherty S.E., Carty M.D., Henderson M.A., Soeller W.C., Gibbs E.M., James D.E., Best J.D. (2002). Identification of a novel glucose transporter-like protein-GLUT-12. *American Journal of Physiology - Endocrinology and Metabolism,* 282, 733–738.

Ronti T., Lupattelli G., Mannarino E. (2006). The endocrine function of adipose tissue: an update. *Clinical Endocrinology*, 64, 355–365.

Rorsman P., Salehi S.A., Abdulkader F., Braun M., Macdonald P.E. (2008). K(ATP)-channels and glucose-regulated glucagon secretion. *Trends in Endocrinology and Metabolism,* 19, 277–284.

Rossetti L., Giaccari A., Barzilai N. (1993). Mechanism by which hyperglycemia inhibits hepatic glucose production in conscious rats. Implications for the pathophysiology of fasting hyperglycemia in diabetes. *The Journal of Clinical Investigation*, 92, 1126–1134.

Sabino-Silva R., Freitas H.S., Lamers M.L., Okamoto M.M., Santos M.F., Machado U.F. (2009). Na^+-glucose cotransporter SGLT1 protein in salivary glands: potential involvement in the diabetes-induced decrease in salivary flow. *Journal of Membrane Biology*, 228, 63-69.

Sabino-Silva R., Mori R.C., David-Silva A., Okamoto M.M., Freitas H.S., Machado U.F. (2010). The Na^+/glucose cotransporters: from genes to therapy. *Brazilian Journal of Medical and Biological Research*, 43, 1019-1026.

Saltiel A.R., Kahn, C.R. (2001). Insulin signalling and the regulation of glucose and lipid metabolism. *Nature,* 414, 799–806.

Saltiel A.R., Pessin J.E. (2002). Insulin signaling pathways in time and space. *Trends in Cell Biology*, 12, 65–71.

Santer R., Schneppenheim R., Dombrowski A., Gotze H., Steinmann B., Schaub J. (1997). Mutations in GLUT2, the gene for the liver-type glucose transporter, in patients with Fanconi-Bickel syndrome. *Nature Genetics,* 17, 324–326.

Santer R., Steinmann B., Schaub J. (2002). Fanconi-Bickel syndrome - A congenital defect of facilitative glucose transport. *Current Molecular Medicine,* 2, 213–227.

Sasaki T., Minoshima S., Shiohama A., Shintani A., Shimizu A., Asakawa S., Kawasaki K., Shimizu, N. (2001). Molecular cloning of a member of the facilitative glucose transporter gene family GLUT11 (SLC2A11) and identification of transcription variants. *Biochemical and Biophysical Research Communications,* 289, 1218–1224.

Scazzocchio B., Varì R., Filesi C., D'Archivio M., Santangelo C., Giovannini C. (2011). Cyanidin-3-O-b-glucoside and protocatechuic acid exert insulin-like effects by upregulating PPARγ activity in human omental adipocytes. *Diabetes,* 60, 2234–2244.

Scheepers A., Joost H.G., Schürmann A. (2004). The Glucose Transporter Families SGLT and GLUT: Molecular Basis of Normal and Aberrant Function. *Journal of Parenteral and Enteral Nutrition*, 28, 365–372.

Scheepers A., Schmidt S., Manolescu A., Cheeseman C.I., Bell A., Zahn C., Joost H.G., Schurmann A. (2005). Characterization of the human SLC2A11 (GLUT11) gene: alternative promoter usage, function, expression, and subcellular distribution of three isoforms, and lack of mouse orthologue. *Molecular Membrane Biology*, 22, 339–351.

Schmid C., Ernst M. (1992). Insulin-like growth factors. In: Gowen M (ed) Cytokines and bone metabolism. CRC Press, Boca Raton, pp 229–265.

Schmid C., Steiner T., Froesch E.R. (1982). Parathormone promotes glycogen formation from [^{14}C] glucose in cultured osteoblast-like cells. *FEBS Letters,* 148, 31–34.

Schmid C., Steiner T., Froesch E.R. (1983). Insulin-like growth factors stimulate synthesis of nucleic acids and glycogen in cultured calvaria cells. *Calcified Tissue International*, 35, 578–585.

Schmidt S., Gawlik V., Holter S.M., Augustin R., Scheepers A., Behrens M., Wurst W., Gailus-Durner V., Fuchs H., Hrabe de Angelis M., Kluge R., Joost H.G., Schurmann A. (2008a). Deletion of glucose transporter GLUT8 in mice increases locomotor activity. *Behavior Genetics*, 38, 396–406.

Schmidt S., Joost H.G., Schurmann A. (2009). GLUT8, the enigmatic intracellular hexose transporter. *American Journal of Physiology - Endocrinology and Metabolism,* 296, 614–618.

Schmidt S., Richter M., Montag D., Sartorius T., Gawlik V., Hennige A.M., Scherneck S., Himmelbauer H., Lutz S.Z., Augustin R., Kluge R., Ruth P., Joost H.G., Schurmann A. (2008b). Neuronal functions, feeding behavior, and energy balance in Slc2a31/- mice. *American Journal of Physiology - Endocrinology and Metabolism*, 295, 1084–1094.

Schmidt U., Briese S., Leicht K., Schurmann A., Joost H.G., Al-Hasani H. (2006). Endocytosis of the glucose transporter GLUT8 is mediated by interaction of a dileucine motif with the beta2-adaptin subunit of the AP-2 adaptor complex. *Journal of Cell Science*, 119, 2321–2331.

Seidner G., Alvarez M.G., Yeh J.I., O'Driscoll K.R., Klepper J., Stump T.S., Wang D., Spinner N.B., Birnbaum M.J., De Vivo D.C., (1998). GLUT-1 deficiency syndrome caused by haploinsufficiency of the blood-brain barrier hexose carrier. *Nature Genetics*, 18, 188–191.

Sesti, G. (2006). Pathophysiology of insulin resistance. *Best Practice and Research Clinical Endocrinology and Metabolism,* 20, 665–679.

Shepherd P.R., Gould G.W., Colville C.A., McCoid S.C., Gibbs E. M., Kahn B.B. (1992). Distribution of GLUT3 glucose transporter protein in human tissues. *Biochemical and Biophysical Research Communications*, 188, 149–154.

Shin B.C., McKnight R.A., Devaskar S.U. (2004). Glucose transporter GLUT8 translocation in neurons is not insulin responsive. *Journal Neuroscience Research,* 75, 835–844.

Shulman G.I., Lacy W.W., Liljenquist J.E. (1980). Effect of glucose, independent of changes in insulin and glucagon secretion, on alanine metabolism in the conscious dog. *Journal of Clinical Investigation*, 65, 496–505.

Shulman R.G., Bloch G., Rothman D.L. (1995). *In vivo* regulation of muscle glycogen synthase and the control of glycogen synthesis. *Proceedings of the National Academic Science USA*, 92, 8535–8542.

Silva B.C., Kousteni S. (2012). Role of Osteoblasts in Regulation of Energy Metabolism. *Clinical Reviews in Bone and Mineral Metabolism*, doi: 10.1007/s12018-012-9128-8.

Sirikul B., Gower B.A., Hunter G.R., Larson-Meyer D.E., Newcomer B.R. (2006). Relationship between insulin sensitivity and *in vivo* mitochondrial function in skeletal muscle. *American Journal of Physiology - Endocrinology and Metabolism,* 291, 724–728.

Solberg D.H., Diamond J.M. (1987). Comparison of different dietary sugars as inducers of intestinal sugar transporters. *American Journal of Physiology,* 252, 574-584.

Stayrook K.R., Bramlett K.S., Savkur R.S., Ficorilli J., Cook T., Christe M.E. (2005). Regulation of carbohydrate metabolism by the farnesoid X receptor. *Endocrinology,* 146, 984–991.

Stuart C.A., Howell M.E., Zhang Y., Yin D. (2009). Insulin stimulated translocation of glucose transporter (GLUT) 12 parallels that of GLUT4 in normal muscle. *The Journal of Clinical Endocrinology and Metabolism,* 94, 3535–3542.

Stumvoll M., Meyer C., Mitrakou A. (1997). Renal glucose production and utilization: new aspects in humans. *Diabetologia*, 40, 749–757.

Szkudelski T., Szkudelska K. (2011). Anti-diabetic effects of resveratrol. *Annals of the New York Academy of Sciences*, 1215, 34–39.

Tappy L., Cayeux M.C., Schneiter P. (1995). Effects of lactate on glucose metabolism in healthy subjects and in severely injured hyperglycemic patients. *American Journal of Physiology*, 268, 630–635.

Thomas D.M., Maher F., Rogers S.D., Best J.D. (1996). Expression and regulation by insulin of GLUT3 in UMR 106-01, a clonal rat osteosarcoma cell line. *Biochemical and Biophysical Research Communications,* 218, 789–793.

Thomas D.M., Rogers S.D., Sleeman M.W., Pasquini G.M., Bringhurst F.R., Ng K.W., Zajac J.D., Best J.D. (1995). Modulation of glucose transport by parathyroid hormone and insulin in UMR 106-01, a clonal rat osteogenic sarcoma cell line. *Journal of Molecular Endocrinology*, 14, 263–275.

Triplitt C.L. (2012). Examining the Mechanisms of Glucose Regulation. *The American Journal of Managed Care,* 18, 4-10.

Turk E., Wright E.M. (1997). Membrane topology motifs in the SGLT cotransporter family. *Journal of Membrane Biology*, 159, 1-20.

Turner N., Heilbronn L.K. (2008). Is mitochondrial dysfunction a cause of insulin resistance? *Trends in Endocrinology and Metabolism*, 19, 324–330.

Uldry M., Ibberson M., Horisberger J.D., Chatton J.Y., Riederer B.M., Thorens B. (2001). Identification of a mammalian H^+-*myo*-inositol symporter expressed predominantly in the brain. *The EMBO Journal*, 20, 4467–4477.

Uldry M., Ibberson M., Hosokawa M., Thorens B. (2002). GLUT2 is a high affinity glucosamine transporter. *FEBS Letters,* 524, 199–203.

Uldry M., Steiner P., Zurich M.G., Beguin P., Hirling H., Dolci W., Thorens, B. (2004). Regulated exocytosis of an H(1)/myoinositol symporter at synapses and growth cones. *The EMBO Journal*, 23 531–540.

Uldry M., Thorens B. (2004). The SLC2 family of facilitated hexose and polyol transporters. *Pflügers Archiv - European Journal of Physiology,* 447, 480–489.

Van Cromphaut S.J. (2009). Hyperglycemia as part of the stress response: the underlying mechanisms. *Best Practice and Research Clinical Anaesthesiology,* 23, 375–386.

Vayro S., Wood I.S., Dyer J., Shirazi-Beechey S.P. (2001). Transcriptional regulation of the ovine intestinal Na^+/glucose cotransporter SGLT1 gene. Role of HNF-1 in glucose activation of promoter function. *European Journal of Biochemistry*, 268, 5460-5470.

Vestri S., Okamoto M.M., de Freitas H.S., Aparecida Dos Santos R., Nunes M.T., Morimatsu M. (2001). Changes in sodium or glucose filtration rate modulate expression of glucose transporters in renal proximal tubular cells of rat. *Journal of Membrane Biology*, 182, 105-112.

Vigilanza P., Aquilano K., Baldelli S., Rotilio G., Ciriolo M.R. (2011). Modulation of intracellular glutathione affects adipogenesis in 3T3-L1 cells. *Journal of Cellular Physiology*, 226, 2016–2024.

Visser W.E., Heemstra K.A., Swagemakers S.M., Ozgür Z., Corssmit E.P., Burggraaf J., van Ijcken W.F., van der Spek P.J., Smit J.W., Visser T.J. (2009). Physiological thyroid hormone levels regulate numerous skeletal muscle transcripts. *Journal of Clinical Endocrinology and Metabolism*, 94, 3487–3496.

Vitart V., Rudan I., Hayward C., Gray N.K., Floyd J., Palmer C.N., Knott S.A., Kolcic I., Polasek O., Graessler J., Wilson J.F., Marinaki A., Riches P.L., Shu X., Janicijevic B., Smolej-Narancic N., Gorgoni B., Morgan J., Campbell S., Biloglav Z., Barac-Lauc L., Pericic M., Klaric I.M., Zgaga L., Skaric-Juric T., Wild S.H., Richardson W.A., Hohenstein P., Kimber C.H., Tenesa A., Donnelly L.A., Fairbanks L.D., Aringer M., McKeigue P.M., Ralston S.H., Morris A.D., Rudan P., Hastie N.D., Campbell H., Wright A.F. (2008). SLC2A9 is a newly identified urate transporter influencing serum urate concentration, urate excretion and gout. *Nature Genetics,* 40, 437–442.

Wan C., Tempel W., Liu Z.J., Wang B.C., Rose R.B. (2005). Structure of the conserved transcriptional repressor enhancer of rudimentary homolog. *Biochemistry,* 44, 5017-5023.

Wang D., Pascual J.M., Yang H., Engelstad K., Mao X., Cheng J., Yoo J., Noebels J.L., De Vivo, D.C. (2006). A mouse model for Glut-1 haploinsufficiency. *Human Molecular Genetics*, 15, 1169–1179.

Watson R.T., Pessin J.E. (2007). GLUT4 translocation: the last 200 nanometers. *Cellular Signaling,* 19, 2209–2217.

Wei J., Ducy P. (2010). Co-dependence of bone and energy metabolisms. *Archives of Biochemistry and Biophysics*, 503, 35–40.

Wells R.G., Pajor A.M., Kanai Y., Turk E., Wright E.M., Hediger M.A. (1992). Cloning of a human kidney cDNA with similarity to the sodium-glucose cotransporter. *Americal Journal of Physiology*, 263, 459-465.

Wheeler M.A., Townsend M.K., Yunker L.A., Mauro L.J. (2002). Transcriptional activation of the tyrosine phosphatase gene, OST-PTP, during osteoblast differentiation. *Journal of Cellular Biochemistry*, 87, 363–376.

Widmer M., Uldry M., Thorens B. (2005). GLUT8 subcellular localization and absence of translocation to the plasma membrane in PC12 cells and hippocampal neurons. *Endocrinology,* 146, 4727–4736.

Wilcox G. (2005). Insulin and Insulin Resistance. *The Clinical Biochemist Reviews,* 26, 19-39.

Wilson-O'Brien, A.L., Patron, N., Rogers, S. (2010). Evolutionary ancestry and novel functions of the mammalian glucose transporter (GLUT) family. *BMC Evolutionary Biology*, 10, 152.

Wong G.L., Luben R.A., Cohn D.V. (1977). 1,25-dihydroxycholecalciferol and parathormone: effects on isolated osteoclastlike and osteoblast-like cells. *Science.* 197, 663–665.

Wood I.S., Trayhurn P. (2003). Glucose transporters (GLUT and SGLT): expanded families of sugar transport proteins. *British Journal of Nutrition,* 89, 3-9.

Wood I.S., Hunter L., Trayhurn P. (2003). Expression of Class III facilitative glucose transporter genes (GLUT-10 and GLUT-12) in mouse and human adipose tissues. *Biochemical and Biophysical Research Communications,* 308, 43–49.

Wright E.M., Hirayama B.A., Loo D.F. (2007). Active sugar transport in health and disease. *Journal of Internal Medicine*, 261, 31-43.

Wright E.M., Loo D.D., Hirayama B.A., Turk E. (2004). Surprising versatility of Na^+-glucose cotransporters: SLC5. *Physiology,* 19, 370-376.

Wu X., Freeze H.H. (2002). GLUT14, a duplicon of GLUT3, is specifically expressed in testis as alternative splice forms. *Genomics,* 80, 553–557.

Wu X., Li W., Sharma V., Godzik A., Freeze, H.H. (2002). Cloning and characterization of glucose transporter 11, a novel sugar transporter that is alternatively spliced in various tissues. *Molecular Genetics and Metabolism,* 76, 37–45.

Xiao H., Massaro D., DeCarlo Massaro G., Biadasz Clerch L. (2004). Expression of lung uncoupling protein-2 mRNA is modulated developmentally and by caloric intake. *Experimental Biology and Medicine,* 229, 479–485.

Yoshida H., Takamura N., Shuto T., Ogata K., Tokunaga J., Kawai K. (2010). The citrus flavonoids hesperetin and naringenin block the lipolytic actions of TNF-alpha in mouse adipocytes. *Biochemical and Biophysical Research Communications,* 394, 728–732.

Yoshizawa T., Hinoi E., Jung D.Y., Kajimura D., Ferron M., Seo J., Graff J.M., Kim J.K., Karsenty G. (2009). The transcription factor ATF4 regulates glucose metabolism in mice through its expression in osteoblasts. *Journal of Clinical Investigation*, 119, 2807–2817.

Zeuthen T., Meinild A.K., Loo D.D., Wright E.M., Klaerke D.A. (2001). Isotonic transport by the Na^+-glucose cotransporter SGLT1 from humans and rabbit. *Journal of Physiology*, 531, 631-644.

Zhao F.Q., Keating A.F. (2007). Functional properties and genomics of glucose transporters. *Current Genomics*, 8, 113-128.

Zhao Y., Fung C., Shin D., Shin B.C., Thamotharan S., Sankar R., Ehninger D., Silva A., Devaskar S.U. (2010). Neuronal glucose transporter isoform 3 deficient mice demonstrate features of autism spectrum disorders. *Molecular Psychiatry*, 15, 286–299.

Zhou L., Cryan E.V., D'Andrea M.R., Belkowski S., Conway B.R., Demarest K.T. (2003). Human cardiomyocytes express high level of Na^+/glucose cotransporter 1 (SGLT1). *Journal of Cellular Biochemistry*, 90, 339-346.

Zoidis E., Ghirlanda-Keller C., Schmid C. (2011). Stimulation of glucose transport in osteoblastic cells by parathyroid hormone and insulin-like growth factor I. *Molecular and Cellular Biochemistry*, 348, 33-42.

Zoidis E., Ghirlanda-Keller C., Schmid C. (2012). Triiodothyronine stimulates glucose transport in osteoblastic cells. *Endocrine,* 41, 501-511.

Zygmunt K., Faubert B., MacNeil J., Tsiani E. (2010). Naringenin, a citrus flavonoid, increases muscle cell glucose uptake via AMPK. *Biochemical and Biophysical Research Communications,* 398, 178–183.

In: Glucose Uptake
Editors: Carter C. Johnson and Davis B. Williams
ISBN: 978-1-62618-670-5

Chapter 2

Glucose Uptake: Knowledge from *In Vivo, In Situ* and *In Vitro* Studies and Health Implications

Fátima Regina Mena Barreto Silva[1,*]*, Marisa Jádna Silva Frederico[1], Allisson Jhonatan Gomes Castro[1], Gabrielle Da Luz[1], Delsi Altenhofen[1] and Luisa Helena Cazarolli[2]

[1]Departamento de Bioquímica, Centro de Ciências Biológicas, Universidade Federal de Santa Catarina, Campus Universitário, Bairro Trindade, Florianópolis, SC, Brazil

[2]Universidade Federal da Fronteira Sul, Campus Universitário Laranjeiras do Sul, Bairro Vila Alberti, PR, Brazil

Abstract

The cellular energy from glucose is obtained through glycogen breakdown and also from skeletal muscle through intermediary metabolites. This waste of whole-body energy is replenished by digested and absorbed dietary glucose on intestine and distributed among muscle, adipose tissue, liver, kidney, brain and others tissues involved in glycemia by means of glucose uptake, metabolism, storage and excretion. Glucose is unable to readily diffuse through the plasma membranes due to its hydrophilic nature. So, at least, two main types of assistance transport proteins are needed: facilitative transporters (glucose carriers (GLUTs) from the SLC2 gene family and active transport proteins (glucose carriers (SGLTs) from the SCL5 gene family. Among twelve known GLUT molecules, GLUT4 is the major glucose transporter for skeletal muscle, adipose tissue (insulin dependent for these both tissues) and cardiac muscle. Other ones, such as GLUTs 1, 2, 3, and 8, facilitate the glucose entry into the brain and liver, for example. Significant amounts of glucose ($\geq$ 80%) are transported into the skeletal muscle where it may be used for energy supply or for glycogen stores. On the other hand, the glucose uptake on adipose tissue is around 5% although it plays an important influence on insulin sensitivity

* Email: mena.barreto@ufsc.br.

in muscle and liver. Furthermore, several transporters are involved on glucose flux from cells to the blood and from organelles to the cytosol. The glucose homeostasis is maintained by a complex network among pancreatic cells, intestine, kidney, skeletal muscle, adipose tissue, liver and lately also the central glucose regulation by brain has been carefully concerned. The reabsorption of glucose by kidney occurs through the sodium glucose cotransporter SGLT2 (the majority) and the remaining ones by SGLT1 (around 10%). SGLT1 is also extensively expressed in the small intestine. The SGLT active glucose transport system takes advantage of sodium electrochemical potential gradient across the brush border membrane and also utilize sodium:potassium adenosine triphosphatase pumps at the basolateral membrane to the suitable glucose uptake. This review provides a discussion about the benefits and what is going on with glucose uptake and health implications, the mechanisms that contribute to glucose homeostasis and the potential impact of glucose transporters as pharmacological targets able to correct glucose imbalance.

Keywords: Glucose uptake, GLUTs, SGLTs, Glycemia, Diabetes

Abbreviations

ADP	*Adenosine* diphosphate
AMPK	5'-AMP-activated protein kinase
ATP	*Adenosine* triphosphate
CCK	Cholecystokinin
cDNA	Complementary deoxyribonucleic acid
cDNA	Complementary deoxyribonucleic acid
CNS	Central nervous system
COOH	Carboxylic acid
ER	Endoplasmic reticulum
ER	Endoplasmic reticulum
G6Pase	Glucose-6-phosphatase
GIP	glucose-dependent insulinotropic polypeptide
GLP-1	glucagon-like peptide-1
GLUTs	Protein symbol of the family of transport facilitators
HepG2	Erythrocyte sugar transporter
HGP	Stimulating hepatic glucose production
HMIT	Myo-inositol transporter H^+/myo-inositol transporter
IP3	Inositol-3-phosphate
K_{ATP} channels	ATP-dependent K^+ channels
K_m	Michaelis constant
MFS	Major facilitator superfamily
mRNA	Messenger ribonucleic acid
NH_2	Amino
PDX-1	Pancreatic and duodenal homeobox 1
PEPCK	Phosphoenolpiruvate carboxykinase
PESPR	Proline, glutamate, serine, and arganine
Pgp	P-glycoprotein

PI-3K	Phosphatidylinositol 3-kinase
PKC	Protein kinase C
SAAT1	Na^+-dependent neutral-amino acid transporter
SGLT	Na^+/glucose cotransporter or symporter
siRNA	Small interfering ribonucleic acid
SLC2, SCL5	Name gene from nomenclature of the sugar transporter genes
SMIT2	Na^+/myo-inositol cotransporter
TM	transmembrane domains

Glucose Uptake and Its Regulation

The maintenance of glycemia at the normal range is important to human health. D-Glucose is the major fuel for most cells that are sensitive to glucose presence and its homeostasis requires an integrated control by the whole organism. The regulation of blood glucose involves a wide communication with diverse tissues, enzymes and receptors via hormones, neural pathways and/or changes in the utilization of substrate. In the physiological state, glycemia is tightly regulated within a narrow range of glucose, despite the fact that substantial shifts in glucose flux occurs between the fasting to the fed state and back again (Maggs et al., 2008; Wasserman, 2009).

In the postabsorptive state (fasting), plasma glucose concentration is determined by the rate that endogenous glucose is released into the circulation and the rate that glucose is taken up by body tissues (Aronoff et al., 2004; Maggs et al., 2008; Wasserman, 2009). In the fasting state and during physical activity, glucagon, secreted from pancreatic α-cells, prevents an abnormal fall in plasma glucose by stimulating hepatic glucose production (HGP), while basal rates of insulin secretion from pancreatic β-cells act primarily to suppress HGP and prevent an abnormal rise in plasma glucose. Glucagon acts primarily in the liver to enhance glycogen breakdown as well as to stimulate gluconeogenesis (by increasing the uptake of gluconeogenic amino acids and by activating the gluconeogenic enzymes). The liver is the primary organ responsible for gluconeogenesis, but after a prolonged fast, the kidney can also contribute to endogenous glucose production (Gerich, 2010; Mitrakou, 2011; Ramnanan et al., 2011; Moore et al., 2012). Although the kidney is unable to release glucose through glycogenolysis, it is involved in the glucose homeostasis by at least two different mechanisms: release of glucose into the circulation via gluconeogenesis and reabsorption of glucose from glomerular filtrate (Gerich, 2010).

During the postabsorptive period, besides glucagon there are other regulators of glycemia such as the growth hormone, cortisol and catecholamine which stimulate lipolysis in the adipose tissue, leading to increased circulating concentrations of free fatty acids and subsequent reductions in insulin-mediated glucose uptake peripherally. As fasting progresses, muscle proteolysis supplies glucogenic amino acids for hepatic glucose production. Also, as muscle glucose uptake is insulin dependent, during fasting (low insulin levels) muscle cells primarily rely on free fatty acids and ketones as fuels, allowing the sparing of glucose in order to keep the glycemia. This is also observed in the adipose tissue, whose glucose metabolism is dependent of high insulin levels. The relationship between fat and carbohydrate oxidation is reciprocal, the oxidation of free fatty acids inhibits the uptake of glucose by insulin-sensitive tissues and enhances hepatic glucose output. Therefore, muscles and adipose tissue, which are major users of glucose in the fed state, become minor users in the fasting state (Jensen, 1997; Beardsall et al., 2003; Giorgino et al., 2005; Shrayyef and Gerich, 2010).

Fasting glucose utilization occurs in both, insulin-dependent (skeletal muscle, adipose tissue, and myocardium) and insulin-independent tissues; among the latter, the brain, an obligate glucose consumer, alone accounts for half of the amount of the hepatic glucose output. This is explained by the fact that the brain is unable to either store or produce glucose and alternative sources of energy are in short supply or are unable to pass the blood–brain barrier (Gerich, 2000). After about few days of starvation, the metabolic profile is set to conserve protein and to supply greater quantities of alternative fuels. In particular, free fatty acids and ketone bodies are utilized to maintain the energy needs (Jensen, 1997; Beardsall et al., 2003; Giorgino et al., 2005; Shrayyef and Gerich, 2010).

In the fed state, glucose homeostasis is more complex since the gastrointestinal tract becomes an important source of exogenous glucose. Marked and rapid changes in glucose flux occur as a result of the considerable inflow of meal-derived glucose into the circulation. Insulin secretion rises sharply during the early meal-time and it stimulates glucose uptake into peripheral tissues and contributes to the suppression of HGP. Through these mechanisms, insulin is a key factor limiting abnormal postprandial glycemia. In parallel, glucagon is suppressed, also contributing to the suppression of HGP (Aronoff et al., 2004; Maggs et al., 2008; Wasserman, 2009).

Importantly, besides insulin and glucagon, the influx of glucose into the circulation from gastrointestinal tract is accompanied by the secretion of several other glucoregulatory hormones including amylin from β-cells in the pancreas and gastrointestinal glucose-dependent insulinotropic polypeptide (GIP), glucagon-like peptide-1 (GLP-1) and cholecystokinin (CCK) from gastrointestinal endocrine cells. These hormones collectively influence glucose homeostasis via several mechanisms of action including the regulation of insulin and glucagon responses as well as the modulation of nutrient passage from the gastrointestinal tract to the appropriate tissues to be stored (Lam, 2010; Marty et al., 2007). Additionally, another key determinant to glucose inflow into the circulation at the time of feeding is the rate at which food and drink enter the small intestine from the stomach, the gastric emptying (Aronoff et al., 2004; Maggs et al., 2008; Wasserman, 2009).

Moreover, the nervous system has been implicated as an insulin target tissue and plays a critical role in the regulation of glucose homeostasis (Donato Jr., 2012). This process involves the regulation of energy balance at the whole body, for example, influencing the gastrointestinal tract and the digestion, absorption, secretion and motility of the nutrients through neural multiple and complex inhibitory and stimulatory reflexes (enteric nervous system). Also, the regulatory role of the central nervous system (CNS) is based on insulin effects and also the action of hormones at CNS such as leptin and amylin that alter food intake, energy expenditure and endogenous glucose production (Morton, 2007; Mourad and Saadé, 2011; Donato Jr., 2012; Ghasemi et al., 2012). Furthermore, the neuronal regulation of homeostasis by nutrient sensing has been deeply discussed. Some studies indicate that nutrient-sensing mechanisms activate negative-feedback systems to regulate energy and glucose homeostasis through a neuronal network. For example, direct metabolic signaling within the intestine activates gut-brain and gut-brain-liver axes to regulate energy and glucose homeostasis, respectively. In parallel, direct metabolism of nutrients within the hypothalamus regulates food intake and blood glucose levels. These findings provide a physiological and neuronal framework by which enhancing or restoring nutrient sensing in the intestine and the brain could normalize energy and glucose homeostasis in diseases such as diabetes and obesity (Lam, 2010). Furthermore, Marty et al. (2007) reported that glucose, besides being a nutrient, is also a signal detected by several glucose-sensing units that are located at different anatomical sites and converge to the hypothalamus to cooperate with leptin and insulin in controlling the melanocortin pathway. These data highlight the importance of the central nervous system in mediating a neuronal circuit with a critical role in orchestrating the control of glucose and energy homeostasis.

Beyond the CNS, other tissues such as liver, adipose tissue, pancreas, kidney, intestine, skeletal and cardiac muscle are important components of the machinery involved in the glucose homeostasis since they are responsible for the glucose uptake, metabolism, storage and excretion (Triplitt, 2012). Intestinal glucose absorption is a rapid and high-capacity phenomenon. Potential glucose sensing systems expressed in the enteroendocrine cells include sweet taste receptors and glucose transporters. In addition, the influence of the enteric nervous system and neuronal pathways has a significant role on glucose sensing and absorption. The absorption process involves the transfer of sugar molecules across plasma membranes, which occurs via integral transport proteins through active and diffusive pathways (Dyer et al., 2007; Araújo and Martel, 2009; Mourad and Saadé, 2011).

Following glucose ingestion and its absorption on small intestine, the increase in plasma glucose concentration triggers the insulin release from β-cells. Glucose is the primary stimulus for insulin secretion, although there are many other metabolic, endocrine and neural control mechanisms (Henquin, 2009; Jitrapakdee et al., 2010; Henquin, 2011; Wiederkehr and Wollheim, 2012). β-cells sense nutrients such as glucose via uptake and intracellular metabolism, resulting in the formation of downstream metabolites, which act as messengers stimulating the hormone secretion. The stimulation of oxidative phosphorylation elevates the cytosolic ATP/ADP ratio. This change favors the closure of the plasma membrane ATP-dependent K^+ channels (K_{ATP} channels) causing plasma membrane depolarization. Once a threshold potential is reached, voltage-gated Ca^{2+} channels open giving rise to cytosolic Ca^{2+} signals that are essentials for insulin granule exocytosis (Henquin, 2009; Henquin, 2011; Wiederkehr; Wollheim, 2012).

In the hierarchy, insulin is, by far, the chief regulating element of plasma glucose levels. Several other factors play a secondary role, glucagon being the next in charge. Postprandial glucose homeostasis is basically related to insulin, which stimulates splanchnic and peripheral glucose uptake, facilitates its conversion to storage compounds and suppresses endogenous (primarily hepatic) glucose production (Shrayyef and Gerich, 2010). In terms of whole-body glucose economy, normally approximately 45% of ingested glucose is thought to be converted into glycogen in the liver, 30% is taken up by skeletal muscle and later converted into glycogen, 15% is taken up by the brain, 5% is taken up by the adipose tissue and 10% is taken up by the kidneys (Gerich, 2000; Meyer et al., 2002; Gerich, 2010).

In the postprandial state, the liver removes glucose from the circulation and stores it in the form of glycogen or metabolizes it through glycolysis to minimize the fluctuation of glycemia (Moore et al., 2012). When glucose enters in hepatocytes through a facilitated diffusion mechanism, liver glucokinase, activated by insulin, promotes glucose phosphorylation and traps glucose on liver cells for conversion into glycogen. Insulin-dependent activation of the insulin receptor tyrosine kinase initiates a signal transduction pathway that results in increased glycogen synthesis (glycogen synthase activation). The proportion of glucose used for glycogen synthesis is dependent on the extent of glycogen stocks; once these are replete, increased amounts of glucose are metabolized through glycolysis or converted to triglyceride. On the other hand, insulin also inhibits liver glycogen phosphorylase *b* hence reducing glucose release from glycogen. Moreover, insulin suppresses gluconeogenesis through both, direct effects on the liver and indirect extra hepatic mechanisms. In the hepatocyte, insulin inhibits the expression of specific enzymes such as phosphoenolpiruvate carboxykinase (PEPCK) and glucose-6-phosphatase (G6Pase), while indirectly, promotes the suppression of glucagon release and the reduction of gluconeogenic precursors (Beardsall et al., 2003; Klover and Mooney, 2004; Moore et al., 2012; Bechmann et al., 2012).

Considering the metabolic fate of glucose in kidneys it has been observed that the uptake and release of glucose are distinct processes occurring at different locations within the kidney and undergoing distinct regulation. Glucose utilization occurs predominantly in the renal medulla, whereas glucose production occurs mainly at the renal cortex. Considering the presence or absence of specific enzymes in different regions of the kidneys, the release of glucose by the normal kidney is mainly, if not exclusively, a result of renal cortical gluconeogenesis, whereas glucose uptake and utilization occur in other parts of the kidney (Stumvoll et al., 1997; Meyer et al., 2002; Gerich, 2010; Mitrakou, 2011; Shrayyef and Gerich, 2010).

Until recently, it was known that in the postprandial period, insulin directly suppresses glucose release from kidneys by affecting the enzymes activation/deactivation and indirectly through the alteration of the gluconeogenic substrate availability as well as the levels of gluconeogenic activators (e.g., suppression of free fatty acids and glucagon release). However, during the past few years a considerable body of evidence indicates that in the postprandial state, although overall endogenous glucose release decreases substantially, renal gluconeogenesis actually increases by approximately 2-fold and accounts for 60% of endogenous glucose release. This happens to compensate for restricted (physiologic) hepatic glucose release and also to facilitate efficient repletion of glycogen stores in the liver. This is based on evidences that the release of glucose by the liver and kidney are interrelated so that a reduction in release by one organ is associated by an increase by the other to further maintain optimal glucose homeostasis. These differences in regulation and reciprocal change in renal and hepatic glucose release have led to the concept of hepatorenal glucose reciprocity. At the same time, insulin stimulates the glucose uptake, its metabolism and conversion into glycogen on renal medulla (Gerich, 2010; Mitrakou, 2011; Meyer et al., 2002; Shrayyef and Gerich, 2010).

The kidney can also influence glucose homeostasis by returning glucose to the circulation via the reabsorption of glucose from glomerular filtrate. Glucose is freely filtered in the glomerulus, so that, as the glycemia increase, the amount of glucose in the glomerular filtrate increases linearly. Reabsorption of filtered glucose also increases linearly until the maximal reabsorptive capacity is exceeded. These processes of reabsorption of glucose from glomerular filtrate and the glucose release into the circulation are mediated by different types of transport: an active (against the gradient concentration) and passive (in favor of the gradient concentration) transport systems (Gerich, 2010; Mitrakou, 2011; Meyer et al., 2002; Shrayyef and Gerich, 2010).

In healthy adults, glycemia is tightly regulated within a range of 70 to 99 mg/dL, and maintained by specific hormones, being insulin the most important one in the postprandial state, period where the blood glucose concentration increases rapidly because of the ingested and absorbed food (Shrayyef and Gerich, 2010; Triplitt, 2012). Insulin regulates glucose metabolism by direct and indirect actions. This hormone stimulates the conversion of simple units such as monosaccharides (including glucose) into complex macromolecules such as glycogen. Through binding to its receptors in different tissues, insulin activates its signaling pathways which involves a complex cascade of protein kinases and regulatory proteins that conduce to the metabolic effects of insulin. This is accomplished in part, by increased glucose uptake in muscle and adipose tissue. The glucose transport is an important step in cell metabolism because it controls the rate of glucose utilization (Beardsall et al., 2003; Shrayyef and Gerich, 2010; Dimitriadis et al., 2011; Leto and Saltiel, 2012). The stimulatory effect of insulin on glucose uptake in muscle and adipose tissue is a consequence of the rapid translocation of glucose transporters from an intracellular site to the plasma membrane of cells (Dimitriadis et al., 2011; Bogan, 2012).

The majority of insulin-stimulated glucose uptake in peripheral tissues occurs in skeletal muscle, where glucose may be either used immediately for energy supply (glycolysis) or energy storage (glycogen), which is mobilized when fuel demands are high (for example during exercise) or when glucose is not abundant. About 50% of glucose taken up by the muscle is oxidized, 35% is stored as glycogen and about 15% is released as lactate and alanine (Beardsall et al., 2003). As stated, skeletal muscle is insulin-dependent, and thus requires insulin for activation of glucose transport as well as for activation of the major glycolysis enzymes (hexokinase and 6-phosphofructokinase) and glycogenic enzyme (glycogen synthase). At the same time, insulin also inhibits the activity of the main glycogenolysis enzyme, glycogen phosphorylase decreasing glycogen breakdown in the fed state (Saltiel and Kahn, 2001; Ferrer et al., 2003; Dimitriadis et al., 2011).

Insulin also promotes glucose uptake and utilization in the adipose tissue. It increases the rate of glycolysis by stimulating hexokinase and 6-phosphofructokinase activity as well as glycogen synthesis. Additionally, it stimulates fatty acid and triacylglycerol synthesis (triglyceride storage – that are released as free fatty acids and utilized as an energy source by other tissues when the fuel availability is low) while inhibits lipolysis (decrease the activation of hormone sensitive lipase). Thus, insulin reduces free fatty acid levels in plasma and stimulates a simultaneous increase in free fatty acids clearance from the circulation (Frayn, 2002; Beardsall et al., 2003; Dimitriadis et al., 2011; Leto and Saltiel, 2012).

Glycemia is tightly regulated and maintained by insulin, glucagon, incretins, CNS and peripheral nervous system among other factors, to meet metabolic requirements at different physiological conditions. Various cells and tissues such as brain, muscle, gastrointestinal tract, liver, kidney, and adipose tissue are also involved in the regulation of glycemia via glucose transport stimulated or not by insulin (Gerich, 2000; Meyer et al., 2002; Gerich, 2010; Lam, 2010).The glucose transporters are of crucial physiologic relevance since they are responsible for the glucose transport, the key for glucose homeostasis.

Facilitative Glucose Transporters - GLUTs

The GLUT protein family (SLC2 gene family) belongs to the Major Facilitator Superfamily (MFS) of membrane transporters (Pao et al., 1998). Most GLUTs catalyze the facilitative (energy-independent) bidirectional transfer of substrates across plasma membranes. In fact, GLUTs can move sugars into or out of cells using the diffusion gradient of glucose (and other sugars) across plasma membrane depending on the type and the cell metabolic state, besides the metabolic state of the organism. Glucose transporters are expressed in every cell of the body, as it might be anticipated from the key role of glucose in providing metabolic energy and building blocks for the synthesis of biomolecules. Theoretically, the appropriate distribution of the whole-body glucose could be controlled by the tissue-specific expression and regulation of several glucose transporter isoforms with distinct kinetic properties. This scenario appears to account, at least in part, for the mechanism by which glucose is properly distributed among cells and organs under different metabolic conditions (Uldry and Thorens, 2004; Zhao and Keating, 2007; Augustin, 2010).

Thirteen members of this family have been identified, GLUT 1-13 and a proton-coupled myo-inositol transporter (HMIT) (Joost et al., 2002). Common structural features of the GLUT family members are the presence of 12 transmembrane domains (TM) with both the amino and carboxy-terminal ends present on the cytosolic side (intracellular) and a unique N-linked oligosaccharide side-chain present in extracellular loops. Also, the amino acid sequence comparisons of the human GLUT family members range from 14–63% identical and 30–79% conserved. The transporters present several conserved residues and motifs designated as sugar transporter signatures which consist of numerous conserved glycine and tryptophan amino acid residues. In a whole, the data suggest that the conserved amino acids can play important roles in substrate binding and/or in the conformational change during the transport process (Joost and Thorens, 2001; Joost et al., 2002; Zhao and Keating, 2007; Augustin, 2010).

Based on the sequence, functional and predicted structural similarities the GLUT family has been divided into three classes: I, II and III (Joost and Thorens, 2001; Joost et al., 2002; Uldry and Thorens, 2004; Augustin, 2010). The class I comprise the classical transporters GLUT1–4 including the gene duplication of GLUT3, which is named as GLUT14. These were the first transporters to be identified, extensively characterized and deeply studied (Augustin, 2010). In a whole, they can be distinguished according to the tissue distribution and the hormonal regulation. The class II of transporters contains the isoforms GLUT5, 7, 9, and 11 which are 36-40% structurally identical and engaged mainly in the active transport of fructose although more functions for the members of this class are not yet firmly established. Class III is composed of isoforms GLUT6, 8, 10, 12 and HMIT (GLUT13). Among them, GLUT8 and 10 are characterized as glucose transporters and HMIT is an H^+/myo-inositol transporter. The function of GLUT6 and 12 is not well defined yet. One particular structural characteristic of the members of class III is the presence of a larger loop 9 (between helices 9 and 10) containing the N-linked glycosylation site and a shorter extracellular loop 1. On the other hand, in the transporters of classes I and II, the structure consists of the N-linked glycosylation site in a larger loop 1 between transmembrane helices 1 and 2. Also, common to all the isoforms is the presence of an internalization signal (dileucine or YSRI motifs in case of GLUT10) that retains these transporters in an intracellular localization under steady state conditions (Joost and Thorens, 2001; Joost et al., 2002; Uldry and Thorens, 2004; Augustin, 2010).

It should be emphasized, however, that the designation of the family does not necessarily reflect the substrate specificity of its members, which may transport sugars or polyols others than glucose or even more than one specific sugar. The expression of several sugar transporter isoforms in individual tissues, and indeed in cells, is a reflection of the different characteristics of each transporter, and provides a high degree of specificity to the control of glucose uptake under a wide range of glucose concentrations. Clearly, there are still much to be learned about the fundamental roles of the GLUT protein family members (Joost and Thorens, 2001; Joost et al., 2002; Wood and Trayhurn, 2003; Uldry and Thorens, 2004; Augustin, 2010).

GLUT1

GLUT1 was the first transporter characterized by molecular cloning. Human GLUT1 is a strongly hydrophobic membrane-spanning glycoprotein comprising 492 amino acids organized as 12 transmembrane domains with a single N-glycosylation site and with a molecular weight of approximately 54 kDa. It is the most conserved isoform and exhibits approximately 74-98% of sequence identity among species. It presents cytoplasmic NH_2 and COOH termini, and despite its hydrophobicity, GLUT1 is readily accessible to solvent water because of the amphipatic characteristic of the transmembrane domains (eight among the 12 domains) (Mueckler et al., 1985, Zhao, Keating, 2007; Carruthers et al., 2009). This transporter provides an efficient pathway for cellular import and export of glucose. In addition, GLUT1 also transports galactose, mannose, glucosamine and ascorbic acid (Uldry et al., 2002; Carruthers et al., 2009).

GLUT1 is found in almost every tissue with different levels of expression in several cell types and is considered one of the most ubiquitously expressed isoform of GLUT. The expression level usually correlates with the rate of cellular glucose metabolism. GLUT1, also known as the HepG2 or erythrocyte sugar transporter, is highly abundant in erythrocyte membranes making up 3–5% of the total erythrocyte proteins but it was not observed in hepatocyte membranes (Zhao and Keating, 2007; Augustin, 2010).

Besides erythrocytes, GLUT1 is highly expressed at the plasma membrane of proliferating cells forming the early developing embryo, between the oocyte to the blastocyst stages (Hogan et al., 1991). Also, GLUT1 is unique in mediating glucose transfer across the blood-brain barrier, catalyzing the rate-limiting step on glucose supply to the central nervous system, an essential fuel for these cells. It was found in astrocytes, endothelial and epithelial-like cell barriers in the brain (Maher et al., 1994; Uldry and Thorens, 2004; Carruthers et al., 2009).

The transporter is also present in eye, peripheral nerve, placenta, lactating mammary gland and in certain tumor cell lines. Additionally, moderate levels of expression are also observed in adipose tissue, skeletal and cardiac muscle, where GLUT1 acts in a cooperative way with GLUT4. In these insulin-sensitive tissues GLUT1 is present constitutively at the plasma membrane, where it presumably provides the glucose basal level required for cellular activity. Additionally, in the basal state, the tissue-specific GLUT4 is distributed in intracellular compartments. However, during the response to insulin signaling, the GLUT4 is translocated to the plasma membrane increasing significantly the glucose uptake (Augustin, 2010; Zhao and Keating, 2007; Carruthers et al., 2009).

GLUT2

GLUT2 is a facilitative glucose transporter and is expressed in high levels in the liver, pancreas, intestine, kidney and brain. GLUT2 is a low-affinity, high capacity transporter of glucose (K_m in the range of 17 mM) and it has the highest K_m for glucose among the known members of the GLUT family. GLUT2 ensures large bidirectional fluxes of glucose in and out the cell due to its low affinity and high capacity. It also transports other dietary sugars, such as mannose, fructose and galactose within the range of physiological concentrations and recently, the transport of dehydroascorbic acid was also reported (Uldry et al., 2002; Leturque et al., 2009). In the pancreas, GLUT2 is the major glucose transporter expressed in β-cells plasma membrane. Since GLUT2 is a high capacity transporter, the intracellular glucose concentration is directly proportional to the extracellular glucose levels. Also, GLUT2 works as a glucose sensor for insulin secretion, thereby being directly involved in the signaling cascade of glucose-induced insulin secretion (Schuit et al., 2001; Uldry and Thorens, 2004; Zhao and Keating, 2007). In hepatocytes, GLUT2 is present in the sinusoidal membrane where it is involved in both, glucose uptake and glucose release into the blood. In addition, it is involved in the glucose-sensing mechanisms on hepato-portal vein. GLUT2 is also expressed on hypothalamus and on brain stem where it may participate in the mechanisms of glucose sensing involved in the control of glucose homeostasis through the control of counter-regulation, food intake and stimulation of glucose uptake by peripheral tissues (Burcelin and Thorens, 2001; Zhao and Keating, 2007; Uldry and Thorens, 2004). On the proximal renal tubules and on enterocytes, GLUT2 participates in the release of absorbed or reabsorbed glucose into the blood stream. It is located at the basolateral membrane of the respective epithelial cells, where GLUT2 forms part of the transcellular pathway for glucose and fructose transport. Also, under conditions of a high-sugar meal ingestion (increased glucose concentrations), GLUT2 is recruited acutely and transiently to the apical membrane of the epithelium in order to increase glucose absorption as well as reabsorption. The accumulated sugars into the enterocytes and proximal tubules cells are subsequently released into the capillaries via the facilitated transporters, GLUT2 and GLUT5, localized on the basolateral surface of the cells (intestine and kidney). Additionally, there are increasing evidences showing that GLUT2 may also mediate a diffusive component of intestine glucose absorption in the brush-border membrane and in the proximal renal tubules (Zhao and Keating, 2007; Uldry and Thorens, 2004; Kellett et al., 2008; Leturque et al., 2009; Augustin, 2010).

GLUT3

GLUT3 is considered a neuron-specific glucose transporter since it is mainly detected in the brain, more specifically, in dendrites and axons, and its level of expression in different regions of the brain correlates with regional cerebral glucose utilization (Haber et al., 1993; Zhao and Keating, 2007; Augustin, 2010; Simpson et al., 2008). Additionally, besides the brain, GLUT3 is also expressed in testis (spermatozoa - glucose uptake and metabolism necessary for motility and maturation), placenta, pre-implantation embryos or certain cancer cells (Augustin, 2010). GLUT3 is also expressed in lymphocytes, monocytes/macrophages, and platelets. In these cells, it is present in intracellular vesicles that can translocate and fuse to the plasma membrane upon cellular activation to ensure increased glucose uptake and metabolism (Simpson et al., 2008; Thorens and Mueckler, 2010).

GLUT3 transporter has a high affinity for glucose (K_m 1.5 mM) and has the highest turnover number of the GLUT isoforms, thus ensuring an efficient glucose uptake in tissues where the demand for glucose as a fuel is considerable, particularly, in the brain. GLUT3 was also shown to transport mannose, galactose, xylose and dehydroascorbic acid but is unable to transport fructose (Colville et al., 1993; Zhao and Keating, 2007). Thus GLUT3 has both, a higher affinity for glucose than GLUT1, -2, or -4 and at least fivefold greater transport capacity than GLUT1 and -4. This is particularly significant for the role of GLUT3 in neurons where the ambient glucose levels are only around 1–2 mM compared with the 5–6 mM in blood glucose levels. Furthermore, as there are approximately the same number of GLUT1 glucose transporters on glial and endothelial cells, the combination of lower K_m and higher capacity enables the neuron preferential access to available glucose (Simpson et al., 2008).

GLUT4

Besides GLUT1, GLUT4 has been widely studied specially due to its important physiological role which is the rate-limiting step in insulin-stimulated glucose uptake.

Beyond, its expression is higher in the insulin sensitive tissues including brown and white adipose tissue, skeletal and cardiac muscle (Huang and Czech, 2007; Augustin, 2010).

GLUT4 comprises 509 amino acids with 12 helices traversing through the lipid bilayer, a glycosylation site is present in the extracellular loop connecting TM1 and TM2 and a large cytoplasmic loop located between transmembrane helices 6 and 7 (Birnbaum, 1989; Mueckler et al., 1985). Both N and C termini are cytoplasmic and these end regions together with the loop regions possess the distinct transporter specific signature sequences involved in the characteristic membrane trafficking capability of GLUT4. These include a distinctive N-terminus sequence with a potentially critical phenylalanine residue, as well as dileucine and acidic motifs in the COOH terminus. These motifs are likely drive kinetic aspects of both, endocytosis and exocytosis in a continuously recycling trafficking system (Corvera et al., 1994; Piper et al., 1993; Al-Hasani et al., 2002; Huang and Czech, 2007; Augustin, 2010). In terms of kinetic properties, GLUT4 is an insulin sensitive glucose transporter and presents a similar affinity for glucose as GLUT1 with a K_m of 5 mM. It is also capable to transport dehydroascorbic acid and glucosamine (Kasahara, Kasahara, 1997).

In the absence of stimuli, GLUT4 is sequestered in intracellular tubulo-vesicular structures clustered either in *trans*-golgi reticulum or in cytoplasm. At steady-state (basal blood glucose levels) only 5–10% of the total cellular GLUT4 content is exposed at the cell surface in either adipocytes or muscle cells. Insulin and exercise are able to rapidly and acutely stimulate GLUT4 translocation to the plasma membrane. GLUT4 translocation represents the rate-limiting step in insulin controlled glucose uptake on skeletal and heart muscle as well as adipose tissue. Insulin changes the net distribution of GLUT4 within minutes, attaining a new steady-state whereby 20–50% of the cellular GLUT4 content becomes exposed at the surface of muscle or fat cells (Huang and Czech, 2007; Antonescu, 2009). When insulin stimulation is withdrawn the GLUT4 proteins are removed from the plasma membrane by endocytosis and recycled back to the intracellular storage compartments. The insulin-stimulated GLUT4 translocation is achieved through multiple highly organized membrane trafficking events, orchestrated by the insulin receptor signals. Several key molecules linking insulin receptor signals and GLUT4 translocation have been recently identified. The cascade of events triggered by insulin involves the phosphatidylinositol 3-kinase (PI3K) mechanism and the activation of the Rho family small GTP-binding protein TC10. Both pathways involve actin remodeling beneath of plasma membrane and around intracellular membranes. Moreover, GLUT4 translocation to the plasma membrane can also be stimulated by exercise. This involves a signaling pathway different from that activated by insulin; triggering the activation of 5'-AMP-activated protein kinase (AMPK) (Huang and Czech, 2007; Leto and Saltiel, 2012; Rowland et al., 2011; Stöckli et al., 2011).

GLUT5

GLUT5 appears to be the only GLUT protein with a high specificity for fructose (K_m 6 mM) without any noticeable glucose transport activity. Its main site of expression is the apical membrane of intestinal epithelial cells (jejunum), where it provides a major route for the absorption of dietary fructose. Low levels of its mRNA are also detected in human kidney, testes, skeletal muscle, and adipocytes. In humans, GLUT5 is expressed only in mature spermatids, suggesting a selective upregulation of GLUT5 expression during germ cell development, from immature spermatogonia to elongated spermatids. In skeletal muscle and adipose tissue its expression is strictly confined to the plasma membrane of adipocytes and to the sarcolemma of muscle. In brain, it has been found in microglia cells, in the human blood-brain barrier, in mouse cerebellum and in rat hippocampus. Regional expression patterns, GLUT5 gene expression appears to be tightly regulated by developmental, nutritional, hormonal, and circadian influences (Burant et al., 1992; Rand et al., 1993; Uldry and Thorens, 2004; Douard and Ferraris, 2008; Augustin, 2010). In addition, GLUT5 plays a role in reabsorption of fructose in the kidney. This protein is present in the apical plasma membrane of S3 segment of renal proximal straight tubules cells, where GLUT5 can potentially recapture the fructose lost from glomerular filtration. Differences in fructose concentration between filtrate and intracellular compartment determine the direction of fructose flux mediated by GLUT5 (Douard and Ferraris 2008).

GLUT5 plays fructose absorption on intestine where it is located at both apical and basolateral membrane of the absorptive epithelial cells (Blakemore et al., 1995; Zhao and Keating, 2007). It seems that GLUT5 has a collaborative action with GLUT2 on fructose absorption. Fructose is transported across the apical membrane of intestinal epithelial cells by GLUT5 (facilitated diffusion) and exits the enterocyte across the basolateral membrane by GLUT2. Some reports propose that GLUT2 is also involved in the apical transport of fructose and also it was demonstrated that GLUT5 is located on basolateral membrane in human jejunum (Blakemore et al., 1995; Kellet et al., 2008; Douard and Ferraris 2008; Jones et al., 2011).

GLUT6

GLUT6/SLC2A6, originally described as GLUT9 has been reassigned as GLUT6. The human isoform consists of 507 amino acids and approximately 48 kDa. It is highly expressed in brain, spleen and peripheral leukocytes. It appears to exhibit glucose transport activity, but its primary physiological substrate has not been definitively identified (Doege et al., 2000, Lisinski et al., 2001). GLUT6 contains an N-terminus dileucine motif, which helps to retain it intracellular and therefore prevents glucose transport. GLUT6 is only detected at the plasma membrane when the dileucine residues are mutated to alanine or when clathrin dependent endocytosis is blocked by overexpression of a dominant-negative dynamin mutant. Whether mechanisms exist to promote cell-surface translocation of these transporters is not yet known, but it has clearly been established that insulin does not promote GLUT6 cell-surface translocation (Lisinski et al., 2001; Augustin, 2010).

GLUT7

GLUT7 is the most recently cloned GLUT member. The human encoded protein is comprised of 524 amino acids. It is more closely related to GLUT5 with 53% identity and it has a high affinity for both, glucose (K_m 0.3 mM) and fructose, but exhibits a low level of transport activity for these hexoses and does not transport galactose, 2-deoxy-D-glucose and xylose (Li et al., 2004; Manolescu et al., 2005; Cheeseman, 2008). Similarly to the other GLUTs, GLUT7 presents 12 membrane-spanning domains. The likely glycosylation site appears on the first extracellular loop between TMs 1 and 2. The comparison of the alignment sequences of the other GLUT proteins with GLUT7 sequence indicated that in the predicted region to form TM7 there is a hydrophobic sequence that might correlate with the substrate specificity. All GLUTs that transport fructose express an isoleucine residue close to the exo-facial end of TM7, whereas the non-fructose-transporting GLUTs have a valine residue at the same position. Mutations of the isoleucine to valine in GLUT7 resulted in the loss of fructose transport activity, whereas glucose transport remained completely unaffected (Li et al., 2004; Manolescu et al., 2005). GLUT7 is primarily expressed in the apical membrane of the brush-border membrane of enterocytes (small intestine) and in the colon. Additionally, it has been shown that the GLUT7 expression in the small intestine can be increased with increasing dietary carbohydrate contents, correlating the sensing luminal carbohydrate mechanisms with GLUT7 transcription and translation process. Also, the GLUT7 mRNA has been detected in the testes and prostate as well although the expression of the protein was not measured yet (Li et al., 2004; Manolescu et al., 2005; Cheeseman, 2008; Zhao and Keating, 2007). To date, the functional activity of GLUT7 and its ability to mediate hexose transport as well as its physiological regulation in intestinal epithelial cells still needs to be assessed. The very low capacity for glucose and fructose transport exhibited by GLUT7 does raise the possibility that the physiological substrate for this transporter can be something other than a hexose, for example this transporter could exchange another organic compound by glucose or fructose (Cheeseman, 2008).

GLUT8

The amino acid sequence of GLUT8 (formerly GLUTX1) has 29.4% of sequence identity with GLUT1 and exhibits all the elements and motifs (sugar transporter signatures) that are characteristic of the GLUT family (Doege, 2000b; Schmidt et al., 2009). GLUT8 is a member of the class III transporters, which is characterized by a larger and presumably glycosylated extracellular loop 9. The molecular mass of human GLUT8 is 51.5 kDa and it is mainly expressed in the testes, with lower levels found in the brain (cerebellum, brain stem, hippocampus and hypothalamus), adrenal gland, liver, spleen, adipose tissue, skeletal and heart muscle, lung, blastocytes and intestine (Carayannopoulos et al., 2000; Ibberson et al., 2000; Schmidt et al., 2009; DeBosch et al., 2012).

The measurement of its transport activity reveals a relatively high affinity for glucose with a K_m of ~2 mM. However, fructose and galactose compete with glucose while cytochalasin B inhibits GLUT8 indicating a multifunctional activity for this sugar (Ibberson et al., 2000). The functional characterization of GLUT8 revealed a dileucine motif in the N-terminus which is believed to target the protein on intracellular compartments. The mutation of the two leucine residues to alanine leads to the expression of GLUT8 at the cell surface and also to the transport activity in mammalian cells (Ibberson et al., 2000; Lisinski et al., 2001). A similar dileucine motif is present in the COOH terminus of GLUT4, which is located in an intracellular compartment and translocates to the cell surface in response to insulin. However, the induction of GLUT8 translocation by insulin, as observed with GLUT4, has been controversial (Sandoval et al., 2000; Schmidt et al., 2009). Thus, recent evidences indicate that GLUT8 is primarily associated with intracellular membranes. It has been detected in several intracellular organelles, such as the rough endoplasmic reticulum, *trans*-Golgi, late endosomes and lysosomes. Taking these into account, GLUT8 seems to be an intracellular transporter potentially catalyzing transport of sugars or sugar derivatives into or out of intracellular organelles (Schmidt et al., 2009).

GLUT9

GLUT9 is a novel, facilitative glucose transporter isoform that exists as two alternative splice variants encoding two proteins (GLUT9a and GLUT9b). The human GLUT9 gene encodes two isoforms through the use of alternative promoters. The major isoform GLUT9a is comprised of 540 amino acids while the GLUT9 ΔN (b) isoform is comprised of 512 amino acids differing at the N-terminus that is shorter. GLUT9 shares 44% and 38% amino acid sequence similarity to GLUT7-GLUT5 and GLUT1, respectively, and also shows several motifs characteristics of the class II transporter (Phay et al., 2000; Augustin et al., 2004; Doblado and Moley, 2009). The substrate specificity of GLUT9 is unique since it is a high-capacity uric acid transporter (K_m of 0.9 and 0.6 mM for the human and mouse protein, respectively) in addition to transport hexose sugars (high affinity transport for glucose and fructose with K_m values of 0.6 and 0.4 mM, respectively). GLUT9 supposedly exchanges glucose or fructose by urate (Augustin et al., 2004; Manolescu et al., 2007, Vitart et al., 2008; Augustin, 2010). In terms of expression, both forms of GLUT9, protein and mRNA are present in the epithelia of various tissues however the two splice variants are expressed differentially in polarized cells. In these cells GLUT9a is localized predominantly on the basolateral surfaces while GLUT9b is expressed on apical surfaces (Doblado and Moley, 2009). In human, GLUT9a is expressed mainly in the kidney and liver, whereas GLUT9b is detected only in the kidney and placenta. In the liver, GLUT9 is expressed on basolateral membrane of hepatocytes. Additionally, in kidney GLUT9 is localized on proximal tubule of epithelial cells and the GLUT9a isoform is localized exclusively on basolateral membrane, whereas GLUT9b is localized on apical membrane (Keembiyehetty et al., 2006; Phay et al., 2000; Augustin et al., 2004). Also, GLUT9 is localized in the β-cells of human and mouse islets and in human chondrocytes. Therefore, GLUT9 represents a functional urate transporter *in vivo* since it is the major player to urate homeostasis by handling urates from kidney and liver (Phay et al., 2000; Evans et al., 2009; Mobasheri et al., 2002; Doblado and Moley, 2009).

GLUT10

GLUT10, composed of 541 amino acids in humans, is a transporter with very high affinity for 2-deoxy-D-glucose (K_m ~0.3 mM). Galactose and glucose compete with 2-deoxy-D-glucose uptake which is inhibited by phloretin (Dawson et al., 2001). The amino acid sequence presents all sugar transporter signatures with the exception of the PESPR (proline, glutamate, serine, and arganine) motif just after the helix 6 that is conserved for all the GLUT isoforms and the absence of the dileucine motif (it contains an alternative C-terminus motif which seems to be involved in the potential intracellular localization of the transporter) (McVie-Wylie, 2001; Dawson et al., 2001; Coucke et al., 2006; Augustin, 2010).

Analysis of the human GLUT10 mRNA distribution suggests that GLUT10 is expressed in a wide variety of glucose-metabolizing tissues such as heart, lung, brain, liver, skeletal muscle, pancreas, placenta and kidney. The expression of GLUT10 is also detected in human and mouse white adipose tissue (McVie-Wylie, 2001; Dawson et al., 2001). Interestingly, the chromosome localization of GLUT10 is in a locus associated with type-2 diabetes and thus, it is a possible candidate for a susceptibility gene involved in this disease (Dawson et al., 2001).

GLUT11

Human GLUT11 (encoded by SLC2A11gene) is a novel sugar transporter which exhibits significant sequence similarity with the members of the GLUT family. The amino acid sequence deduced from its cDNAs predicts 12 putative membrane-spanning helices and all the motifs (sugar-transporter signatures) that have previously been shown to be essential for sugar-transport activity. The closest relative of GLUT11 is the fructose transporter GLUT5 (sharing 41.7% amino acid identity with GLUT11). GLUT11 exhibits detectable glucose and fructose transport activity but not for galactose (Doege et al., 2001; Sasaki et al., 2001; Scheepers et al., 2005; Zhao and Keating, 2007).

In human, three isoforms of GLUT11 (GLUT11-A, GLUT11-B, and GLUT11-C) have been cloned, which differ only at their N-termini sequences (coding 7, 14 and 10 amino acids, respectively). The three isoforms are expressed in a tissue specific manner: GLUT11-A (493 amino acid residues and low-affinity for glucose) is present in heart, skeletal muscle, and kidney; GLUT11-B (503 amino acid residues and it was shown to increase fructose transport in Chinese hamster ovary cells) is expressed in placenta, liver, brain, lung, trachea, adipose tissue, and kidney; and GLUT11-C found in adipose tissue, heart, skeletal muscle, and pancreas (Scheepers et al., 2005; Sasaki et al., 2001; Wu et al., 2002). To note, the northern blot analysis, which represents the cumulated expression of the three spliced forms, has shown GLUT11 mRNA expressed in various tissues, most abundantly in skeletal muscle and heart, at an intermediate level in the brain, small intestine, lung and peripheral blood leukocytes and at low levels in the liver, kidney and placenta (Wu et al., 2002).

GLUT12

GLUT12 is a larger protein containing 617 amino acids and contains several similar features to GLUT4 and GLUT8. As a class III isoform, GLUT12 also contains dileucine motifs — both at the N- and C-terminal ends of the protein, suggesting a predominant intracellular expression site (Rogers et al., 2002; Stuart et al., 2009; Flessner et al., 2009). GLUT12 has been shown to have preferential affinity for glucose and 2-deoxy-D-glucose over other hexoses (Rogers et al., 2003b). GLUT12 is mainly expressed in skeletal muscle, heart, small intestine, adipose tissue, prostate, pregnant and lactating rat mammary gland (Rogers et al., 2002; Macheda et al., 2003; Ware et al., 2011). Furthermore, GLUT12 mRNA is highly detected in breast and prostate cancer cells (Rogers et al., 2003a; Chandler et al., 2003).

As stated, GLUT12 is present in human skeletal muscle and is distributed similarly to GLUT4 between intracellular and cell surface in the basal state. In response to euglycemic insulin infusions in humans (euglycemic clamp experiment), substantial amounts of GLUT12 moved to the muscle plasma membrane-enriched fractions (obtained from muscle biopsies) in a manner that was quantitatively very similar to the effects on GLUT4. GLUT12 represents about 12% of the insulin-translocatable GLUTs in normal muscle (Stuart et al., 2006; Stuart et al., 2009; Purcell et al., 2011). Thus, the presence of target motifs similar to GLUT4 and GLUT8 as well as the localization in insulin-sensitive tissues, postulate that the GLUT12 can be a second insulin-responsive glucose transporter (Purcell et al., 2011; Stuart et al., 2009; Rogers et al., 2002; Rogers et al., 2003b).

GLUT13/HMIT

GLUT13, also called proton-coupled myo-inositol transporter (HMIT), is a myo-inositol transporter and is the only GLUT protein that appears to work as a proton-coupled symporter (Uldry et al., 2001; Bourgeois et al., 2005). As for other class III GLUT members, HMIT is restricted to an intracellular location since it is a glycosylated protein that contains three conserved internalization signals: an endoplasmic reticulum (ER) retention signal in the N-terminus region, a dileucine internalization signal and a tyrosine based internalization motif at the C-terminal end (Uldry et al., 2001). Although the HMIT amino acid sequence contains all motifs known to be important for glucose transport activity, surprisingly no sugar transport activity has been found for HMIT. Instead, HMIT has been identified as a lower-affinity/higher-capacity H^+-coupled myo-inositol symporter with a K_m of about 100 μM. More recently, HMIT has been shown to transport inositol-3-phosphate (IP3) (Di Daniel et al., 2009a; Di Daniel et al., 2009b). This transport activity is strongly activated by decreasing the extracellular pH, which increases the V_{max} without changing the K_m (~100 μM) (Uldry et al., 2001, 2004; Bourgeois et al., 2005; Fu et al., 2012).

Despite the low level of expression in adipose tissue and kidney, HMIT is predominantly expressed in the brain, with higher expression in hippocampus, hypothalamus, cerebellum and brain stem. In neurons, HMIT is generally expressed in the Golgi apparatus. However, in brain astrocytes it is detected both on the plasma membrane and in the cytosol. When HMIT is present in intracellular vesicles its movement to the cell surface can be induced following cell depolarization, activation of protein kinase C or increased intracellular calcium concentrations (Di Daniel et al., 2009a; Uldry et al., 2001, 2004). The predominant expression of HMIT in the CNS suggests that it has a key role in the control of myo-inositol and phosphoinositides metabolism in the brain. In this tissue myo-inositol is a major osmolyte and also serves as the precursor for phosphatidylinositol, the major inositol-containing phospholipid that are crucial to the brain signal transduction (Cremona and De Camilli, 2001; Di Daniel et al., 2009a; Di Daniel et al., 2009b).

GLUT14

GLUT14 belongs to the class I transporters and consists of 11 exons with a genomic organization similar to that of GLUT3 probably it resulted from a duplication of GLUT3 and thus it is most likely a glucose transporter although its specific role in glucose metabolism is unknown. GLUT14 has two alternatively spliced forms: the shorter form of GLUT14 (GLUT14-S) consists of 10 exons and produces a 497 amino acid protein that is 94.5% identical to GLUT3. The longer form (GLUT14-L) has an additional exon and codes for a protein with 520 amino acids that differs from GLUT14-S only at the N-terminus. GLUT14-S/L contains 12 putative membrane-spanning helices along with sugar-transporter signature motifs that have previously been shown to be essential for sugar transport activity. The putative glycosylation sites of GLUT14-S/L are present in loop 1. In contrast to the expression of GLUT3 in many tissues, both isoforms of GLUT14 are specifically expressed in testes (Wu and Freeze, 2002).

Na^+/Glucose Cotransporters - SGLTs

The Na^+/glucose cotransporters (SGLTs isoforms), from the SCL5 gene family to active transport proteins, transport monosaccharides only in the presence of a sodium-driven electrochemical gradient (Augustin, 2010; Sabino-Silva et al., 2010). Each isoform has a tissue-specific distribution and more than one isoform can be expressed in the same tissue simultaneously or not. The expressions of the transporter genes, the kinetic properties as well as the proposed functions for each transporter are differentially regulated. These differences reveal that each transporter isoform plays a specific role in glucose uptake in various tissues and in glucose homeostasis (Zhao and Keating, 2007).

The SGLTs are extremely important for survival since they assure energy substrate taken from food (glucose absorption) and also prevent its loss in urine and/or saliva (glucose reabsorption). There are 12 members of the human family (SLC5) that code for 60- to 80-kDa membrane proteins containing 580 –718 amino acids. These proteins are involved in the transport of glucose (Na^+- cotransporters for sugars), myo-inositol, iodide, amino acids, short-chain fatty acids, vitamins, osmolytes, choline and some ions across the brush border membrane of the small intestine epithelium and the renal proximal tubules. Although 6 isoforms of Na^+/glucose cotransporters have been described, the SGLT1 and SGLT2 isoforms (encoded by the genes SLC5A1 and SLC5A2, respectively) are believed to be the most important ones and have been extensively explored in studies focusing on glucose fluxes (revised in: Wright and Turk, 2004; Sabino-Silva et al., 2010; Wright et al., 2011).

SGLT1

Hediger et al., 1987, used a novel expression cloning strategy to isolate cDNA encoding the sodium-dependent glucose transporter expressed in the small intestinal mucosa. SGLT1 is a 75 kDa membrane protein with a Na^+/glucose stoichiometry of 2:1 and shows no sequence homology to GLUT proteins. It contains 664 amino acid residues arranged in 14 transmembrane α-helices, a single glycosylation site between transmembrane helices 5 and 6 and two phosphorylation sites between 6 and 7 as well as between 8 and 9 transmembrane helices (Hediger et al., 1987; Turk and Wright, 1997; Wright et al., 2004). The NH_2 and COOH termini are extracellular and intramembrane and the glucose binding and translocation domains involve five transmembrane segments, which are next to the COOH terminus (Wright et al., 2004; Sabino-Silva et al., 2010; Wright et al., 2011).

The SGLT1 protein, encoded by the SLC5A1 gene, is located mainly in the apical membranes of the small intestine (enterocytes) but it has also been detected in the kidney (renal proximal straight tubules - S3 segment), trachea, parotid and submandibular salivary glands as well as in the heart. SGLT1 has high affinity for glucose, but low capacity to transport it and is specifically inhibited by phlorizin. Furthermore, the cotransporter affinity and the kinetics are the same for glucose and galactose, the preferred substrates, whereas mannose (or 2-deoxy-D-glucose) is barely accepted. Through the SGLT1, two sodium ions are transported for each molecule of glucose. This cotransport is a secondary Na^+/K^+ ATPase-dependent process, able to transport glucose into the cell against its concentration gradient. Crane et al. (1961) were pioneering researchers who postulated that the active glucose transport across the intestinal epithelium is driven by the sodium gradient across the membrane, by means of a Na^+/glucose cotransport. It is believed that Na^+ and glucose transport by the SGLT1 occur by a sequential mechanism in which two Na^+ ions on the extracellular side bind to the transporter right before glucose inducing a conformational change in the glucose binding site and increasing the transporter affinity for this substrate. Then, another conformational change determines the simultaneous shift of the Na^+ and glucose binding sites to the inner surface of the membrane with the glucose molecule and immediately after the two Na^+ ions being released within the cell. The low affinity of the cytosolic sites for glucose and Na^+ and the low intracellular Na^+ concentration relative to the extracellular concentration (10 versus 140 mequiv/L) promote these dissociations. Within this process, the SGLT1 protein is able to recycle around one thousand times per second at 37°C. The sodium electrochemical potential gradient across the brush border membrane promotes sodium influx and provides the energy which drives glucose into the enterocytes. Thus, the energy for the whole process comes from the adenosine-5'-triphosphate (ATP) consumed by Na^+/K^+ ATPase. The process is reversible, i.e., the rate and direction of Na^+/glucose cotransport depend on the size and direction of the sodium electrochemical potential gradient and the glucose concentrations. Glucose that accumulated inside the enterocyte is then transported across the basolateral membrane by facilitative diffusion transporters (Wood and Trayhurn, 2003; Wright et al., 2003; Wright et al., 2004; Sabino-Silva et al., 2010; Wright et al., 2011).

It has been demonstrated that besides the two Na^+ ions plus the one glucose molecule, the SGLT1 protein is also able to transport H_2O and urea molecules through a water and urea channel in the protein. The coupled water/urea transport occurs due to changes induced by ligands (sodium and glucose) in the protein structure during the transport cycle. The significance of the SGLT1 channel is that aquaporins and urea transporters are absent from the intestinal brush border and therefore SGLT1 may play an important role in fluid and urea transport across the intestine. It is also similar for other cotransporters as well, that act as water pumps. There is a stoichiometric relationship between the rate of Na^+/glucose cotransport and the rate of water transport (2 Na^+: 1 glucose ~ 260 water molecules). Coupled-water transport is independent of the osmotic gradient and even can occur against it. Also, together with the Na^+ and glucose transport SGLT1 promotes the depolarization of plasma membrane, which can serve as a signal and in the absence of glucose SGLT1 can support a Na^+ current (Wright et al., 2004; Wright et al., 2011).

SGLT2

The SGLT2 was originally described as an amino acid cotransporter called Na^+-dependent neutral-amino acid transporter (SAAT1), but later it was recognized as a glucose transporter (Mackenzie et al., 1994). This protein contains 672 amino acids and its NH_2 and COOH termini are extracellular. The SGLT2 transports glucose and galactose (low affinity) with different affinities via a secondary active transport mechanism. Differently from SGLT1, SGLT2 is a low-affinity and high-capacity glucose transporter, which transports 1 Na^+ ion for each glucose molecule (stoichiometry of 1 Na^+:1 sugar). The Na^+-electrochemical gradient provided by the Na^+–K^+ ATPase pump is used to transport glucose into cells against its concentration gradient (Wood and Trayhurn, 2003; Sabino-Silva et al., 2010; Wright et al., 2011). The SGLT2 encoding gene (SLC5A2) is predominantly expressed in kidneys, but low mRNA expression has also been demonstrated in mammary glands, liver, lungs, intestine, skeletal muscle, heart, thyroid, cerebellum, salivary glands and spleen (Turk and Wright, 1997; Wood and Trayhurn, 2003; Sabino-Silva et al., 2010).

Usually, all filtered glucose is reabsorbed into blood by the renal proximal tubules (S1, S2 and S3 segments). The mechanisms of glucose reabsorption across the renal tubules are quite similar to that in the small intestine, mediated by an active process in the brush-border membrane of the tubular epithelium and a facilitated process in the basolateral membrane. The sodium electrochemical gradient generated by the active sodium transport provides the energy required for glucose transport. The SGLT2 is highly expressed in the luminal membrane of the S1 and S2 segment of the renal proximal tubule. This isoform is the main responsible for glucose reabsorption from the glomerular filtrate. In fact, more than 90% of the filtered glucose is reabsorbed in the initial segments of the proximal tubule via SGLT2. Any remaining glucose is recovered by SGLT1 (also expressed in the kidney – S3 segment) thus preventing glucose loss in the urine (Wood and Trayhurn, 2003; Wright et al., 2004; Sabino-Silva et al., 2010; Wright et al., 2011).

Other Na^+/Glucose Cotransporters

Six among twelve known SGLT family members have the capacity to bind and/or transport monosaccharides and besides SGLT1 and SGLT2, there are other four isoforms (SGLT3 to 6) that have been described (Wright et al., 2011).

SGLT3 was firstly identified as a renal SAAT1 in pigs. However, studies with the human analogue of SGLT3 (hSGLT3, SLC5A4), reclassified it as a Na^+-glucose cotransporter (Wright et al., 2011). SGLT3 has functional characteristics intermediate between SGLT1 and SGLT2. So, whereas SGLT3 stoichiometry is the same as that for SGLT1 (2 Na^+:1 sugar), sugar affinity and specificity are similar to SGLT2 (SGLT3 has low affinity for sugars, but is highly selective for D-glucose) (Díez-Sampedro et al., 2003). SGLT3 has been detected in a wide variety of tissues such as small intestine, testes, spleen, uterus, brain, kidney, lung and skeletal muscle. Also, it has been described in cholinergic neurons in the enteric (intestinal) nervous system and in the neuromuscular junctions in the skeletal muscle (Díez-Sampedro et al., 2003; Wright et al., 2004; Wright et al., 2011). It is noteworthy that in some species (for example, in humans) SGLT3 is not a sugar transporter, instead it works as a glucose-sensor in the plasma membrane. It conveys information to the cell about the external glucose concentration acting as Na^+-glucose cotransporter or a glucose-sensitive ion channel, directly through the membrane potential (increases in glucose concentration depolarize plasma membrane through Na^+-glucose cotransport or by an increase of Na^+ conductance) or indirectly coupled through plasma membrane G proteins (Díez-Sampedro et al., 2003; Wright et al., 2004; Wright et al., 2011).

Limited studies have been carried out with other cotransporters (SGLTs 4, 5 and 6). There is one report on SGLT4 (SLC5A9) showing that it is a low-affinity glucose transporter. Also, it exhibits unusual sugar selectivity since it transports D-mannose and fructose but not galactose or 3-O-methyl-D-glucoside. SGLT4 is expressed predominantly in small intestine and kidney, liver, brain, lung, trachea, uterus and pancreas (Tazawa et al., 2005; Wright et al., 2011).

In addition, SGLT5 (SLC5A10) is expressed at kidney cortex and exhibits Na^+-dependent, phlorizin-sensitive transport of mannose, fructose and galactose. SGLT5 is one of the specific transporters that contribute to the plasma mannose levels by regulating its reabsorption in the kidney. Interestingly, whereas SGLT4 is predominantly expressed in the intestine, SGLT5 is exclusively found in the kidney, suggesting a complementary role for these two transporters on regulation of mannose absorption and reabsorption, respectively

(Leicht et al., 2006; Grempler et al., 2012; Chen et al., 2010; Wright et al., 2011). Finally, SGLT6 (SMIT2 - Na^+/myo-inositol cotransporter), encoded by the SLC5A11 gene, is a 675 amino acid cotransporter that specifically facilitates the transport of two myo-inositols per sodium ion. SGLT6 has low affinity for D-glucose in the small intestine and kidney and its preferred substrate is D-chiro-inositol. SGLT6 is expressed in many different organs and tissues including the intestine, kidney, brain and lung (Coady et al., 2002; Wright et al., 2011).

Perspectives

The glucose uptake into skeletal muscle and adipose tissue is one of the most important metabolic pathways and both tissues are the major sites of glucose disposal, thereby regulating glucose homeostasis. Facilitated glucose uptake into fat and muscle cells is mediated by an enriched-plasma membrane concentration of GLUT4 via insulin and/or contraction-mediated GLUT4 translocation in the case of muscle cells (Furtado et al., 2002; Watson et al., 2004).

Insulin, the most potent physiological anabolic agent, is able to promote the synthesis and storage of lipids, proteins and carbohydrates as well as to inhibit their catabolism and metabolites releasing into the blood. The first step by which insulin regulates the transport of glucose into the cells involves the traffic of GLUT4 in both, skeletal muscle and fat cells (Leto and Saltiel, 2012).

Although the GLUT4 translocation is a complex process, glucose uptake is accomplished via insulin-responsive GLUT4 traffic rather than by increasing the intrinsic activity of the transporter (Watson et al., 2004; Leto and Saltiel, 2012). Recent advances indicate that the intestinal sugar absorption is one of the major central mechanisms that determine postprandial glycemia.

The activity of disaccharidases, including sucrase and isomaltase, is one of the important factors that seem to protect postprandial hyperglycemia (Liu et al., 2011). It has been showed high activity of these enzymes both in small intestine of diabetic patients and experimental diabetic animals (Dyer et al., 2002; Liu et al., 2011). In addition the classical active absorption mediated by the Na^+/glucose cotransporter (SGLT1), for GLUT2 is attributed a paracellular flow of glucose that depends on high glucose concentration during meal intake and, both transporters are regulated by enterocytes (Kellet and Brot-Larroche, 2005; Wright et al., 2011).

In essence, a multiplicity of mechanisms in intestine (glucose transporters and disaccharidases), skeletal muscle, adipose tissue, kidney, CNS and β–cells interact to mediate glucose homeostasis under different conditions (Kellet et al., 2008; Pereira et al., 2011; Pereira et al., 2012; Wright et al., 2011). Several exogenous natural or synthetic derivative compounds have been contributed to the advances of new models to identify potential therapeutic targets to study and understand the glycemic control (Cazarolli et al., 2006; Zanatta et al., 2007; Cazarolli et al., 2008a, b; Cazarolli et al., 2009a, b; Damazio et al., 2009; Folador et al., 2010; Pereira et al., 2011; Kappel et al., 2012; Frederico et al., 2012; Kappel et al., 2013a, b). However, despite this progress, many questions remain to be understood. So, there are some issues that should be addressed to enable the characterization of glucose transporters as specific and critical target to the diabetes and obesity therapy. For example, these transporters should be assayed using *in silico* dynamic approach, analyzed by X-ray diffraction.

Also, studies of protein-ligand interaction using circular dichroism, alteration on the structure of protein by nuclear magnetic resonance, studies on biophysical analysis by fluorescence and surface plasmon resonance in the presence of different compounds would help to clarify the GLUTs network. Furthermore, particularly the nanotechnology, patch clamp, proteins transfection and siRNAs techniques can dramatically helps to biopharmaceutical knowledge and improve the quality of life of patients, in the near future.

Acknowledgments

The studies carried out in the authors' laboratory were supported by Conselho Nacional de Desenvolvimento e Tecnológico (CNPq), Coordenação de Pessoal de Nível Superior (CAPES), CAPES/COFECUB (Project 554/07), Fundação de Amparo à Pesquisa do Estado de Santa Catarina (FAPESC) and Financiadora de Estudo e Projetos (FINEP).

References

Al-Hasani, H; Kunamneni, R; Dawson, K; Hinck, C.S; Muller-Wieland, D; Cushman, S.W. Roles of the N- and C-termini of GLUT4 in endocytosis. *Journal of Cell Science,* 2002 115:131-140.

Antonescu, C.N; Foti, M; Sauvonnet, N; Klip, A. Ready, set, internalize: mechanisms and regulation of GLUT4 endocytosis. *Bioscience Reports,* 2009 29(1):1-11.

Araújo, J.R; Martel, F. Regulação da Absorção Intestinal de Glicose - Uma Breve Revisão. *Arquivos de Medicina,* 2009 23, (2):35-43.

Aronoff, S.L; Berkowitz, K; Schreiner, B; Want, L. Glucose metabolism and regulation: beyond insulin and glucagon. *Diabetes Spectrum: A Publication of American Diabetes Association,* 2004 17:183–190.

Augustin, R. The protein family of glucose transport facilitators: it's not only about glucose after all. *IUBMB Life,* 2010 62(5):315–333.

Augustin, R; Carayannopoulos, MO; Dowd, LO; Phay, JE; Moley, JF; Moley, KH. Identification and characterization of human glucose trans-porter-like protein-9 (GLUT9): alternative splicing alters trafficking. *The Journal of Biological Chemistry,* 2004 279:16229–16236.

Beardsall, K; Yuen, K; Williams, R; Dunger, D. Applied physiology of glucose control. *Current Paediatrics,* 2003,13:543-548.

Bechmann, L.P; Hannivoort, R.A; Gerken, G; Hotamisligil, G.S; Trauner, M; Canbay, A. The interaction of hepatic lipid and glucose metabolism in liver diseases. *Journal of Hepatology,* 2012 56:952–964.

Birnbaum, M.J. Identification of a novel gene encoding an insulin-responsive glucose transporter protein. *Cell,* 1989 57:305–15.

Blakemore, SJ; Aledo, JC; James, J; Campbell, FC; Lucocq, JM; Hundal, HS. The GLUT5 hexose transporter is also localized to the basolateral membrane of the human jejunum. *The Biochemical Journal,* 1995 309:7–12.

Bogan, J.S. Regulation of Glucose Transporter Translocation in Health and Diabetes. *Anual Review of Biochemistry,* 2012 81:507–32.

Bourgeois, F; Coady, M.J; Lapointe, J.Y. Determination of transport stoichiometry for two cation-coupled myo-inositol cotransporters: SMIT2 and HMIT. *The Journal of Physiology,* 2005 563:333-343.

Burant, C.F; Takeda, J; Brot-Laroche, E; Bell, G.I; Davidson, NO. Fructose transporter in human spermatozoa and small intestine is GLUT5. *The Journal of Biological Chemistry,* 1992 267:14523-14526.

Burcelin, R; Thorens, B. Evidence that extrapancreatic GLUT2 dependent glucose sensors control glucagon secretion. *Diabetes,* 2001 50:1282–1289.

Carayannopoulos, M.O; Chi, M.M; Cui, Y; Pingsterhaus, J.M; McKnight, RA; Mueckler, M; Devaskar, SU; Moley, KH. GLUT8 is a glucose transporter responsible for insulin-stimulated glucose uptake in the blastocyst. *The Procedings of National Academy of Sciences of the United States of America,* 2000 97:7313-7318.

Carruthers, A; DeZutter, J; Ganguly, A; Devaskar, S.U. Will the original glucose transporter isoform please stand up! *American Journal of Physiology: Endocrinology and Metabolism,* 2009 297:836–848.

Cazarolli, L.H; Zanatta, L; Jorge, A.P; de Sousa, E; Horst, H; Woehl, V.M; Pizzolatti, M.G; Szpoganicz, B; Silva, F.R. Follow-up studies on glycosylated flavonoids and their complexes with vanadium: their anti-hyperglycemic potential role in diabetes. *Chemical Biological of Interactions,* 2006 163:177-191.

Cazarolli, L.H; Zanatta, L; Alberton, E.H; Figueiredo, M.S; Folador, P; Damazio, R.G; Pizzolatti, M.G; Silva, F.R. Flavonoids: Prospective Drug Candidates. *Mini Reviews in Medicinal Chemistry,* 2008a 13:1429-1440.

Cazarolli, L.H; Zanatta, L; Alberton, E.H; Figueiredo, M.S; Folador, P; Damazio, R.G; Pizzolatti, M.G; Silva, F.R. Flavonoids: Cellular and Molecular Mechanism of Action to Glucose Homeostasis. *Mini Reviews in Medicinal Chemistry,* 8, 2008b:1032-1038.

Chandler, J.D; Williams, E.D; Slavin, J.L; Best, J.D; Rogers, S. Expression and localization of GLUT1 and GLUT12 in prostate carcinoma. *Cancer,* 2003 97:2035–2042.

Cheeseman, CI. GLUT7: a new intestinal facilitated hexose transporter. *American Journal of Physiology: Endocrinology and Metabolism,* 2008 295:238–241.

Chen, J; Wiliams, S; Ho, S; Loraine, H; Hagan, D; Whaley, J; Feder, J. Quantitative PCR tissue expression profiling of the human SGLT2 gene and related family members. *Diabetes Therapy: Research, Treatment and Education of Diabetes and Related Disorders,* 2010 1(2):57–92.

Coady, M.J; Wallendorff, B; Gagnon, D.G; Lapointe, J.Y. Identification of a novel Na/myo-inositol cotransporter. *The Journal of Biological Chemistry,* 2002 277:35219 –35224.

Colville, C.A; Seatter, M.J; Jess, T.J; Gould, G.W; Thomas, H.M. Kinetic analysis of the liver-type (GLUT2) and brain-type (GLUT3) glucose transporters in Xenopus oocytes: substrate specificities and effects of transport inhibitors. *The Biochemical Journal,* 1993 290(3):701–706.

Corvera, S; Chawla, A; Chakrabarti, R; Joly, M; Buxton, J; Czech MP. A double leucine within the GLUT4 glucose transporter COOH-terminal domain functions as an endocytosis signal. *The Journal of Cell Biology,* 1994 126:979–989.

Coucke, P.J; Willaert, A; Wessels, M.W; Callewaert, B; Zoppi, N; De Backer, J; Fox, JE; Mancini, GM; Kambouris, M; Gardella, R; Facchetti, F; Willems, PJ; Forsyth, R; Dietz, HC; Barlati, S; Colombi, M; Loeys, B; De Paepe, A. Mutations in the facilitative glucose transporter GLUT10 alter angiogenesis and cause arterial tortuosity syndrome. *Nature Genetics,* 2006 38:452–457.

Crane, R.K; Miller, D; Bihler, I. (1961). The restrictions on possible mechanisms of intestinal transport of sugars. In Kleinzeller, A.; Kotyk, A. *Membrane Transport and Metabolism. Proceedings of a Symposium held in Prague, August 22–27, 1960.* Prague: Czech Academy of Sciences. 439–449.

Cremona, O; De Camilli, P. Phosphoinositides in membrane traffic at the synapse. *The Journal of Science,* 2001 114:1041–1105.

Damazio, R.G; Zanatta, A.P; Cazarolli, L.H; Mascarello, A; Chiaradia, L.D; Nunes, R.J; Yunes, R.A; Silva F.R. Nitrochalcones: potential in vivo insulin secretagogues. *Biochimie,* 2009 91, 1493-1498.

Dawson, P.A; Mychaleckyj, J.C; Fossey, SC; Mihic, J; Craddock, A.L; Bowden, D.W. Sequence and functional analysis of GLUT10: a glucose transporter in the type 2 diabetes-linked region of chromosome 20q12–13.1. *Molecular Genetics Metabolism,* 2001 74:186 –199.

DeBosch, B.J; Chi, M; Moley, K.H. Glucose Transporter 8 (GLUT8) Regulates enterocyte fructose transport and global mammalian fructose utilization. *Endocrinology,* 2012 153:4181–4191.

Di Daniel, E.; Kew, J.N; Maycox, PR. Investigation of the H+–myo-inositol transporter (HMIT) as a neuronal regulator of phosphoinositide signalling. *Biochemical Society Transactions,* 2009b 37(5):1139-1143.

Di Daniel, E; Mok, M.H; Mead, E; Mutinelli, C; Zambello, E; Caber-lotto, L.L; Pell, T.J; Langmead, C.J; Shah, A.J; Duddy, G; Kew, J.N; Maycox, PR. Evaluation of expression and function of the H+/myo-inositol transporter HMIT. *BMC Cell Biology,* 2009a 10, 54.

Díez-Sampedro, A; Hirayama, B.A; Osswald, C; Gorboulev, V; Baumgarten, K; Volk, C; Wright; E.M; Koepsell, H. A glucose sensor hiding in a family of transporters. *PNAS,* 2003 100 20:11753–11758.

Dimitriadis, G; Mitrou, P; Lambadiari, V; Maratou, E; Raptis, SA. Insulin effects in muscle and adipose tissue. *Diabetes Reserach of Clinical Practice,* 2011 93:52–59.

Doblado, M; Moley, K.H. Facilitative glucose transporter 9, a unique hexose and urate transporter. *American Journal of Physiology: Endocrinology and Metabolism,* 2009 297, 831–835.

Doege, H; Bocianski, A; Joost, H.G; Schurmann, A. Activity and genomic organization of human glucose transporter 9 (GLUT9), a novel member of the family of sugar-transport facilitators predominantly expressed in brain and leucocytes. *The Biochemical Journal,* 2000 350:771-776.

Doege, H; Bocianski, A; Scheepers, A; Axer, H; Eckel, J; Joost, H.G; Schurmann, A. Characterization of human glucose transporter (GLUT) 11 (encoded by SLC2A11), a novel sugar-trans-port facilitator specifically expressed in heart and skeletal muscle. *The Biochemical Journal,* 2001 359:443–449.

Doege, H; Schurmann, A; Bahrenberg, G; Brauers, A; Joost, H.G. GLUT8, a novel member of the sugar transport facil-itator family with glucose transport activity. *The Journal of Biological Chemistry,* 2000b 275:16275 –16280.

Donato, J. The central nervous system as a promising target to treat diabetes mellitus. *Current Topics in Medicinal Chemistry*, 2012 12:2070-2081.

Douard, V; Ferraris, R.P. Regulation of the fructose transporter GLUT5 in health and disease. *American Journal of Physiology: Endocrinology and Metabolism,* 2008 295, 227–237.

Dyer, J; Daly, K; Salmon, K.S.H; Arora, D.K; Kokrashvili, Z; Margolskee, R.F, Shirazi-Beechey, S.P. Intestinal glucose sensing and regulation of intestinal glucose absorption. *Biochemical Society Transactions,* 2007 35(5):1191-1194.

Dyer, J.; Wood, I.S; Palejwala, A; Ellis, A; Shirazi-Beechey, S.P. Expression of monosaccharide transporters in intestine of diabetic humans. *American The Journal of Physiology Gastrointestinal and Liver Physiology,* 2002 282:G241-248.

Evans, S.A; Doblado, M; Chi, M.M; Corbett, J.A; Moley, K.H. Facilitative Glucose Transporter 9 (GLUT9) expression affects glucose sensing in pancreatic beta cells. *Endocrinology,* 2009 150:5302–5310.

Ferrer, J.C; Favre, C; Gomis, R.R; Fernández-Novell, J.M; García-Rocha, M; de la Iglesia, N; Cid, E; Guinovart, J.J. Control of glycogen deposition. *FEBS Letters,* 2003 546:127-132.

Flessner, L. B.; Moley, K. H. Similar [DE]XXXL[LI] motifs differentially target GLUT8 and GLUT12 in Chinese hamster ovary cells. *Traffic,* 2009 10, 324–333

Folador, P; Cazarolli, L.H; Gazola, A.C; Reginatto, F.H; Schenkel, E.P; Silva, F.R. Potential insulin secretagogue effects of isovitexin and swertisin isolated from Wilbrandia ebracteata roots in non-diabetic rats. *Fitoterapia,* 2010 81, 1180-1187.

Frayn, K.N. Adipose tissue as a buffer for daily lipid flux. *Diabetologia,* 2002 45:1201–1210.

Frederico, M.J; Castro, A.J; Mascarello, A; Mendes, C.P; Kappel, V.D; Stumpf, T.R; Leal, P.C; Nunes, RJ; Yunes, R.A; Silva, F.R. Acylhydrazone contribute to sérum glucose homeostasis through dual physiological targets. *Current Topics in Medicinal Chemistry,* 2012 12, 2049-2058.

Fu, H; Li, B; Hertz, L; Peng, L. Contributions in astrocytes of SMIT1/2 and HMIT tomyo-inositol uptake at different concentrations and pH. *Neurochemistry International,* 2012 61:187–194.

Furtado, L.M; Somwar, R; Sweeney, G; Niu, W; Klip, A. (2002) Activation of the glucose transporter GLUT4 by insulin. *Biochem. Cell Biol.,* 80:569-78.

Gerich, J.E. Physiology of glucose homeostasis. *Diabetes Obes. Metab.,* 2000 2: 345–350.

Gerich, J.E. Role of the kidney in normal glucose homeostasis and in the hyperglycaemia of diabetes mellitus: therapeutic implications. *Diabetic Medicine: A Journal of the Bristish Diabetic,* 2010 27:136–142.

Ghasemi, R; Haeri, A; Dargahi, L; Mohamed, Z; Ahmadiani, A. Insulin in the brain: sources, localization and functions. *Molecular Neurobiology*, 2013 47(1):145-171.

Giorgino, F; Laviola, L; Eriksson, J.W. Regional differences of insulin action in adipose tissue: insights from in vivo and in vitro studies. *Acta Physiolica Scandnavica,* 2005,183:13–30.

Grempler, R; Augustin, R; Froehner, S; Hildebrandt, T; Simon, E; Marka, M; Eickelmann, P. Functional characterisation of human SGLT-5 as a novel kidney-specific sodium-dependent sugar transporter. *FEBS Letters,* 2012 586 248–253.

Haber, R.S; Weinstein, S.P; O'Boyle, E; Morgello, S. Tissue distribution of the human GLUT3 glucose transporter. *Endocrinology,* 1993 132: 2538-2543.

Hediger, M.A; Coady, M; Ikeda, T.S; Wright, E.M. Expression cloning and cDNA sequencing of the Na+/glucose co-transporter. *Nature,* 1987 330:379-381.

Henquin, J.C. Regulation of insulin secretion: a matter of phase control and amplitude modulation. *Diabetologia,* 2009 52:739–751.

Henquin, J.C. The dual control of insulin secretion by glucose involves triggering and amplifying pathways in β-cells. *Diabetes Reserach of Clinical Practice,* 2011 93(1):27-31.

Hogan, A; Heyner, S; Charron, M.J; Copeland, N.G; Gilbert, D.J; Jenkins, NA; Thorens, B; Schultz, GA. Glucose transporter gene expression in early mouse embryos. *Development,* 1991 113: 363-372.

Huang, S.; Czech, M. P. (2007). The GLUT4 glucose transporter. *Cell Metab., 5,* 237–252.

Ibberson, M; Uldry, M; Thorens, B. GLUTX1, a novel mammalian glucose transporter expressed in the central nervous sys-tem and insulin-sensitive tissues. *The Journal of Biological Chemistry,* 2000 275:4607–4612.

Jensen, M.D. Lipolysis: contribution from regional fat. *Annual Review of Nutrition,* 1997 17: 127–139.

Jitrapakdee, S; Wutthisathapornchai, A; Wallace, J.C; MacDonald, M.J. Regulation of insulin secretion: role of mitochondrial signalling. *Diabetologia,* 2010 53:1019–1032.

Jones, H.F; Butler, R.N; Brooks, D.A. Intestinal fructose transport and malabsorption in humans. *American Journal of Physiology: Endocrinology and Metabolism,* 2011 300: 202–206.

Joost, H.G; Thorens, B.The extended GLUT-family of sugar/polyol transport facilitators: nomenclature, sequence characteris-tics, and potential function of its novel members. *Molecular Membrane Biology,* 2001 18, 247–256.

Joost, H.G; Bell, G.I; Best, J.D; Birnbaum, M.J; Charron, M.J; Chen, Y.T; Doege, H; James, D.E; Lodish, H.F; Moley, K.H; Moley, J.F; Mueckler, M; Rogers, S; Schurmann, A; Seino, S; Thorens, B. Nomenclature of the GLUT/SLC2A family of sugar/polyol transport facilitators. *American Journal of Physiology: Endocrinology and Metabolism,* 2002 282: 974–976.

Kappel, V.D; Zanatta, L; Postal, B.G; Silva, F.R. Rutin potentiates calcium uptake via voltage-dependent calcium channel associated with stimulation of glucose uptake in skeletal muscle. *Archives of Biochemistry Biophysics,* 2013a.

Kappel, V.D; Frederico, M.J; Postal, B.G; Mendes, C.P; Cazarolli, L.H; Silva, FR. The role of calcium in intracellular pathways of rutin in rat pancreatic islets: potential insulin secretagogue. *The European Journal of Pharmacology,* 2013b 702: 264-268.

Kappel, V.D; Pereira, D.F; Cazarolli, L.H; Guesser, S.M; da Silva, C.H; Schenkel, E.P; Reginatto, F.H; Silva, F.R. Short and long-term effects of *baccharis articulata* on glucose homeostasis. *Molecules,* 2012, 17:6754-6768.

Kasahara, T; Kasahara, M. Characterization of the rat Glut4 glucose transporter expressed in the yeast Saccharomyces cerevisiae: comparison with the Glut1 glucose transporter. *Biochimica and Biophysica Acta,* 1997 *1324,* 111-119.

Keembiyehetty, C; Augustin, R; Carayannopoulos, M.O; Steer, S; Manolescu, A; Cheeseman, C.I; Moley, K.H. Mouse glucose transporter 9 splice variants are expressed in adult liver and kidney and are up-regulated in diabetes. *Molecular Endocrinology,* 2006 20:686–697.

Kellett, G.L; Brot-Laroche, E. Apical GLUT2. A major pathway of intestinal sugar absorption. *Diabetes,* 2005 54:3056-3062.

Kellett, G.L; Brot-Laroche, E; Mace, O.J; Leturque, A. Sugar absorption in the intestine: the role of GLUT2. *Annual Review of Nutrition,* 2008 28:35–54.

Klover, P.J; Mooney, R.A. Hepatocytes: critical for glucose homeostasis. *Internatinal Journal of Biochemestry Cell,* 2004 36, 753–758.

Leicht, S; Grohmann, S; Page, K; Streicher, R; Mark, M; Eichelmann, P. hSGLT5 (SLC5A10) is a sodium-dependent sugar transporter exclusively expressed in the kidney. *American Diabetes Association,* 2006 66:1512.

Leto, D; Saltiel, A.R. Regulation of glucose transport by insulin: traffic control of GLUT4. *Nature Reviews. Molecular Cell Biology,*2012 13, 383-396.

Leturque, A; Brot-Laroche, E; Le Gall, M. GLUT2 mutations, translocation, and receptor function in diet sugar managing. *American Journal of Physiology: Endocrinology and Metabolism,* 2009 296:985–992.

Li, Q; Manolescu, A; Ritzel, M; Yao, S; Slugoski, M; Young, J.D; Chen, XZ; Cheeseman, C.I. Cloning and functional char-acterization of the human GLUT7 isoform SLC2A7 from the small intestine. *American Journal of Physiology. Gastrointestinal and Liver Physiology,* 2004 287:G236–G242.

Lisinski, I; Schurmann, A; Joost, HG; Cushman, S.W; Al-Hasani, H. Targeting of GLUT6 (formerly GLUT9) and GLUT8 in rat adipose cells. *The Biochemical Journal,* 2001 358:517–522.

Liu, L; Yu, Y-L; Liu, C; Wang, X-T; Liu, X-D; Xie, L. Insulin deficiency induces abnormal increase in intestinal disaccharidase activities and expression under diabetic states, evidences from in vivo and in vitro study. *Biochemical Pharmacology,* 2011 82:1963-1970.

Macheda, M.L; Williams, E.D; Best, J.D; Wlodek, ME; Rogers, S. Expression and localisation of GLUT1 and GLUT12 glucose transporters in the pregnant and lactating rat mammary gland. *Cell Tissue Research,* 2003 311:91–97.

Mackenzie, B; Panayotova-Heiermann, M; Loo, D.D; Lever, J.E; Wright, E.M. SAAT1 is a low affinity Na+/glucose cotrans-porter and not an amino acid transporter. A reinterpretation. *The Journal of Biological Chemistry,* 1994 269:22488-22491.

Maggs, D; MacDonald, I; Nauck, M.A. Glucose homeostasis and the gastrointestinal tract: insights into the treatment of diabetes. *Diabetes Obesity and Metabolism,* 2008 10(1):18-33.

Maher, F; Vannucci, S.J; Simpson, I.A. Glucose transporter proteins in brain. *FASEB Journal,* 1994 8:1003–1011.

Manolescu, A; Salas-Burgos, AM; Fischbarg, J; Cheeseman, C.I. Identification of a hydrophobic residue as a key determinant of fructose transport by the facilitative hexose transporter SLC2A7 (GLUT7). *The Journal of Biological Chemistry,* 2005 280:42978–42983.

Manolescu, A.R; Augustin, R; Moley, K; Cheeseman, C. A highly conserved hydrophobic motif in the exofacial vestibule of fructose transporting SLC2A proteins acts as a critical determinant of their substrate selectivity. *Molecular Membrane Biology,* 2007 24:455–463.

McVie-Wylie, A.J; Lamson, DR ; Chen, Y.T. Molecular cloning of a novel member of the GLUT family transporters, SLC2A10 (GLUT10), localized on chromosome 20q13.1: a candidate gene for NIDDM susceptibility. *Genomics,* 2001 72:113 –117.

Meyer, C; Dostou, J.M; Welle, S.L; Gerich, J.E. Role of human liver, kidney, and skeletal muscle in postprandial glucose homeostasis. *American Journal of Physiology: Endocrinology and Metabolism,* 2002 282:419–427.

Mitrakou, A. Kidney: Its impact on glucose homeostasis and hormonal regulation. *Diabetes Reserach of Clinical Practice,* 2011 93:66–72.

Mobasheri, A; Neama, G; Bell, S; Richardson, S; Carter, S.D. Human articular chondrocytes express three facilitative glucose transporter isoforms: GLUT1, GLUT3 and GLUT9. *Cell Biology International,* 2002 26:297–3000.

Moore, M.C; Coate, K.C; Winnick, JJ; An, Z; Cherrington, A.D. Regulation of hepatic glucose uptake and storage *in vivo*. *Advaces in Nutrition*, 2012 3:286–294.

Morton, G.J. Hypothalamic leptin regulation of energy homeostasis and glucose metabolism. *The Journal of Physiology.,* 2007 583(2):437–443.

Mourad, F.H; Saadé, N.E. Neural regulation of intestinal nutrient absorption. *Progress in Neurobiology,* 2011 95:149–162.

Mueckler, M; Caruso, C; Baldwin, S.A; Panico, M; Blench I, Morris H.R, Allard WJ, Lienhard GE, Lodish HF. Sequence and structure of a human glucose transporter. *Science,* 1985 229:941–945.

Pao, S.S; Paulsen, I.T; Saier, MHJr. Major facilitator superfamily. *Micro-biology Molecular Biology Reviews,* 1998 62:1–34.

Pereira, D.F; Cazarolli, L.H; Lavado, C; Mengatto, V; Figueiredo, MS; Guedes, A; Pizzolatti, MG; Silva, FR. Effects of flavonoids on alpha-glucosidase activity: potential targets for glucose homeostasis. *Nutrition,* 2011:1161-1167.

Pereira, D.F; Kappel, V.D; Cazarolli, L.H; Boligon, A.A; Athayde, ML; Guesser, S.M; Da Silva, E.L; Silva, F.R. Influence of the traditional Brazilian drink Ilex paraguariensis tea on glucose homeostasis. *Phytomedicine,* 2012 19:868-877.

Phay, J.E; Hussain, H.B; Moley, J.F. Cloning and expression analysis of a novel member of the facilitative glucose transporter family, SLC2A9 (GLUT9). *Genomics,* 2000 66:217-220.

Piper, R.C; Tai, C; Kulesza, P; Pang, S; Warnock, D; Baenziger, J; Slot, J.W; Geuze, H.J; Puri, C; James, D.E. GLUT-4 NH2 terminus contains a phenylalanine-based targeting motif that regulates intracellular sequestration. *The Journal of Cell Biology,* 1994 121:1221–1232.

Purcell, S.H; Aerni-Flessner, L.B; Willcockson, A.R; Diggs-Andrews, K.A; Fisher, S.J; Moley, K.H. Improved insulin sensitivity by GLUT12 overexpression in mice. *Diabetes,* 2011 60:1478–1482.

Ramnanan, C.J; Edgerton, D.S; Kraft G; Cherrington A.D. Physiologic action of glucagon on liver glucose metabolism. *Diabetes Obesity and Metabolism,* 2011 13(1):118 –125.

Rand, E.B; Depaoli, A.M; Davidson N.O; Bell, G.I; Burant, C.F. Sequence, tissue distribution, and functional characterization of the rat fructose transporter GLUT5. *American Journal of Physiology,* 1993 264:1169–G1176.

Rogers, S; Chandler, J.D; Clarke, A.L; Petrou, S; Best, J.D. Glucose transporter GLUT12-functional characterization in Xenopus laevis oocytes. *Biochemical and Biophysical Research Communications,* 2003b 308:422-426.

Rogers, S; Docherty, S.E; Slavin, J.L; Henderson, M.A; Best, J.D. Differential expression of GLUT12 in breast cancer and normal breast tissue. *Cancer Letters,* 2003a 193:225–233.

Rogers, S; Macheda, M.L; Docherty, S.E; Carty, M.D; Henderson, M.A; Soeller, W.C; Gibbs, E.M; James, D.E; Best, J.D. Identification of a novel glucose transporter-like protein-GLUT-12. *American Journal of Physiology: Endocrinology and Metabolism,* 2002 282: 733–738.

Rowland, A.F; Fazakerley, D.J; James, D.E. Mapping Insulin/GLUT4 Circuitry. *Traffic,* 2011 12:672–681.

Sabino-Silva, R; Mori, R.C; David-Silva, A; Okamoto, M.M; Freitas H.S; Machado, U.F. The Na/glucose cotransporters: from genes to therapy. *Brazilian Journal of Medical and Biological Research,* 2010 43(11):1019-1026.

Saltiel, A.R; Kahn, C.R. Insulin signalling and the regulation of glucose and lipid metabolism. *Nature,* 2001 414, 799–806.

Sandoval, IV; Martinez-Arca, S; Valdueza, J; Palacios, S; Holman, G.D. Distinct reading of different structural determinants modulates the dileucine-mediated transport steps of the lysosomal membrane protein LIMPII and the insulin-sensitive glucose transporter GLUT4. *The Journal of Biological Chemistry,* 2000 275:39874–39885.

Sasaki, T; Minoshima, S; Shiohama, A; Shintani, A; Shimizu, A; Asakawa, S; Kawasaki, K; Shimizu, N. Molecular cloning of a member of the facilitative glucose transporter gene family GLUT11 (SLC2A11) and identification of transcription variants. *Biochemical and Biophysical Research Communications,* 2001 289:1218–1224.

Scheepers, A; Schmidt, S; Manolescu, A; Cheeseman, C.I; Bell, A; Zahn, C; Joost, H.G; Schurmann, A. Characterization of the human SLC2A11 (GLUT11) gene: alternative promoter usage, function, expression, and subcellular distribution of three isoforms, and lack of mouse orthologue. *Molecular Membrane Biology,* 2005 22:339–351.

Schmidt S, Joost H.G, Schurmann A. GLUT8, the enigmatic intracellular hexose transporter. *American Journal of Physiology: Endocrinology and Metabolism,* 2009 296:614–618.

Schuit, F.C; Huypens, P; Heimberg, H; Pipeleers, D.G. Glucose sensing in pancreatic beta-cells: a model for the study of other glucose-regulated cells in gut, pancreas, and hypothalamus. *Diabetes,* 2001 50:1-11.

Shrayyef, M.Z; Gerich, JE. Normal glucose homeostasis. In: Poretsky L. 2nd editor. Principles of diabetes mellitus. Spring Street, New York, NY 10013, USA: Springer Science-Business Media, LLC; 2010.

Simpson, IA; Dwyer, D; Malide, D; Moley, K.H; Travis, A; Vannucci, S.J. The facilitative glucose transporter GLUT3: 20 years of distinction. *American Journal of Physiology: Endocrinology and Metabolism,* 2008 295:242–253.

Stöckli, J; Fazakerley, D.J; James, D.E. GLUT4 exocytosis. *The Journal of Science,* 2011 124: 4147–4159.

Stuart, C.A; Howell, M.E; Zhang, Y; Yin, D. Insulin-stimulated translocation of glucose transporter (GLUT) 12 parallels that of GLUT4 in normal muscle. *Journal Clinical Endocrinology and Metabolism,* 2009 94:3535–3542.

Stuart, C.A; Yin, D; Howell, M.E; Dykes, R.J; Laffan, JJ; Ferrando, A.A. Hexose transporter mRNAs for GLUT4, GLUT5, and GLUT12 predominate in human muscle. *American Journal of Physiology: Endocrinology and Metabolism,* 2006 291:1067–1073.

Stumvoll, M; Meyer, C; Mitrakou, A; Nadkarni, V; Gerich, J.E. Renal glucose production and utilization: new aspects in humans. *Diabetologia,* 1997 40:749–757.

Tazawa, S; Yamato, T; Fujikura, H; Hiratochi, M; Itoh, F; Tomae, M; Takemura, Y; Maruyama, H; Sugiyama, T; Wakamatsu, A; Isogai, T; Isaji, M. SLC5A9/SGLT4, a new Na-dependent glucose transporter, is an essential transporter for mannose, 1,5-anhydro-D-glucitol, and fructose. *Life Sciences,* 2005 76:1039 –1050.

Thorens, B; Mueckler, M. Glucose transporters in the 21st Century. *American Journal of Physiology: Endocrinology and Metabolism,* 2010 298:141–145.

Triplitt, C.L. Examining the mechanisms of glucose regulation. *The American Journal of Managed Care*, 2012 18:4-10.

Turk, E; Wright, E.M. Membrane topology motifs in the SGLT cotransporter family. *Journal Membrane Biology,* 1997 159, 1-20.

Uldry, M; Thorens, B. The SLC2 family of facilitated hexose and polyol transporters. *Pflugers Archives – Eurpean Journal of Physiology,* 2004 447, 480–489.

Uldry, M; Ibberson, M; Horisberger, J.D; Chatton, J.Y; Riederer, BM; Thorens, B. Identification of a mammalian H(+)-myo-inositol symporter expressed predominantly in the brain. *EMBO Journal*, 2001 20: 4467-4477.

Uldry, M; Ibberson, M; Hosokawa, M; Thorens, B. GLUT2 is a high affinity glucosamine transporter. *FEBS Letters,* 2002 524:199–203.

Uldry, M; Steiner, P; Zurich, M.G; Beguin, P; Hirling, H; Dolci, W; Thorens, B. Regulated exocytosis of an H+/myo-inositol sym-porter at synapses and growth cones. *EMBO Journal,* 2004 23:531-540.

Vitart, V; Rudan, I; Hayward, C; Gray, NK; Floyd J; Palmer, C.N; Knott, SA; Kolcic, I; Polasek O; Graessler, J; Wilson, J.F; Marinaki, A; Riches, P.L; Shu, X; Janicijevic, B; Smolej-Narancic, N; Gorgoni, B., Morgan, J., Campbell, S; Biloglav, Z; Barac-Lauc, L., Pericic, M; Klaric, I.M; Zgaga, L; Skaric-Juric, T; Wild, S.H; Richardson, W.A; Hohenstein, P; Kimber, C.H; Tenesa, A; Donnelly, L.A; Fairbanks, L.D; Aringer, M; McKeigue, P.M; Ralston, S.H; Morris, A.D; Rudan, P; Hastie, N.D; Campbell, H; Wright, A.F. SLC2A9 is a newly identified urate transporter influencing serum urate concentration, urate excretion and gout. *Nature Genetics,* 2008 40:437–442.

Ware, B; Bevier, M; Nishijima, Y; Rogers, S; Carnes, C.A; Lacombe, VA. Chronic heart failure selectively induces regional heterogeneity of insulin-responsive glucose transporters. *American Journal of Physiology Regulatory Integrative and Comparative Physiology,* 2011 301:1300–1306.

Wasserman, D.H. Four grams of glucose. *American Journal of Physiology: Endocrinology and Metabolism,* 2009 296:11–21.

Watson, R.T; Kanzaki, M; Pessin, J.E. Regulated membrane trafficking of the insulin-responsive glucose transporter 4 in adipocytes. *Endocrine Reviews,* 2004 25:177-204.

Wiederkehr, A.; Wollheim, C. B. Mitochondrial signals drive insulin secretion in the pancreatic b-cell. *Molecular Cell Endocrinology,* 2012 353, 128–137.

Wood, I.S.; Trayhurn, P. Glucose transporters (GLUT and SGLT): expanded families of sugar transport proteins. *The British Journal of Nutrition,* 2003 89, 3–9.

Wright, E.M.; Turk, E. The sodium/glucose cotransport family SLC5. *Pflügers Archives,* 2004 447, 510 –518.

Wright, E.M; Loo; D.D; Hirayama, B.A; Turk, E. Surprising versatility of Na+-glucose cotransporters: SLC5. *Physiology,* 2004 19:370-376.

Wright, E.M; Loo, D.D.F; Hirayama, B.A. Biology of human sodium glucose transporters. *Physiology Reviews,* 2011 91:733–794.

Wright, E.M; Martín, M.G; Turk, E. Intestinal absorption in health and disease—sugars. *Best Practice & Research Clinical Gastroenterology,* 2003 17(6):943 – 956.

Wu, X; Freeze, H.H. GLUT14, a duplicon of GLUT3, is specifically expressed in testis as alternative splice forms. *Genomics*, 2002 80, 553–557.

Wu, X; Li, W; Sharma, V; Godzik, A; Freeze, H.H. Cloning and characterization of glucose transporter 11, a novel sugar transporter that is alternatively spliced in various tissues. *Molecular Genetics Metabolism,* 2002 76:37–45.

Zanatta, L; de Sousa, E; Cazarolli, L.H; Junior, A.C; Pizzolatti, M.G; Szpoganicz, B; Silva, F.R. Effect of crude extract and fractions from vitex megapotamica leaves on hyperglycemia in alloxan-diabetic rats. *Journal of Ethnopharmacology,* 2007 109, 151-155.

Zhao, F.Q; Keating, A.F. Functional properties and genomics of glucose transporters. *Current Genomics,* 2007 8, 113-128.

In: Glucose Uptake
Editors: Carter C. Johnson and Davis B. Williams
ISBN: 978-1-62618-670-5

Chapter 3

Expression and Regulation of Neuronal Glucose Transporters in Health and Disease

Aníbal I. Acuña, Angara Zambrano, Ilona I. Concha and Maite A. Castro*

Instituto de Bioquímica y Microbiología,
Facultad de Ciencias, Universidad Austral de Chile,
Valdivia, Chile

Abstract

The brain is an energy demanding organ and relies on glucose as its main energy source. Glucose is taken up by brain cells through glucose transporters, which are expressed in every cell of the central nervous system. Maintaining a constant influx of energetic substrates is necessary for the survival and function of every cell in the body. However, glucose uptake can be modified under different physiological and pathological conditions.

In this chapter we will discuss neuronal glucose transporter expression and function, with descriptions of neuronal glucose transport regulation during periods of both rest and activity.

We will give some insight into the signalling pathways involved in the regulation of glucose transporter function and availability. And finally, by focusing on neurodegenerative disorders, with special emphasis on Alzheimer's disease, we will describe glucose transporter function, regulation and signalling under disease conditions.

* Corresponding author: Dr. Maite A. Castro, Instituto de Bioquímica y Microbiología, Facultad de Ciencias, Universidad Austral de Chile, casilla 547, Valdivia, Chile. Tel: 56-63-221474, E-mail: macastro@uach.cl.

1. Introduction

While the brain represents only 2% of the body mass in an adult human, it accounts for 25% of the total (resting) body energy consumption (Siesjèo 1978). Brain activity is sustained on the energy provided by glucose. Most of the energy consumed in the brain is attributable to restoration of the resting membrane potential following neuronal depolarisation. Neuronal activity is responsible for 80% of brain energy consumption (Sibson et al., 1998; Rothman et al., 1999). Neuronal functions such as neurotransmitter recycling and axonal and dendritic transport also contribute to brain energy consumption (Ames 2000). When blood glucose levels decrease to 2-3 mM, some cognitive impairment can be observed. Below a level of 1 mM, mental confusion is evident (Cox et al., 2008; Dwyer 2002).

Glucose enters the brain through the blood-brain barrier (BBB, Huber et al., 1997; Boado and Pardridge 1994; Nualart et al., 1999). Glucose transport across the BBB is not affected by acute hyperglycaemia or hyperinsulinaemia (Hasselbach et al., 1999, 2005). The BBB permeability may be altered under starvation conditions, permitting entrance of other metabolic substrates in substantial quantities. Starvation induces higher concentrations of ketone bodies in the blood that correlates with an increased expression of monocarboxylate transporters, MCTs in the BBB (Hasselbach et al., 1995).

Neurons have a highly oxidative metabolism (Hertz 2008) and are able to utilise glucose because they possess glucose transporters and have glycolytic and tricarboxylic acid (TCA) cycle enzymes. Metabolism of glucose is tightly regulated to meet two main needs: the generation of ATP and the provision of carbon for biosynthetic reactions in conjunction with local functional activities. Glucose can also follow an alternative oxidative route, the pentose phosphate pathway (PPP), providing NADPH that is mainly used for regeneration of glutathione (Kletzien et al., 1994), thus conferring an antioxidant role upon neurons (Ben-Yoseph et al. 1996, Herrero-Mendez et al., 2009).

Maintaining an adequate glucose uptake is essential for the brain and for neurons. However, glucose transporter function and localisation can be modified under certain physiological and pathological conditions. In the following sections we provide an up-to-date review of: glucose transport in brain and neurons; neuronal glucose transport under conditions of synaptic activity; pathways involved in regulating the function and subcellular localisation of neuronal glucose transporters. For each section we will discuss the related pathological conditions.

2. Glucose Transport in Brain and Glucose Transporters

Glucose from the bloodstream can enter the brain through the blood brain barrier (BBB) because BBB cells possess GLUTs, facilitative glucose transporters (Agus et al., 1997, Pardridge et al., 1990, Nualart et al., 1999). All brain cells can uptake glucose because all of them express GLUTs (Table 1). Thus, glucose can reach intracellular compartments in brain and in this way be metabolised.

The *SLC2 family* of *GLUTs* currently has 14 members (Joost et al., 2002; Doege et al., 2001; Ibberson et al., 2000; Joost y Thorens, 2001; Lisinski et al., 2001; Phay et al., 2000; Rogers et al., 2002; Wu y Freeze, 2002), many of which are expressed in brain (Table 1). GLUTs have molecular sizes ranging from 45 to 55 kDa and they are only able to translocate glucose D-enantiomers across the plasma membrane. GLUTs are responsible for the facilitative transport of glucose across the plasma membrane (Muekler, 1994; Joost and Thorens, 2001).

Table 1. Glucose transporter expression in brain

TRANSPORTER	TISSUE EXPRESSION	FUNCTION	REFERENCES
GLUT1			
55 kDa	Endothelial cells of the BBB.	Supplies basal glucose. Glucose entry in brain.	Agus et al., 1997; Boado and Pardridge, 1994; Bolz et al., 1996; Devraj et al., 2011; Pardridge et al., 1990; Nualart et al., 1999.
45 kD	Epithelial cells of the choroid plexus, ependymal cells, hypothalamic glial cells, glial cells, vascular feet of astrocytes.	Supplies basal glucose. Glucose entry in brain.	Aller et al., 1997; Chari et al., 2011; Nualart et al., 1999; Silva-Alvarez et al., 2005; Simpson et al., 1994.
GLUT2	Found in embryonic brain and hypothalamic ependymal cells.	Glucose sensing.	García et al., 2003; Nualart et al., 1999.
GLUT3	Predominant isoform in axons and dendrites of neurons.	Mediates glucose delivery to neurons.	Aller et al., 1997; Castro et al., 2007; Leino et al., 1997; Maher et al., 1999; Vanucci et al., 1997.
GLUT4	Slightly expressed in hippocampus and cerebellum.		Choeiri et al., 2002; Vannucci et al., 2000.
GLUT5	Embryonic cerebellum. Microglial cells.		Maher et al., 1994; Mueckler et al., 1994; Nualart et al., 1999.
GLUT6	Only GLUT6 mRNA has been localised in the brain.		Doege et al., 2000; Godoy et al., 2006.
GLUT8	Some neurons in the hippocampus and other brain areas.		Ibberson et al., 2000; Ibberson et al., 2002; Lisinski et al., 2001; Reagan et al., 2002.
GLUT13 (HMIT)	Primarily expressed in neurons.	Myoinositol and proton-coupled transport.	Augustin et al., 2010; Uldry et al., 2001.
SGLT1	Expressed in BBB. Expression is inducible by ischaemia.		Elfeber et al., 2004; Wright et al., 2011.
SGLT3	Expressed in hippocampus and cerebral cortex.		Enerson and Drenes, 2006; Vemula et al., 2008; Wright et al., 2011.
SGLT4	Expressed in hippocampus and cerebral cortex.		Wright et al., 2004.

Considering their primary structure, 12 transmembrane domains have been proposed for the GLUT transporter family (Uldry et al., 2001). GLUT1 is ubiquitous and thus this isoform should be responsible for basal glucose transport in many different cell types (Thorens et al., 1988).

In brain, GLUT1 exists as two different molecular weight forms (45 and 55 kDa). Both forms have similar kinetic properties (Birnbaum et al., 1986). The 55 kDa form has been described at the BBB, in endothelial cells from cerebral microvasculature (Nualart et al., 1999, Agus et al., 1997, Pardridge et al., 1990; Boado and Pardridge, 1994, Bolz et al., 1996). This form of GLUT1 is glycosylated (Maher et al., 1992).

Devraj and colleagues (2011) demonstrated that GLUT1 in luminal (blood-facing) and abluminal (brain-facing) endothelial cells is differentially phosphorylated.

A differential phosphorylation of the 55 kDa form has also been demonstrated, suggesting differences between GLUT1 expressed at the luminal and abluminal membranes of ependymal cells (Devraj et al., 2011). The 45 kDa form of GLUT1 is expressed in epithelial cells from choroid plexus, ependymal cells (Nualart et al., 1999; Silva-Alvarez et al., 2005), hypothalamic glial cells (Chari et al., 2011), astrocytes and neurons (Aller et al., 1997). Immunohistochemical analysis of GLUT1 suggests a correlation between the presence of GLUT1 and local cerebral glucose utilisation (Zeller et al., 1997). GLUT2 however has a low affinity for glucose and its expression is very limited in brain. It has been described in embryonic cerebellum (Nualart et al., 1999) and hypothalamus (García et al., 2003). GLUT3 is a major neuronal glucose-transporter isoform (Maher et al., 1996). It is a high-affinity transporter with a catalytic constant higher than that of GLUT1 (Nagamatsu et al., 1993).

Immunohistochemistry studies generally detect the most intense staining of GLUT3 in the neuropil, reflecting high concentrations in axons and dendrites (Aller et al., 1997; Leino et al., 1997; Vannucci et al., 1997, Maher et al., 1999; Castro et al., 2007). Inducible astrocytic GLUT3 expression during ischaemia has recently been reported (Iwabuchi and Kaeahara, 2011). GLUT4 has been described mainly in hippocampus and cerebellum (Choeiri et al., 2002). Neuronal GLUT4 expression was detected in Purkinje and granule cells from cerebellum, in granule cells from the olfactory bulb and in hippocampus (Vannucci et al., 2000). GLUT5 is predominantly a fructose transporter (Mueckler, 1994). Immunohistochemical methods detected GLUT5 in microglial cells (Maher et al., 1994) and embryonic cerebellum (Nualart et al., 1999). GLUT6 in brain has not been fully characterised. GLUT6 mRNA has been found in brain (Doege et al., 2000; Godoy et al., 2006), but protein expression has not yet been reported. GLUT8 is found in hippocampal neurons (Reagan et al., 2002) and in neurons from other brain areas (Ibberson et al., 2000; Lisinski et al., 2001; Ibberson et al., 2002).

This transporter has a dileucine motif, which targets the protein to insulin-regulated trafficking pathways (Manolescu et al., 2008). GLUT13 (HMIT), the proton-coupled myo-inositol transporter, is the only GLUT reported to use a proton gradient to energise substrate movement (Uldry et al., 2001). In brain, myo-inositol is a precursor for phosphatidylinositol, a key regulator for several signalling pathways. Because HMIT is predominantly expressed in brain, there is increasing interest in a possible role for HMIT in myo-inositol/ phosphatidylinositol physiology in neurons (Uldry et al., 2001; Augustin et al., 2010).

The sodium-dependent glucose co-transporters (SGLT) are responsible for the transport of glucose (and galactose) and sodium across the plasma membrane via a secondary active transport mechanism (Wright et al., 2011). SGLTs are mainly important for their role at epithelial barriers such as the brush border membrane of mature enterocytes in the small intestine and the brush border membrane in renal tubules (Wright et al., 2004; Wright et al., 2011). Expression of five SGLT isoforms has been documented in brain (Table 1).

However, SGLT function in the nervous system is poorly understood. Ischaemia-induced SGLT1 expression was determined at the BBB (Elfeber et al., 2004). The presence of SGLT2 mRNA has been shown in rat brain (Enerson y Drenes, 2006) and SGLT2 activity has been measured at the BBB (Vemula et al., 2008). In cultured hypothalamic neurons, mRNA corresponding to SGLT1, SGLT3a and SGLT3b has been detected (O`Malley et al., 2006), while protein expression of SGLT4 and SGLT5 has been described in brain (Wright et al., 2004).

2.1. GLUT3 Localisation in the Plasma Membrane

Neurons are able to take up glucose from the extracellular space because they possess glucose transporters, mainly the neuronal glucose transporter, GLUT3 (Carruthers 1990; Maher et al., 1992; Nagamatsu et al., 1993; Simpson et al., 2007). The neuronal glucose transporter GLUT3 can be observed in intracellular compartments as well as at the cell surface. This transporter has been shown to be targeted to the apical membranes of polarised cells by the presence of a targeting signal in the C-terminal cytosolic tail domain (Harris et al., 1992; Inukai et al., 2004). Intracellular localisation of GLUT3 is proposed to be within synaptic-like vesicles that are distinct from classic synaptic vesicles (Thoidis et al., 1999). A similar localisation was observed in platelets, where GLUT3 vesicles translocate after thrombin treatment (Heijnen et al., 1997). GLUT3 was observed to be localised in fluid membrane domains (Sakyo et al., 2002; Sakyo et al., 2007) and caveolin-rich detergent-resistant membrane domains (Rauch et al., 2007). Therefore it is possible that GLUT3 membrane localisation depends on cell type and cell requirements. Using Total Internal Reflection Microscopy, we have observed that treatment of a neuronal cell line with the cholesterol sequestering compound methyl-β-cyclodextrin and cholesterol loading methods have different effects on cell surface localisation of GLUT1 and GLUT3 (unpublished data, Figure 1). In the absence of cholesterol, GLUT3 appears to be misplaced and even loses its surface localisation. This effect can be reversed by loading with cholesterol. On the other hand, GLUT1 surface localisation seems to increase in cells depleted of cholesterol, an effect that is unaltered by later addition of cholesterol.

Although GLUT1 has been observed in detergent-resistant domains (Sakyo et al., 2007; Rauch et al., 2006), cholesterol depletion triggers GLUT1 translocation (Caliceti et al., 2012) and stimulates glucose uptake (Barnes et al., 2004) in leukaemia cells and liver epithelial cells respectively. Further studies are required to investigate glucose transporter trafficking as well as the subcellular and membrane localisation of transporters in neurons. However, glucose uptake and glucose transporters have been studied under resting and active neuronal conditions because of the high energetic requirements of neurons.

3. Neuronal Glucose Uptake during Periods of Rest and Activity

Under resting conditions neuronal cells take up glucose mainly through GLUT3. Extracellular brain glucose concentrations are considerably lower than normal blood glucose, reported to be around 2-3 mM (Silver and Erecińska 1994). A high affinity for glucose and a higher turnover number for GLUT3 transporters compared to astrocytic GLUT1 transporters ensures a constant supply of glucose (Maher et al., 1996). Glycolytic and oxidative metabolism are both active in neurons. Glucose taken up is rapidly phosphorylated by hexokinase-I in the first step of the glycolytic pathway.

Since hexokinase-I has an affinity constant for glucose that is one order of magnitude lower than that for the extracellular glucose concentration, glucose is readily phosphorylated and committed to neuronal metabolism (Lowry and Passonneau 1964).

Glucose follows different routes that are relevant for neuronal physiology. Glycolysis and the tricarboxylic acid cycle, coupled to oxidative phosphorylation, help to maintain ATP levels. The pentose phosphate pathway generates five-carbon sugars and NADPH for use in biosynthetic reactions. NADPH also restores the reduced form of the antioxidant glutathione and is favoured in neurons (Herrero-Mendez 2009).

Different approaches have been used to study neuronal glucose uptake in cell cultures and brain slices. Uptake of a fluorescent glucose analogue, 6-NBDG, has been observed to be higher in neurons than in astrocytes, although intracellular glucose levels measured with a FRET nanosensor increased faster in astrocytes than in neurons.

It has been suggested that kinetic differences between glucose transporter isoforms may account for the observed differences in glucose analogue uptake (Jakoby et al., 2012). A computer simulation study proposes an inverse correlation between 6-NBDG uptake and glycolysis rate (calculated for GLUT1 and hexokinase-I; Dinuzzo et al., 2013).

Radiolabelled glucose analogues have also been used to analyse glucose uptake in cultured neurons and to describe the kinetic properties of facilitative glucose transporters (Maher et al., 1996; Castro et al. 2007). Neuronal glucose uptake and usage was also measured indirectly, with electrophysiological recordings used to evaluate neuronal activity (Castro et al., 2007).Neuronal activity requires large quantities of ATP, with most energy being needed for processes that follow synaptic transmission, such as restoring the membrane gradient after neuronal depolarisation, neurotransmitter recycling and intracellular signalling (Attwell and Laughlin 2001; Ames 2000).

Neuronal energy substrate uptake and metabolism has to meet the increased energy demands that accompany brain activity and since neurons are unable to synthesise glycogen, substrates need to be taken up from the extracellular space. The increased blood flow observed with brain activity is accompanied by an increase in glucose utilisation (Sokoloff et al., 1977). Glutamatergic synaptic activity, which accounts for 80% of cortical synapse activity, increases glucose uptake via an up-regulation of surface GLUT3 expression, though functional glucose transport activity has not been evaluated (Weisová et al., 2009; Ferreira 2011). Although a previous study showed an increased ^{3}H-2-deoxyglucose uptake in 20 minute uptake assays after stimulation with glutamate, the fact that this increase might represent glycolytic activity, as 2-deoxyglucose is phosphorylatable by hexokinase, can not be ruled out (Ward et al., 2007).

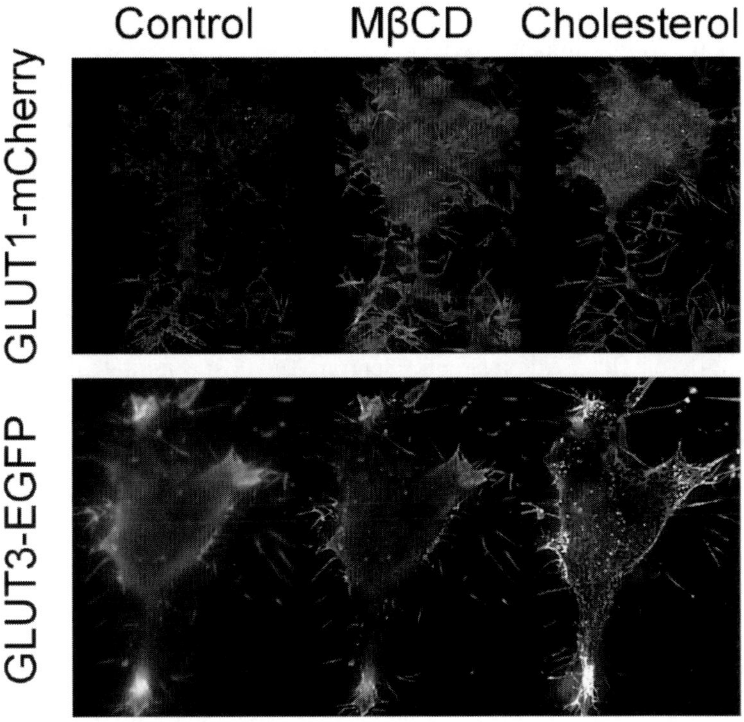

a

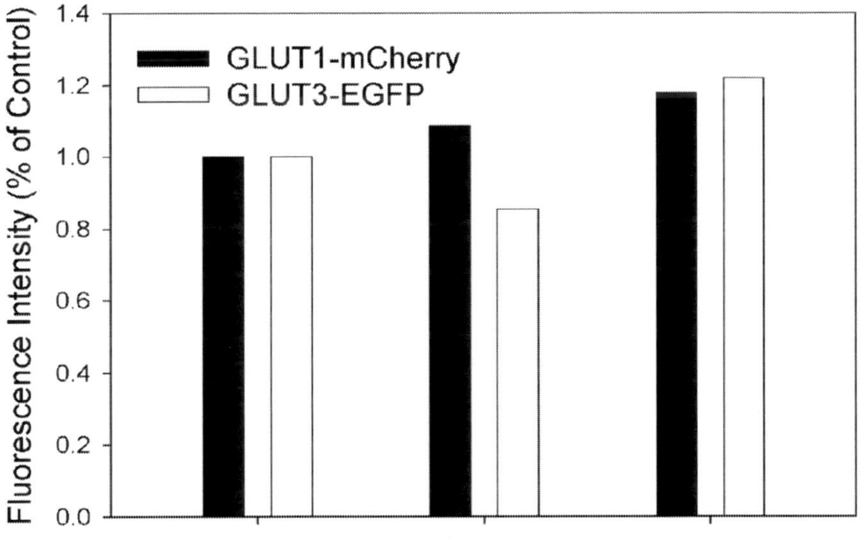

b

Figure 1. Cholesterol depletion induces a decrease in GLUT3 localisation at the plasma membrane. *Top:* Total internal reflection microscopy (TIRM) images from the neuronal cell line NG108 expressing GLUT1-mCherry and GLUT3-EGFP. Cells were incubated in serum-free media for 15 min (control). Following this, the same cells were treated with 10 mM methyl-β-cyclodextrin (MβCD), the cholesterol sequestering compound, for 15 min. Finally, cells were washed and treated with 30μg/ml Synthechol (cholesterol) for 20 min. *Bottom*: Bars plot graph for fluorescence intensity measured during the above-mentioned treatments. Data were acquired using an Olympus IX71 microscope equipped with an Olympus TIRF objective (60x, oil, NA 1.49). The evanescent field was 180 nm.

Neuronal lactate uptake has been a controversial topic for over 15 years (Allaman et al., 2010; Castro et al., 2009; Chih and Roberts, 2003; Dienel, 2009; Gjedde, 2002; Hertz, 2004; Hertz et al., 2007; Pellerin and Magistretti 1994; Pellerin and Magistretti 2012; Mangia et al., 2003; Mangia et al., 2009). However, the paradigm of glucose as the essential brain energy substrate remains firm. What is not clear, is whether glucose metabolised by astrocytes and released as lactate serves as an energy substrate for active neurons. Glycolysis and oxidation of lactate into pyruvate are not able to be activated simultaneously. Activation of glycolysis produces an increase in pyruvate levels and a decrease in NAD^+/NADH ratio, thus favouring lactate dehydrogenase catalysis of pyruvate to lactate.

Therefore, if neurons were to use lactate as the major energy source following synaptic activity, glycolysis or glucose uptake into neurons should be inhibited. There are various (coexisting) theories that have been proposed to explain this (For a review see Castro et al., 2009, Beltrán et al., 2012). Two hypotheses that involve glucose uptake inhibition have been proposed.

In the first, glutamate has been shown to inhibit glucose transport in neurons through single-cell time-lapse confocal microscopy in neuron-astrocyte co-cultures (Porras et al., 2008). Since these experiments were performed in co-cultures, the possibility that glutamate-treated astrocytes play a role in neuronal glucose uptake inhibition can not be excluded. Thus the effect of glutamate on neurons may be an indirect one. The second hypothesis demonstrates, using radiolabelled and fluorescent glucose analogues, that intracellular ascorbic acid inhibits glucose utilisation in neurons by inhibiting the neuronal glucose transporter GLUT3 (Castro et al., 2007; Castro et al., 2009, Beltrán et al., 2011). Astrocytes are thought to be involved in ascorbic acid recycling (Astuya et al. 2005).

During synaptic activity, ascorbic acid is released from intracellular reservoirs (O'Neill et al., 1984; Ghasemzadeh et al., 1991; Yusa, 2001) and can be taken up by neurons efficiently because they express the ascorbic acid transporter SVCT2 (Castro et al., 2001). Intracellular ascorbic acid is also able to stimulate lactate transport in neurons and in cells that express GLUT3 (Castro et al., 2008). The specific nature of this mechanism relies on GLUT3 as a necessary component for inhibition. Ascorbic acid is unable to inhibit glucose uptake in neurons treated with a specific shRNA (to block GLUT3 expression) or in cells that do not express GLUT3.

The role of ascorbic acid as a regulator of metabolic substrate preferences has been termed the ascorbic acid metabolic switch (Castro et al., 2009; Beltrán et al., 2011). An *in vivo* study using two-photon microscopy revealed similar glucose uptake in astrocytes and neurons.

Following neuronal activity, glucose uptake was only seen to be increased in astrocytes (Chuquet et al., 2010). Lactate has been shown to sustain the energetic requirements during neuronal activation both *in vitro* and *in vivo* (Ivanov et al., 2011; Wyss et al., 2011). Neuronal glucose transporters are able to efficiently supply glucose to neurons.

However, during periods of brain activity, glucose uptake (or glycolysis) inhibition mechanisms are required in order to allow the uptake and utilisation of lactate as a faster energy source. This effect may occur in two steps. In the first, glycolysis and glucose uptake are increased to try and meet energetic demands.

In a later step, glucose uptake and utilisation are inhibited. Thus, lactate can be taken up by neurons and neuronal ATP levels replenished.

4. Regulation of Glucose Uptake in Neuronal Cells: Signalling Associated with GLUT Trafficking and Activation

Multiple organ systems, such as the central nervous system, liver, kidney, pancreas and intestine, participate in homeostatic feedback loops that work to maintain constant levels of nutrients in the bloodstream partly by altering nutrient transporter expression on the surface of their constituent cells (Edinger 2007, Zambrano et al., 2010 (1)).

The best known example of signal transduction that is related to GLUT expression at the plasma membrane is GLUT4. GLUT4 is not the most abundant GLUT isoform in the brain, but it is strongly insulin-sensitive, while other GLUT isoforms (GLUT1, GLUT3, GLUT8, GLUT12) are only weakly insulin-sensitive (Chua et al., 2012; Piper et al., 1991; Navarrete Santos et al., 2004; Stuart et al., 2009; Wilson et al., 1995).

For this reason, we will describe the regulatory mechanism of GLUT4 expression at the plasma membrane and then compare this mechanism with findings reported for neuronal glucose transporter GLUT3 expression at the cell surface.

4.1. Regulatory Mechanism of Glucose Uptake

Growth signals increase the transcription of nutrient transporter mRNAs. In contrast, much less is known about how signal transduction cascades affect the trafficking and degradation of these proteins, although some progress is being made in this field. For this reason, despite the fact that GLUT4 trafficking in response to insulin signalling has been very well studied, many molecular details are still lacking. Insulin stimulates glucose transport into muscle and adipose tissue 10- to 30-fold with a half time of 2-5 minutes. The major glucose transporter expressed in these tissues is GLUT4, which has received a great deal of attention. Much less is known about the regulation of other glucose transporters. Specifically, the mechanisms that govern insulin-stimulated glucose transport in muscle and fat cells have captured the interest of researchers for decades. New insights into insulin signalling reveal that phosphorylation events initiated by the insulin receptor regulate key GLUT4 trafficking proteins, including small GTPases, tethering complexes and the synaptic vesicle fusion machinery (Leto and Saltiel, 2012). Translocation of GLUT4 to the plasma membrane following insulin stimulation represents the convergence of two complex systems: signal transduction and vesicular transport (Stockli et al. 2011).

Following a meal, increased nutrients in the blood lead to secretion of insulin. In turn, this hormone prevents gluconeogenesis in the liver and promotes glucose uptake into muscle and adipose tissue through regulated trafficking of GLUT4 from intracellular stores to the plasma membrane (Saltiel and Khan, 2001). Glucose uptake is the rate-limiting step in glucose utilisation and/or storage and as such has a key role in the maintenance of glucose homeostasis. Insulin increases the steady-state plasma membrane GLUT4 levels by 5- to 30-fold. Three different features of GLUT4 trafficking have been implicated in this redistribution: an increase in the exocytosis rate constant, a decrease in the endocytosis rate constant and an increased amount of GLUT4 in the cell surface recycling pool (Stockli et al. 2011).

In the absence of insulin, most GLUT4 is distributed between endosomes, the trans-Golgi network (TGN) and heterogeneous tubulovesicular structures comprised of endosomal sorting intermediates and specialised GLUT4 storage vesicles (GSVs). In the absence of insulin, only about 5% of the total transporter pool is found on the cell surface. Exclusion of GLUT4 from the cell surface depends on efficient sorting and sequestration into GSVs that do not readily cycle to the plasma membrane in the absence of stimulation but translocate to the membrane in response to insulin or exercise, which results in a ten-fold increase in glucose uptake (Leto and Saltiel, 2012). Specifically, insulin signals through TBC1D4 (TBC1 domain family member 4) and TBC1D1 (tre-2/USP6, BUB2, cdc16 domain family, member 1), both Akt substrates, to modulate Rab GTPases and through the Rho GTPase TC10α to act on other targets. GLUT4 is sorted into highly specialised store vesicles containing e.g. VAMP2 (vesicle-associated membrane protein 2). In stimulated cells these vesicles are mobilised to the plasma membrane where insulin is able to modulate the trafficking itinerary and endosome process (Bogan 2012).

Insulin is secreted by the pancreas and engages its receptor on the surface of myocytes and adipocytes, thereby activating the canonical PI3K-Akt pathway. Activation of this pathway is necessary and sufficient to trigger exocytosis of GSVs to the plasma membrane (Stockli et al., 2011). Oncogenic alleles of Akt can support growth factor-independent glucose transporter expression (Edinger and Thompson, 2002; Rathmell et al., 2003). Akt is a serine/threonine kinase that lies downstream from many growth factor receptors.

In addition to regulating glucose uptake, growth factors and Akt also affect the rate of glycolysis (Vander Heider et al., 2001; Plas et al., 2001). The ability of Akt to support growth-factor independent cell survival is completely dependent on the availability of extracellular glucose. While activated Akt protects from growth factor withdrawal-induced death, myristoylated Akt-expressing cells are as sensitive as control cells to glucose withdrawal (Edinger 2005).

Akt phosphorylates TBC1D4 (TBC1 domain family, member 4) at multiple sites, which results in 14-3-3 binding to TBC1D4 and the presumed inhibition of TBC1D4 GAP (GTPase-activating proteins) activity. This enables a large number of GSVs to move to, dock at and fuse with the plasma membrane (Sano et al., 2003; Ramm et al., 2006).

Insulin-dependent phosphorylation of a range of other molecules also plays a key regulatory role in GLUT4 translocation: ESYT1 (Extended synaptotagmin-1) and TC10 (Rho-related GTP-binding protein RhoQ) phosphorylation by CDK5 (Cyclin-dependent kinase 5); MYO1C (Unconventional myosin-1c) and MYO5 (Myosin-5) phosphorylation by CamKII (Calcium/calmodulin-dependent protein kinase II); RIP140 (NRIP1, nuclear receptor interacting protein 1) and SEC5 (Exocyst complex component 2) phosphorylation by PKC (Protein kinase C); and PIKFYVE (1-phosphatidylinositol 3-phosphate 5-kinase), AS250 (250kDa substrate of Akt) and MYO5 phosphorylation by Akt.

Akt is the principal insulin-regulated signal transducer for GLUT4 translocation and several steps in the GLUT4-trafficking pathway are regulated by insulin, including the approach, tethering, docking and fusion of vesicles at the plasma membrane (for review see Stockli et al., 2011). Likewise, mTOR appears to lie downstream of Akt in the growth factor signal transduction cascade. Treatment of cells with rapamycin (specific inhibitor of mTOR complex) blocked the growth factor-independent expression of nutrient transporters supported by activated Akt (Edinger 2005).

The PI3K-dependent signalling pathway is also initiated after insulin binds to its cognate Tyr kinase receptor on the cell surface, leading to recruitment and Tyr phosphorylation of the insulin receptor substrate (IRS) family of adaptor proteins (Myers and White, 1996). Tyr-phosphorylated IRS proteins serve as docking sites for the SH2 domain of the p85 regulatory subunit of class I PI3K, and the interaction of IRS proteins and PI3K results in PI3K activation and the subsequent synthesis of phosphatidylinositol-3,4,5-trisphosphate (PtdIns(3,4,5)P3) from PtdIns(4,5)P2 at the plasma membrane. PtdIns(3,4,5)P3 in turn serves as a docking site for several PH domain-containing Ser/Thr kinases that are implicated in glucose uptake, including phosphoinositide-dependent kinase 1 (PDK1) and Akt. PDK1 and mTORC2 activate Akt through dual Ser/Thr phosphorylation (Leto and Saltiel 2012).

AMPK (5'-adenosine monophosphate-activated protein kinase) is a serine/threonine kinase that plays a crucial role in regulating energy homeostasis, acting as an integrator of cellular and organismal energy balance and energy-dependent responses. AMPK also functions at tissue and organism levels to promote context-specific responses to physiological signals of metabolic status.

AMPK modulates many aspects of cellular metabolism. AMPK was first known to be activated by ATP depletion (increased AMP/ATP ratio) and related stimuli (exercise, starvation, hypoxia, cellular pH and redox status, increased creatine/phosphocreatine ratio; Kahn et al., 2005; Hardie 2007).

However, certain drugs, hormones, and cellular stressors that do not alter AMP/ATP ratio also activate AMPK. Thus, AMPK in various cells and tissues senses both physiological and pathophysiological stimuli. AMPK stimulates catabolism by activating glucose uptake (GLUT4 translocation), glycolysis [activating 6-phosphofructo-2-kinase kinase (PFK-2)], glucose oxidation, and fatty acid oxidation (by relieving CPT-1 inhibition by malonyl-CoA; Ramamurthy and Ronnett 2012).

4.2. Regulation of Glucose Uptake in Neuronal Cells

Despite the critical importance of glucose for brain function, little is known about the molecular events regulating normal glucose uptake. It has been shown that the distribution of glucose transporters exposed to the cell surface is a major regulator of glucose uptake, so that trafficking of glucose transporters between the cytosol and the plasma membrane is under strict control (Pessin and Bell, 1992). As described previously, this has been studied extensively for the adipocyte and muscle cell-specific transporter isoform GLUT4, the trafficking and consequent cell glucose uptake of which are under the control of insulin. Neuronal glucose uptake relies on GLUT3 at the plasma membrane but very little is known about regulation of the subcellular localisation of this transporter. Nor is it well understood just how translocation of the neuronal glucose transporter isoform GLUT3 vesicles to the plasma membrane occurs, as well as how they fuse with the membrane. Reports show that GLUT3 in PC12 cells colocalises with SNARE complex proteins SNAP-25 and syntaxin 1, suggesting that fusion of GLUT3-containing vesicles with the plasma membrane may be mediated by these proteins. In addition, GLUT3 vesicle fusion would seem to be regulated, as depolarisation increases GLUT3 insertion into the plasma membrane. In both PC12 cells and neurons, GLUT3 is present in a distinct population of small vesicles that have many vesicular proteins in common with synaptic vesicles (Weisova 2009).

Depolarisation of neurons by high extracellular K^+ acutely increases glucose uptake and increases the amount of GLUT3 exposed to the extracellular surface (Uemura and West Greenlee, 2001). Therefore, regulation of glucose uptake in neurons would seem to rely, in part, on regulated, Ca^{2+}-dependent exocytosis of GLUT3 vesicles at the plasma membrane.

In the brain, glutamate is the major neurotransmitter mediating most of the occurring synaptic transmission. Glutamate excitation produces an increase in the surface expression of GLUT3 that persists for hours following the initial glutamate treatment. Increased GLUT3 surface expression was mediated by AMPK, as inhibition of AMPK, either through knockdown or pharmacological inhibition, blocked the GLUT3 response to glutamate (Amato and Man, 2011). Glutamate excitation induced a rapid alteration in the AMP:ATP ratio associated with activation of the AMPK. Pharmacological activation of AMPK with AICAR (5-aminoimidazole-4- carboxamide riboside) alone also increased GLUT3 surface expression, with a hyperpolarisation of mitochondrial membrane potential that was evident in many neurons (Weisova, 2009).

Additional studies suggest that synaptic activity increases surface expression of GLUT3 leading to an elevation of intracellular glucose. This effect was blocked by the NMDA receptor (NMDAR) and by neuronal nitric oxide synthase (nNOS) inhibition. Interestingly, the Akt inhibitor I (Akt-I) blocked NMDAR-induced GLUT3 surface expression while a nNOS-phosphomimetic mutant (S1412D) enhanced GLUT3 expression at the cell surface. These results suggest that NMDAR/Akt-dependent nNOS phosphorylation is coupled to GLUT3 trafficking. The NMDAR-induced increase in surface GLUT3 may represent a novel pathway for control of energy supply during neuronal activity (Ferreira et al., 2011)

It is important to add that GLUT3 is upregulated in response to many stressors, such as hypoxia, hypoglycaemia, insulin and IGF-1. HIF-1α, a transcription factor that mediates adaptive responses to changes in tissue oxygenation, can stimulate GLUT3 expression. In addition, GLUT3 is regulated by some other transcription factors such as SP1, SP3, MSY-1 and CREB, etc (Rajakumar et al., 2004).

In peripheral tissues, insulin is a key signalling component in glucose uptake, while the regulation of neuronal glucose influx has traditionally been viewed as an insulin-insensitive process. However, there is evidence to suggest that both the insulin and insulin-like growth factor 1 (IGF1) signalling pathways are involved in glucose uptake in the brain. Insulin treatment in cultured hippocampal neurons facilitates GLUT3 membrane translocation, which requires membrane depolarisation for the final fusion of GLUT3-carrying vesicles with the plasma membrane in order to facilitate glucose uptake (Amato and Man, 2011). In addition to its varied role in peripheral tissues, insulin has profound effects in the CNS, where it regulates various key physiological functions, such as food intake, energy homeostasis, reproductive endocrinology, sympathetic activity, peripheral insulin actions, and even learning and memory (Gupta and Dey, 2012).

Insulin markedly potentiates Ca^{2+}-dependent neuronal glucose uptake induced by KCl, an effect that coincides with increased immunolabelling of the exofacial epitope of GLUT3. The increase in glucose uptake is due to increased fusion of GLUT3 vesicles with the plasma membrane, resulting in more GLUT3 being exposed to the extra-cellular surface (Uemura and Greenlee, 2001). Insulin potentiation of glucose uptake stimulated by depolarisation with KCl is probably a result of increased numbers of docked GLUT3 available for fusion with the plasma membrane.

This neuronal glucose uptake is regulated by at least two separable factors: the promotion of GLUT3 translocation to the plasma membrane and a membrane depolarisation that induces fusion of GLUT3 vesicles with the plasma membrane.

High extracellular K^+ is known to double the rate of neuronal glucose oxidation, although this K^+ effect is secondary to cell depolarisation and reflects the high-energy demand for Na^+ extrusion. Thus, insulin promotion of GLUT3 translocation to the plasma membrane may be an important regulatory mechanism that modulates neuronal glucose uptake in response to the metabolic demands of neurons (Uemura et al., 2006).

The insulin-like growth factor (IGF-1) is homologous to insulin and acts through its own, but closely related tyrosine kinase receptor that shares many signalling components and cellular responses with the insulin receptor (Treins et al., 2005). The IGF-1 receptor is widely distributed in the brain and has been shown to induce GLUT3 expression, with SP1 participating in this pathway. Both receptors, IGF-1 and insulin, are homologous with identical signal-transducing domains controlling most of the same intracellular pathways. The IGF-1 receptor plays a critical role in brain glucose metabolism. It activates phosphatidylinositol 3-kinase, PI3K/Akt and extracellular signal-regulated kinase (ERK) pathways which mediate cell proliferation and/or survival (Mairet-Coello et al., 2009). Several cytokines, including IGF-1, have also been reported to modulate HIF-1α protein expression or stability through PI3K/Akt/mTOR and/or ERK pathways (Fukuda et al., 2002; Slomiany and Rosenzweig, 2006). Expression of GLUT3 in response to IGF-1 was dependent on PI3K and mTOR activity and required the transcription factor HIF-1α (Yu et la., 2012).

5. Neuronal Glucose Transporters and Neurodegenerative Diseases

According to the World Health Organization, there were more than 35.6 million people worldwide living with dementia in 2010. The most common form of dementia is Alzheimer's disease, which probably contributes to 60–70% of cases (WHO, 2012).

The impact of neurodegenerative diseases on the society and economy will continue to increase as the population ages. Neurodegenerative diseases are chronic, progressive neurological disorders associated with neuronal loss.

These diseases have many similarities at a sub-cellular level including atypical protein assemblies, failure of normal protein degradation pathways, induced cell death, impaired axonal transport and metabolic failures (Rubinsztein et al., 2006; De Vos et al., 2008; Bredesen et al., 2006; Lin and Beal, 2006).

As discussed above, glucose is an essential substrate for the maintenance of brain and neuronal activity. Neurons are particularly sensitive to fluctuations in energy levels. Energy deficiency is tightly related to neuronal survival and viability and also contributes to age-related disorders. Alterations in uptake and/or utilisation of glucose can thus be associated with pathological conditions.

Several neurodegenerative diseases show metabolic failure that includes alterations in the expression of glucose transporters, altered expression of metabolic enzymes and altered activity of the enzymes involved in energy metabolism. Whether these metabolic alterations contribute to the onset and progression of the disease or whether they are a consequence of the disease is not yet fully understood. Alzheimer's disease (AD), Parkinson's disease (PD), Huntington's disease (PD) and amyotrophic lateral sclerosis (ALS) are neurodegenerative diseases that present metabolic failure.

Parkinson's disease is the most common neurodegenerative movement disorder of the central nervous system. Clinical symptoms arise from a progressive degeneration of dopaminergic neurons in the substantia nigra and other monoaminergic neurons in the brainstem (Braak et al., 2003). Studies have found that mitochondrial dysfunction contributes to the pathogenesis of PD and is thought to be related to mutations in PINK1 and DJ1 genes (Schapira, 2006; Wallace, 2005). Hypometabolism of glucose has been described in PD patients (Huang et al., 2007; Yong et al., 2007; Perneczy et al., 2008; Hosokai et al., 2009; Lee et al., 2008; Lee et al., 2010; Bohnen et al., 2011; Borghammer et al., 2011).

However, there is only limited molecular evidence to support this symptom. Neuronal cell cultures with PINK1 deficiency associated with reduced neuronal glucose uptake had their uptake restored to control levels by scavenging or inhibiting ROS species (Gandhi et al., 2009)

Amyotrophic lateral sclerosis is the third most common human adult-onset neurodegenerative disease. ALS is a progressive and severely disabling neurological disease characterised by muscle atrophy, spasticity, eventual paralysis and death.

A progressive degeneration and elimination of upper motor neurons in the cerebral cortex occurs as well as in the lower motor neurons of the brainstem and spinal cord (Rowland and Shneider, 2001; Sathasivam et al., 2001).

Research has revealed that mitochondrial dysfunction, including abnormalities of mitochondrial morphology and function, are associated with ALS, (Sasaki and Iwata, 1999; Menzies et al., 2002; Martin, 2012). Cu/Zn-SOD in transgenic mice presented decreased levels of glucose transport in cerebral cortex synaptic terminals (Guo et al., 2000). An upregulation of Glut3 was observed *in vivo* in motor neurons from $SOD1^{G93A}$ mice that could be reproduced *in vitro* when non-transgenic motor neurons were grown on $SOD1^{G93A}$ primary astrocytes. This suggests that a dysregulated energy provision by mutant SOD1 astrocytes occurs in response to motor neurons (Ferraiuolo et al., 2011).

Huntington's disease (HD) is a progressive, autosomal dominant, neurodegenerative disorder. Its common symptoms are motor dysfunction and cognitive abnormalities. The disease is caused by an expanded polyglutamine (polyQ) stretch in the Huntingtin gene, resulting in major cell loss in the striatum, a region of the basal ganglia that integrates cortical information for behavioural output. HD is characterised by widespread neurodegeneration with preferential deterioration of medium-sized spiny neurons (MSSNs) in the striatum (Penney and Young, 1998). The major excitatory input to MSSNs comes from the cortex (corticostriatal pathway) and the thalamus. HD also causes dysfunction and subsequent death of neurons in other brain regions, including the cortex. Defects in energy metabolism may even extend to presymptomatic subjects. Decreased GLUT1 and GLUT3 transporter expression has been shown in the caudate of postmortem HD brains (Gamberino and Brennan, 1994). This decrease in glucose transporter expression correlates with positron emission tomography studies that have demonstrated marked reductions in glucose metabolism in basal ganglia (Mazziotta et al., 1987) and in cerebral cortex of symptomatic Huntington's disease patients (Leenders et al., 1986; Kuwert et al., 1993). However, Oláh et al. (2008) described an increase in several glycolytic enzymes and in ATP production in brain from animals with Huntington's disease.

The same authors described a marked contrasting decrease in glyceraldehyde-3-phosphate dehydrogenase activity. Impairment in enzyme activity of the TCA cycle (Lim et al., 2008) and in oxidative phosphorylation has also been described. A key role for proliferator-activated receptor gamma coactivator-1α has been proposed for the control of energy metabolism in the early stages of Huntington's disease (Finck and Kelly, 2006). Metabolic abnormalities are not limited to the nervous system as for many neurodegenerative diseases (AD, PD, HD and ALS) and have been associated with type II diabetes or insulin resistance (Bruce et al., 2001; Sandyk 1993; Podolsky et al., 1972; Ristow 2004; Pradat et al., 2010). Given the higher incidence of AD and increasing research efforts that aim to better understand these disorders, we will further discuss the findings related to alterations seen in glucose uptake in AD.

5.1. Failure of Glucose Uptake Associated with Alzheimer's Disease

Growing evidence supports the notion that Alzheimer's disease could be conceptualised as a metabolic disease with progressive impairment of the brain's capacity to utilise glucose and to respond to insulin and insulin-like growth factor (IGF) stimulation (de la Monte 2012). Inhibition of neuronal glucose utilisation may decrease neuronal activity and precede neuronal degeneration. Restrictions to glucose availability increase the sensitivity of primary neurons to glutamate excitation. It has been suggested that the impairment of neuronal glucose utilisation in the brain of AD patients may precede neuronal degeneration (Ogawa et al., 1996; Ferreira et al., 2010; Zhang et al., 2012). AD is known to be associated with abnormalities in energy metabolism, including decreased insulin sensitivity, compromised mitochondrial activities and impaired lipid metabolism (Amato and Man, 2011).

In AD there is a severe reduction of glucose uptake and metabolism (Hoyer, 2004; Jagust et al. 1991; Minoshima et al., 1995; Nicholson et al., 2010). A decreased brain glucose metabolism has been observed even before the onset of the disease in several groups of at-risk individuals (Petersen et al., 1999; Kennedy et al., 1995; Mosconi et al., 2006; Reiman et al., 2004; Mosconi et al., 2007, Small et al., 2000). This suggests that the impairment of glucose uptake/metabolism is a cause of neurodegeneration or is involved in the pathogenesis of AD. The reduction in glucose uptake is accompanied by a decrease in GLUT1 and GLUT3 in the brains of patients with AD (Simpson et al., 1994; Simpson and Davies, 1994; Harr et al., 1995). This decrease in glucose transporters was found to be correlated with tau hyperphosphorylation, with a higher density of neurofibrillary tangles and a downregulation of HIF-1 in human AD brains (Liu et al., 2008). A recent study reveals a mechanism that might be responsible for the altered GLUT3 expression in AD. The study shows that the promoter of human GLUT3 contains three potential cAMP response elements (CRE)-like elements, two of which (CRE-2 and CRE-3) are required to promote GLUT3 expression. In AD, CRE-binding protein (CREB) would be proteolysed by calpain I thus decreasing GLUT3 expression and impairing glucose uptake and metabolism in AD brain (Jin et al., 2013).

In AD, beta-amyloid (βA) deposits are one of the fundamental causes of the disease. Several growth factors, such as IL-3, that increase glucose uptake, can rescue neuronal cells neuronal glucose uptake. It has been suggested that βA decreases glucose uptake by inhibiting fusion of these vesicles (Uemura and Greenlee, 2001). Likewise, it has been shown that βA impairs glucose uptake in cultured hippocampal neurons and astrocytes. Deposits of βA are correlated with a decreased GLUT1 expression in the hippocampus (Hooijmans et al., 2007). Exposure of neurons to βA also leads to impaired mitochondrial activity, suppressed production of ATP, and ultimately neuronal cell death. It is not known whether βA affects glucose uptake in the brain of AD patients. However, *in vitro* studies suggest that βA inhibition of neuronal glucose uptake precedes neuronal degeneration and may thus contribute to changes in glucose utilisation observed in the brain of AD patients.

Research has demonstrated that βA inhibition of neuronal glucose uptake is mediated by a G protein and cAMP. βA transiently increased intracellular cAMP and either forskolin or dibutyryl cAMP (Bt2cAMP) also inhibited neuronal glucose uptake, suggesting that cAMP plays a role in inhibiting neuronal glucose uptake. It is of interest to note that βA-mediated inhibition of neuronal glucose uptake occurs despite a significant increase in GLUT3 translocation to the plasma membrane and an upregulation of GLUT3 mRNA transcription. Further, βA inhibited a K^+-induced increase in GLUT3 at the extracellular surface, suggesting that βA inhibits neuronal glucose uptake by inhibiting fusion of GLUT3-containing vesicles with the plasma membrane (Prapong et al., 2002).

Several additional studies have indicated that neurons are insulin-resistant in AD and that their glucose uptake is decreased. The expression levels of insulin and insulin-like growth factor (IGF-1) proteins and their receptors, as well as expression of insulin receptor substrate-1/2 proteins, are significantly reduced in AD brains. These studies indicate that the insulin/IGF signalling pathway is clearly compromised and that neurons had become insulin-resistant in AD. Streptozotocin (STZ)-treated mice show decreased levels of GLUT1 and GLUT3 rat brain. This decrease might arise from STZ-induced impaired insulin signalling (Deng et al., 2009).

Further, AD-like alterations in protein kinase B and glycogen synthase kinase-β in hippocampus and frontal cortex were seen in STZ-treated rats (Salkovic-Petrisic et al., 2006). Treatment with flavonoids (hesperetin and hesperidin) reverted the effect of βA on GLUT3 expression and PI3K signalling. Thus an increased GLUT3 expression and Akt phosphorylation were observed (Huang et al., 2012). However, AMPK signalling provides an alternative pathway for regulation of glucose uptake. It seems that AMPK may be a major regulator of energy metabolism in neurons, whereas the insulin/IGF pathway might play a more significant role in regulation of neuronal growth and survival (Salminen 2011).

Conclusion

Glucose is an essential energy source for the adult human brain. The brain constitutes only 25% of the adult body weight but it utilises 50% of the total glucose supply. Neuronal activity accounts for 80% of brain energy consumption. Therefore, regulation of the expression and sub cellular localisation of the neuronal glucose transporter GLUT3 is key to the regulation of neuronal energy supply. GLUT3 is expressed at cholesterol-rich domains in the plasma membrane and synaptic-like vesicles in the intracellular space. GLUT3 translocation to the plasma membrane is induced by at least three signalling pathways (Figure 2).

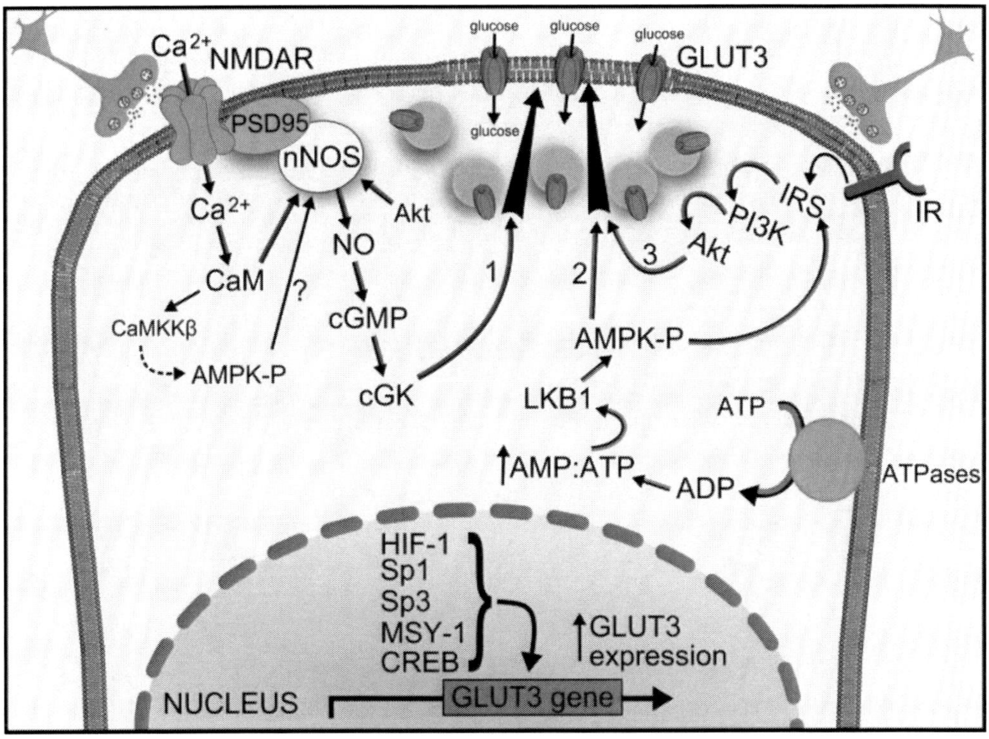

Figure 2. Multiple pathways may contribute to increased GLUT3 expression at the cell surface. GLUT3 is the predominant glucose transporter isoform expressed in neurons. It is expressed at cholesterol-rich domains in the plasma membrane and in synaptic-like vesicles. Expression of GLUT3 at the plasma membrane may be induced by at least four pathways. These pathways should coexist thus assuring efficient glucose uptake in neuronal cells. First (red arrows): An increase in intracellular Ca^{2+} induced by NMDAR activation might lead to nNOS activation and GLUT3 translocation to the plasma membrane. Second (blue arrows): Synaptic activity induces an increase in the AMP:ATP ratio because activation of ATPases is necessary to restore ion gradients. An increase in AMP:ATP ratio induces AMPK activation and GLUT3 translocation to the plasma membrane. Third (purple arrows): Insulin receptor activation leads to GLUT3 translocation to the plasma membrane in an Akt-dependent manner. Fourth: GLUT3 expression is upregulated in response to many stressors mediating adaptive responses to oxygen deprivation and hypoglycaemia, among others. Several transcription factors can induce GLUT3 expression. Akt: protein kinase B; AMPK-P: activated (phosphorylated) AMP kinase; CaM: calmodulin-dependent protein; CaMKKβ: calmodulin-dependent protein kinase kinase β; cGK: cGMP-dependent protein kinase; cGMP: cyclic guanosine mono phosphate; CREB: *cAMP* response element-binding protein; GLUT3: glucose transporter isoform 3; HIF-1: hypoxia-inducible factor-*1*; IR: insulin receptor; IRS: insulin receptor substrate; LKB1: serine/threonine kinase LKB1; MSY-1: transcription factor MSY-1; NMDAR: NMDA receptor; nNOS: neuronal nitric oxide synthase; PI3K: phosphoinositide 3-kinase; PSD95: postsynaptic density protein 95; Sp1: transcription factor SP1; Sp3: transcription factor SP3.

Glutamatergic synaptic activity induces NMDAR activation, increased intracellular Ca^{2+} concentration and nNOS activation causing GLUT3 translocation to the cell surface. Synaptic activity in general induces activation of ATPases because restoration of the ion gradient is essential for successful synaptic transmission. ATPase activation induces an increase in the

AMP:ATP ratio, AMPK activation and thus, GLUT3 translocation to the plasma membrane. Insulin receptor activation can also induce GLUT3 translocation to the cell surface via activation of the PI3K/Akt pathway.

However, neuronal cells can respond to several stressors in an adaptive manner, producing transcription factors which are able to stimulate GLUT3 expression. There are several reports of metabolic impairment in various neurodegenerative disorders such as AD, HD, ALS and PD. Indeed, many of these impairments are related to failures in glucose uptake. Disruptions in energy production may affect neuronal transmission and thus, neuronal survival. Moreover, deregulation of energy metabolism could be implicated in an increased production of oxidative species that also affects neuronal survival.

Acknowledgments

TIRFM experiments were performed at Dr. S. Brauchi's laboratory (Instituto de Fisiología, UACh). This work was supported by Chilean grants FONDECYT 1110571 (M.A.C.) and 1110508 (I.I.C.), DID-UACh.

References

Agus, D. B., Gambhir, S. S., Pardridge, W. M., Spielholz, C., Baselga, J., Vera, J. C., and Golde, D. W. (1997). Vitamin C crosses the blood-brain barrier in the oxidized form through the glucose transporters. *The Journal of clinical investigation* 100, 2842-2848.

Allaman, I., Belanger, M. and Magistretti, P. J. (2011). Astrocyte-neuron metabolic relationships: for better and for worse. *Trends in neurosciences* 34, 76-87.

Aller, C. B., Ehmann, S., Gilman-Sachs, A., and Snyder, A. K. (1997). Flow cytometric analysis of glucose transport by rat brain cells. *Cytometry* 27, 262-268.

Amato, S. and Man, H. Y. (2011). Bioenergy sensing in the brain: the role of AMP-activated protein kinase in neuronal metabolism, development and neurological diseases. *Cell cycle* 10, 3452-3460.

Ames, A., 3rd (2000). CNS energy metabolism as related to function. *Brain research Brain research reviews* 34, 42-68.

Astuya, A., Caprile, T., Castro, M., Salazar, K., Garcia Mde, L., Reinicke, K., Rodriguez, F., Vera, J. C., Millan, C., Ulloa, V., et al. (2005). Vitamin C uptake and recycling among normal and tumor cells from the central nervous system. *Journal of neuroscience research* 79, 146-156.

Attwell, D. and Laughlin, S. B. (2001). An energy budget for signaling in the grey matter of the brain. *Journal of cerebral blood flow and metabolism: official journal of the International Society of Cerebral Blood Flow and Metabolism* 21, 1133-1145.

Augustin, R. (2010). The protein family of glucose transport facilitators: It's not only about glucose after all. *IUBMB life* 62, 315-333.

Barnes, K., Ingram, J. C., Bennett, M. D., Stewart, G. W., and Baldwin, S. A. (2004). Methyl-beta-cyclodextrin stimulates glucose uptake in Clone 9 cells: a possible role for lipid rafts. *The Biochemical journal* 378, 343-351.

Beltrán, F. A., Acuna, A. I., Miro, M. P., Angulo, C., Concha, I. I., and Castro, M. A. (2011). Ascorbic acid-dependent GLUT3 inhibition is a critical step for switching neuronal metabolism. *Journal of cellular physiology* 226, 3286-3294.

Beltrán, F. A., Acuña, A. I., Miro, M. P., and Castro, M. A. (2012). Brain Energy Metabolism in Health and Disease. In: *Neuroscience - Dealing With Frontiers,* C. M. Contreras, ed. (InTech).

Ben-Yoseph, O., Boxer, P. A. and Ross, B. D. (1996). Assessment of the role of the glutathione and pentose phosphate pathways in the protection of primary cerebrocortical cultures from oxidative stress. *Journal of neurochemistry* 66, 2329-2337.

Birnbaum, M. J., Haspel, H. C. and Rosen, O. M. (1986). Cloning and characterization of a cDNA encoding the rat brain glucose-transporter protein. *Proceedings of the National Academy of Sciences of the United States of America* 83, 5784-5788.

Biskup, S., Gerlach, M., Kupsch, A., Reichmann, H., Riederer, P., Vieregge, P., Wullner, U., and Gasser, T. (2008). Genes associated with Parkinson syndrome. *Journal of neurology* 255 Suppl. 5, 8-17.

Boado, R. J. and Pardridge, W. M. (1994). Measurement of blood-brain barrier GLUT1 glucose transporter and actin mRNA by a quantitative polymerase chain reaction assay. *Journal of neurochemistry* 62, 2085-2090.

Bogan, J. S. (2012). Regulation of glucose transporter translocation in health and diabetes. *Annual review of biochemistry* 81, 507-532.

Bohnen, N. I., Koeppe, R. A., Minoshima, S., Giordani, B., Albin, R. L., Frey, K. A., and Kuhl, D. E. (2011). Cerebral glucose metabolic features of Parkinson disease and incident dementia: longitudinal study. *Journal of nuclear medicine: official publication, Society of Nuclear Medicine* 52, 848-855.

Bolz, S., Farrell, C. L., Dietz, K., and Wolburg, H. (1996). Subcellular distribution of glucose transporter (GLUT-1) during development of the blood-brain barrier in rats. *Cell and tissue research* 284, 355-365.

Borghammer, P., Hansen, S. B., Eggers, C., Chakravarty, M., Vang, K., Aanerud, J., Hilker, R., Heiss, W. D., Rodell, A., Munk, O. L., et al. (2012). Glucose metabolism in small subcortical structures in Parkinson's disease. *Acta neurologica Scandinavica* 125, 303-310.

Braak, H., Rub, U., Gai, W. P., and Del Tredici, K. (2003). Idiopathic Parkinson's disease: possible routes by which vulnerable neuronal types may be subject to neuroinvasion by an unknown pathogen. *Journal of neural transmission* 110, 517-536.

Bredesen, D. E., Rao, R. V. and Mehlen, P. (2006). Cell death in the nervous system. *Nature* 443, 796-802.

Bruce, D. G., Harrington, N., Davis, W. A., Davis, T. M., and Fremantle Diabetes, S. (2001). Dementia and its associations in type 2 diabetes mellitus: the Fremantle Diabetes Study. *Diabetes research and clinical practice* 53, 165-172.

Caliceti, C., Zambonin, L., Prata, C., Vieceli Dalla Sega, F., Hakim, G., Hrelia, S., and Fiorentini, D. (2012). Effect of plasma membrane cholesterol depletion on glucose transport regulation in leukemia cells. *PloS one* 7, e41246.

Carruthers, A. (1990). Facilitated diffusion of glucose. *Physiological reviews* 70, 1135-1176.

Castro, M., Caprile, T., Astuya, A., Millan, C., Reinicke, K., Vera, J. C., Vasquez, O., Aguayo, L. G., and Nualart, F. (2001). High-affinity sodium-vitamin C co-transporters

(SVCT) expression in embryonic mouse neurons. *Journal of neurochemistry* 78, 815-823.

Castro, M. A., Angulo, C., Brauchi, S., Nualart, F., and Concha, I. I. (2008). Ascorbic acid participates in a general mechanism for concerted glucose transport inhibition and lactate transport stimulation. Pflugers Archiv: *European journal of physiology* 457, 519-528.

Castro, M. A., Beltrán, F. A., Brauchi, S., and Concha, I. I. (2009). A metabolic switch in brain: glucose and lactate metabolism modulation by ascorbic acid. *Journal of neurochemistry* 110, 423-440.

Castro, M. A., Pozo, M., Cortes, C., Garcia Mde, L., Concha, I. I., and Nualart, F. (2007). Intracellular ascorbic acid inhibits transport of glucose by neurons, but not by astrocytes. *Journal of neurochemistry* 102, 773-782.

Cox, D., Gonder-Frederick, L., McCall, A., Kovatchev, B., and Clarke, W. (2002). The effects of glucose fluctuation on cognitive function and QOL: the functional costs of hypoglycaemia and hyperglycaemia among adults with type 1 or type 2 diabetes. *International journal of clinical practice Supplement*, 20-26.

Chari, M., Yang, C. S., Lam, C. K., Lee, K., Mighiu, P., Kokorovic, A., Cheung, G. W., Lai, T. Y., Wang, P. Y., and Lam, T. K. (2011). Glucose transporter-1 in the hypothalamic glial cells mediates glucose sensing to regulate glucose production in vivo. *Diabetes* 60, 1901-1906.

Chih, C. P. and Roberts Jr, E. L. (2003). Energy substrates for neurons during neural activity: a critical review of the astrocyte-neuron lactate shuttle hypothesis. *Journal of cerebral blood flow and metabolism: official journal of the International Society of Cerebral Blood Flow and Metabolism* 23, 1263-1281.

Choeiri, C., Staines, W. and Messier, C. (2002). Immunohistochemical localization and quantification of glucose transporters in the mouse brain. *Neuroscience* 111, 19-34.

Chua, L. M., Lim, M. L., Chong, P. R., Hu, Z. P., Cheung, N. S., and Wong, B. S. (2012). Impaired neuronal insulin signaling precedes Abeta42 accumulation in female AbetaPPsw/PS1DeltaE9 mice. *Journal of Alzheimer's disease: JAD* 29, 783-791.

Chuquet, J., Quilichini, P., Nimchinsky, E. A., and Buzsaki, G. (2010). Predominant enhancement of glucose uptake in astrocytes versus neurons during activation of the somatosensory cortex. *The Journal of neuroscience: the official journal of the Society for Neuroscience* 30, 15298-15303.

De la Monte, S. M. (2012). Brain insulin resistance and deficiency as therapeutic targets in Alzheimer's disease. *Current Alzheimer research* 9, 35-66.

De Vos, K. J., Grierson, A. J., Ackerley, S., and Miller, C. C. (2008). Role of axonal transport in neurodegenerative diseases. *Annual review of neuroscience* 31, 151-173.

Deng, Y., Li, B., Liu, Y., Iqbal, K., Grundke-Iqbal, I., and Gong, C. X. (2009). Dysregulation of insulin signaling, glucose transporters, O-GlcNAcylation, and phosphorylation of tau and neurofilaments in the brain: Implication for Alzheimer's disease. *The American journal of pathology* 175, 2089-2098.

Devraj, K., Klinger, M. E., Myers, R. L., Mokashi, A., Hawkins, R. A., and Simpson, I. A. (2011). GLUT-1 glucose transporters in the blood-brain barrier: differential phosphorylation. *Journal of neuroscience research* 89, 1913-1925.

Dienel, G. A. and Cruz, N. F. (2009). Exchange-mediated dilution of brain lactate specific activity: implications for the origin of glutamate dilution and the contributions of glutamine dilution and other pathways. *Journal of neurochemistry* 109 Suppl. 1, 30-37.

Dinuzzo, M., Giove, F., Maraviglia, B., and Mangia, S. (2013). Glucose metabolism down-regulates the uptake of 6-(N-(7-nitrobenz-2-oxa-1,3-diazol-4-yl)amino)-2-deoxyglucose (6-NBDG) mediated by glucose transporter 1 isoform (GLUT1): theory and simulations using the symmetric four-state carrier model. *Journal of neurochemistry.*

Doege, H., Bocianski, A., Scheepers, A., Axer, H., Eckel, J., Joost, H. G., and Schurmann, A. (2001). Characterization of human glucose transporter (GLUT) 11 (encoded by SLC2A11), a novel sugar-transport facilitator specifically expressed in heart and skeletal muscle. *The Biochemical journal* 359, 443-449.

Doege, H., Schurmann, A., Bahrenberg, G., Brauers, A., and Joost, H. G. (2000). GLUT8, a novel member of the sugar transport facilitator family with glucose transport activity. *The Journal of biological chemistry* 275, 16275-16280.

Dwyer, D. S. (2002). *Glucose Metabolism in the Brain* (London: Academic Press).

Edinger, A. L. (2005). Growth factors regulate cell survival by controlling nutrient transporter expression. *Biochemical Society transactions* 33, 225-227.

Edinger, A. L. (2007). Controlling cell growth and survival through regulated nutrient transporter expression. *The Biochemical journal* 406, 1-12.

Edinger, A. L. and Thompson, C. B. (2002). Akt maintains cell size and survival by increasing mTOR-dependent nutrient uptake. *Molecular biology of the cell* 13, 2276-2288.

Elfeber, K., Kohler, A., Lutzenburg, M., Osswald, C., Galla, H. J., Witte, O. W., and Koepsell, H. (2004). Localization of the Na+-D-glucose cotransporter SGLT1 in the blood-brain barrier. *Histochemistry and cell biology* 121, 201-207.

Enerson, B. E. and Drewes, L. R. (2006). The rat blood-brain barrier transcriptome. *Journal of cerebral blood flow and metabolism: official journal of the International Society of Cerebral Blood Flow and Metabolism* 26, 959-973.

Ferraiuolo, L., Kirby, J., Grierson, A. J., Sendtner, M., and Shaw, P. J. (2011). Molecular pathways of motor neuron injury in amyotrophic lateral sclerosis. *Nature reviews Neurology* 7, 616-630.

Ferreira, I. L., Resende, R., Ferreiro, E., Rego, A. C., and Pereira, C. F. (2010). Multiple defects in energy metabolism in Alzheimer's disease. *Current drug targets* 11, 1193-1206.

Ferreira, J. M., Burnett, A. L. and Rameau, G. A. (2011). Activity-dependent regulation of surface glucose transporter-3. *The Journal of neuroscience: the official journal of the Society for Neuroscience* 31, 1991-1999.

Finck, B. N. and Kelly, D. P. (2006). PGC-1 coactivators: inducible regulators of energy metabolism in health and disease. *The Journal of clinical investigation* 116, 615-622.

Fukuda, R., Hirota, K., Fan, F., Jung, Y. D., Ellis, L. M., and Semenza, G. L. (2002). Insulin-like growth factor 1 induces hypoxia-inducible factor 1-mediated vascular endothelial growth factor expression, which is dependent on MAP kinase and phosphatidylinositol 3-kinase signaling in colon cancer cells. *The Journal of biological chemistry* 277, 38205-38211.

Gamberino, W. C. and Brennan, W. A., Jr. (1994). Glucose transporter isoform expression in Huntington's disease brain. *Journal of neurochemistry* 63, 1392-1397.

Gandhi, S., Wood-Kaczmar, A., Yao, Z., Plun-Favreau, H., Deas, E., Klupsch, K., Downward, J., Latchman, D. S., Tabrizi, S. J., Wood, N. W., et al. (2009). PINK1-

associated Parkinson's disease is caused by neuronal vulnerability to calcium-induced cell death. *Molecular cell* 33, 627-638.

Garcia, M., Millan, C., Balmaceda-Aguilera, C., Castro, T., Pastor, P., Montecinos, H., Reinicke, K., Zuniga, F., Vera, J. C., Onate, S. A., et al. (2003). Hypothalamic ependymal-glial cells express the glucose transporter GLUT2, a protein involved in glucose sensing. *Journal of neurochemistry* 86, 709-724.

Ghasemzadeh, B., Cammack, J. and Adams, R. N. (1991). Dynamic changes in extracellular fluid ascorbic acid monitored by in vivo electrochemistry. *Brain research* 547, 162-166.

Gjedde, A., Marrett, S. and Vafaee, M. (2002). Oxidative and nonoxidative metabolism of excited neurons and astrocytes. *Journal of cerebral blood flow and metabolism: official journal of the International Society of Cerebral Blood Flow and Metabolism* 22, 1-14.

Godoy, A., Ulloa, V., Rodriguez, F., Reinicke, K., Yanez, A. J., Garcia Mde, L., Medina, R. A., Carrasco, M., Barberis, S., Castro, T., et al. (2006). Differential subcellular distribution of glucose transporters GLUT1-6 and GLUT9 in human cancer: ultrastructural localization of GLUT1 and GLUT5 in breast tumor tissues. *Journal of cellular physiology* 207, 614-627.

Guo, Z., Kindy, M. S., Kruman, I., and Mattson, M. P. (2000). ALS-linked Cu/Zn-SOD mutation impairs cerebral synaptic glucose and glutamate transport and exacerbates ischemic brain injury. *Journal of cerebral blood flow and metabolism: official journal of the International Society of Cerebral Blood Flow and Metabolism* 20, 463-468.

Gupta, A. and Dey, C. S. (2012). PTEN, a widely known negative regulator of insulin/PI3K signaling, positively regulates neuronal insulin resistance. *Molecular biology of the cell* 23, 3882-3898.

Hardie, D. G. (2007). AMP-activated/SNF1 protein kinases: conserved guardians of cellular energy. *Nature reviews Molecular cell biology* 8, 774-785.

Harr, S. D., Simonian, N. A. and Hyman, B. T. (1995). Functional alterations in Alzheimer's disease: decreased glucose transporter 3 immunoreactivity in the perforant pathway terminal zone. *Journal of neuropathology and experimental neurology* 54, 38-41.

Harris, D. S., Slot, J. W., Geuze, H. J., and James, D. E. (1992). Polarized distribution of glucose transporter isoforms in Caco-2 cells. *Proceedings of the National Academy of Sciences of the United States of America* 89, 7556-7560.

Hasselbalch, S. G., Knudsen, G. M., Capaldo, B., Postiglione, A., and Paulson, O. B. (2001). Blood-brain barrier transport and brain metabolism of glucose during acute hyperglycemia in humans. *The Journal of clinical endocrinology and metabolism* 86, 1986-1990.

Hasselbalch, S. G., Knudsen, G. M., Jakobsen, J., Hageman, L. P., Holm, S., and Paulson, O. B. (1995). Blood-brain barrier permeability of glucose and ketone bodies during short-term starvation in humans. *The American journal of physiology* 268, E1161-1166.

Hasselbalch, S. G., Knudsen, G. M., Videbaek, C., Pinborg, L. H., Schmidt, J. F., Holm, S., and Paulson, O. B. (1999). No effect of insulin on glucose blood-brain barrier transport and cerebral metabolism in humans. *Diabetes* 48, 1915-1921.

Heijnen, H. F., Oorschot, V., Sixma, J. J., Slot, J. W., and James, D. E. (1997). Thrombin stimulates glucose transport in human platelets via the translocation of the glucose transporter GLUT-3 from alpha-granules to the cell surface. *The Journal of cell biology* 138, 323-330.

Herrero-Mendez, A., Almeida, A., Fernandez, E., Maestre, C., Moncada, S., and Bolanos, J. P. (2009). The bioenergetic and antioxidant status of neurons is controlled by continuous degradation of a key glycolytic enzyme by APC/C-Cdh1. *Nature cell biology* 11, 747-752.

Hertz, L. (2004). The astrocyte-neuron lactate shuttle: a challenge of a challenge. *Journal of cerebral blood flow and metabolism: official journal of the International Society of Cerebral Blood Flow and Metabolism* 24, 1241-1248.

Hertz, L. (2008). Bioenergetics of cerebral ischemia: a cellular perspective. *Neuropharmacology* 55, 289-309.

Hertz, L., Peng, L. and Dienel, G. A. (2007). Energy metabolism in astrocytes: high rate of oxidative metabolism and spatiotemporal dependence on glycolysis/glycogenolysis. *Journal of cerebral blood flow and metabolism: official journal of the International Society of Cerebral Blood Flow and Metabolism* 27, 219-249.

Hooijmans, C. R., Graven, C., Dederen, P. J., Tanila, H., van Groen, T., and Kiliaan, A. J. (2007). Amyloid beta deposition is related to decreased glucose transporter-1 levels and hippocampal atrophy in brains of aged APP/PS1 mice. *Brain research* 1181, 93-103.

Hosokai, Y., Nishio, Y., Hirayama, K., Takeda, A., Ishioka, T., Sawada, Y., Suzuki, K., Itoyama, Y., Takahashi, S., Fukuda, H., et al. (2009). Distinct patterns of regional cerebral glucose metabolism in Parkinson's disease with and without mild cognitive impairment. *Movement disorders: official journal of the Movement Disorder Society* 24, 854-862.

Hoyer, S. (2004). Glucose metabolism and insulin receptor signal transduction in Alzheimer disease. *European journal of pharmacology* 490, 115-125.

Huang, C., Mattis, P., Tang, C., Perrine, K., Carbon, M., and Eidelberg, D. (2007). Metabolic brain networks associated with cognitive function in Parkinson's disease. *NeuroImage* 34, 714-723.

Huang, S. (2012). Inhibition of PI3K/Akt/mTOR Signaling by Natural Products. Anti-cancer agents in medicinal chemistry.

Huber, J. D., Egleton, R. D. and Davis, T. P. (2001). Molecular physiology and pathophysiology of tight junctions in the blood-brain barrier. *Trends in neurosciences* 24, 719-725.

Ibberson, M., Riederer, B. M., Uldry, M., Guhl, B., Roth, J., and Thorens, B. (2002). Immunolocalization of GLUTX1 in the testis and to specific brain areas and vasopressin-containing neurons. *Endocrinology* 143, 276-284.

Ibberson, M., Uldry, M. and Thorens, B. (2000). GLUTX1, a novel mammalian glucose transporter expressed in the central nervous system and insulin-sensitive tissues. *The Journal of biological chemistry* 275, 4607-4612.

Inukai, K., Shewan, A. M., Pascoe, W. S., Katayama, S., James, D. E., and Oka, Y. (2004). Carboxy terminus of glucose transporter 3 contains an apical membrane targeting domain. *Molecular endocrinology* 18, 339-349.

Ivanov, A., Mukhtarov, M., Bregestovski, P., and Zilberter, Y. (2011). Lactate Effectively Covers Energy Demands during Neuronal Network Activity in Neonatal Hippocampal Slices. *Frontiers in neuroenergetics* 3, 2.

Iwabuchi, S. and Kawahara, K. (2011). Inducible astrocytic glucose transporter-3 contributes to the enhanced storage of intracellular glycogen during reperfusion after ischemia. *Neurochemistry international* 59, 319-325.

Jagust, W. J., Seab, J. P., Huesman, R. H., Valk, P. E., Mathis, C. A., Reed, B. R., Coxson, P. G., and Budinger, T. F. (1991). Diminished glucose transport in Alzheimer's disease: dynamic PET studies. *Journal of cerebral blood flow and metabolism: official journal of the International Society of Cerebral Blood Flow and Metabolism* 11, 323-330.

Jin, N., Qian, W., Yin, X., Zhang, L., Iqbal, K., Grundke-Iqbal, I., Gong, C. X., and Liu, F. (2013). CREB regulates the expression of neuronal glucose transporter 3: a possible mechanism related to impaired brain glucose uptake in Alzheimer's disease. *Nucleic acids research.*

Joost, H. G., Bell, G. I., Best, J. D., Birnbaum, M. J., Charron, M. J., Chen, Y. T., Doege, H., James, D. E., Lodish, H. F., Moley, K. H., et al. (2002). Nomenclature of the GLUT/SLC2A family of sugar/polyol transport facilitators. *American journal of physiology Endocrinology and metabolism* 282, E974-976.

Joost, H. G. and Thorens, B. (2001). The extended GLUT-family of sugar/polyol transport facilitators: nomenclature, sequence characteristics, and potential function of its novel members (review). *Molecular membrane biology* 18, 247-256.

Kahn, B. B., Alquier, T., Carling, D., and Hardie, D. G. (2005). AMP-activated protein kinase: ancient energy gauge provides clues to modern understanding of metabolism. *Cell metabolism* 1, 15-25.

Kennedy, A. M., Frackowiak, R. S., Newman, S. K., Bloomfield, P. M., Seaward, J., Roques, P., Lewington, G., Cunningham, V. J., and Rossor, M. N. (1995). Deficits in cerebral glucose metabolism demonstrated by positron emission tomography in individuals at risk of familial Alzheimer's disease. *Neuroscience letters* 186, 17-20.

Kletzien, R. F., Harris, P. K. and Foellmi, L. A. (1994). Glucose-6-phosphate dehydrogenase: a "housekeeping" enzyme subject to tissue-specific regulation by hormones, nutrients, and oxidant stress. *FASEB journal: official publication of the Federation of American Societies for Experimental Biology* 8, 174-181.

Kuwert, T., Lange, H. W., Boecker, H., Titz, H., Herzog, H., Aulich, A., Wang, B. C., Nayak, U., and Feinendegen, L. E. (1993). Striatal glucose consumption in chorea-free subjects at risk of Huntington's disease. *Journal of neurology* 241, 31-36.

Lee, F. J. and Liu, F. (2008). Genetic factors involved in the pathogenesis of Parkinson's disease. *Brain research reviews* 58, 354-364.

Lee, M. S., Lyoo, C. H., Ryu, Y. H., Lim, H. S., Nam, C. M., Kim, H. S., and Rinne, J. O. (2011). The effect of age on motor deficits and cerebral glucose metabolism of Parkinson's disease. *Acta neurologica Scandinavica* 124, 196-201.

Leenders, K. L., Frackowiak, R. S., Quinn, N., and Marsden, C. D. (1986). Brain energy metabolism and dopaminergic function in Huntington's disease measured in vivo using positron emission tomography. *Movement disorders: official journal of the Movement Disorder Society* 1, 69-77.

Leino, R. L., Gerhart, D. Z., van Bueren, A. M., McCall, A. L., and Drewes, L. R. (1997). Ultrastructural localization of GLUT 1 and GLUT 3 glucose transporters in rat brain. *Journal of neuroscience research* 49, 617-626.

Leto, D. and Saltiel, A. R. (2012). Regulation of glucose transport by insulin: traffic control of GLUT4. *Nature reviews Molecular cell biology* 13, 383-396.

Lim, D., Fedrizzi, L., Tartari, M., Zuccato, C., Cattaneo, E., Brini, M., and Carafoli, E. (2008). Calcium homeostasis and mitochondrial dysfunction in striatal neurons of Huntington disease. *The Journal of biological chemistry* 283, 5780-5789.

Lin, M. T. and Beal, M. F. (2006). Mitochondrial dysfunction and oxidative stress in neurodegenerative diseases. *Nature* 443, 787-795.

Lisinski, I., Schurmann, A., Joost, H. G., Cushman, S. W., and Al-Hasani, H. (2001). Targeting of GLUT6 (formerly GLUT9) and GLUT8 in rat adipose cells. *The Biochemical journal* 358, 517-522.

Liu, Y., Liu, F., Iqbal, K., Grundke-Iqbal, I., and Gong, C. X. (2008). Decreased glucose transporters correlate to abnormal hyperphosphorylation of tau in Alzheimer disease. *FEBS letters* 582, 359-364.

Lowry, O. H. and Passonneau, J. V. (1964). The Relationships between Substrates and Enzymes of Glycolysis in Brain. *The Journal of biological chemistry* 239, 31-42.

Maher, F. (1995). Immunolocalization of GLUT1 and GLUT3 glucose transporters in primary cultured neurons and glia. *Journal of neuroscience research* 42, 459-469.

Maher, F., Davies-Hill, T. M. and Simpson, I. A. (1996). Substrate specificity and kinetic parameters of GLUT3 in rat cerebellar granule neurons. *The Biochemical journal* 315 (Pt 3), 827-831.

Maher, F., Vannucci, S., Takeda, J., and Simpson, I. A. (1992). Expression of mouse-GLUT3 and human-GLUT3 glucose transporter proteins in brain. *Biochemical and biophysical research communications* 182, 703-711.

Maher, F., Vannucci, S. J. and Simpson, I. A. (1994). Glucose transporter proteins in brain. *FASEB journal: official publication of the Federation of American Societies for Experimental Biology* 8, 1003-1011.

Mairet-Coello, G., Tury, A. and DiCicco-Bloom, E. (2009). Insulin-like growth factor-1 promotes G(1)/S cell cycle progression through bidirectional regulation of cyclins and cyclin-dependent kinase inhibitors via the phosphatidylinositol 3-kinase/Akt pathway in developing rat cerebral cortex. *The Journal of neuroscience: the official journal of the Society for Neuroscience* 29, 775-788.

Mangia, S., Garreffa, G., Bianciardi, M., Giove, F., Di Salle, F., and Maraviglia, B. (2003). The aerobic brain: lactate decrease at the onset of neural activity. *Neuroscience* 118, 7-10.

Mangia, S., Simpson, I. A., Vannucci, S. J., and Carruthers, A. (2009). The in vivo neuron-to-astrocyte lactate shuttle in human brain: evidence from modeling of measured lactate levels during visual stimulation. *Journal of neurochemistry* 109 Suppl. 1, 55-62.

Manolescu, A. R., Witkowska, K., Kinnaird, A., Cessford, T., and Cheeseman, C. (2007). Facilitated hexose transporters: new perspectives on form and function. *Physiology* 22, 234-240.

Mazziotta, J. C., Phelps, M. E., Pahl, J. J., Huang, S. C., Baxter, L. R., Riege, W. H., Hoffman, J. M., Kuhl, D. E., Lanto, A. B., Wapenski, J. A., et al. (1987). Reduced cerebral glucose metabolism in asymptomatic subjects at risk for Huntington's disease. *The New England journal of medicine* 316, 357-362.

Menzies, F. M., Ince, P. G. and Shaw, P. J. (2002). Mitochondrial involvement in amyotrophic lateral sclerosis. *Neurochemistry international* 40, 543-551.

Minoshima, S., Frey, K. A., Koeppe, R. A., Foster, N. L., and Kuhl, D. E. (1995). A diagnostic approach in Alzheimer's disease using three-dimensional stereotactic surface projections of fluorine-18-FDG PET. *Journal of nuclear medicine: official publication, Society of Nuclear Medicine* 36, 1238-1248.

Mosconi, L., Brys, M., Switalski, R., Mistur, R., Glodzik, L., Pirraglia, E., Tsui, W., De Santi, S., and de Leon, M. J. (2007). Maternal family history of Alzheimer's disease predisposes to reduced brain glucose metabolism. *Proceedings of the National Academy of Sciences of the United States of America* 104, 19067-19072.

Mosconi, L., Sorbi, S., de Leon, M. J., Li, Y., Nacmias, B., Myoung, P. S., Tsui, W., Ginestroni, A., Bessi, V., Fayyazz, M., et al. (2006). Hypometabolism exceeds atrophy in presymptomatic early-onset familial Alzheimer's disease. *Journal of nuclear medicine: official publication, Society of Nuclear Medicine* 47, 1778-1786.

Mueckler, M. (1994). Facilitative glucose transporters. *European journal of biochemistry / FEBS* 219, 713-725.

Myers, M. G., Jr. and White, M. F. (1996). Insulin signal transduction and the IRS proteins. *Annual review of pharmacology and toxicology* 36, 615-658.

Nagamatsu, S., Sawa, H., Kamada, K., Nakamichi, Y., Yoshimoto, K., and Hoshino, T. (1993). Neuron-specific glucose transporter (NSGT): CNS distribution of GLUT3 rat glucose transporter (RGT3) in rat central neurons. *FEBS letters* 334, 289-295.

Navarrete Santos, A., Tonack, S., Kirstein, M., Kietz, S., and Fischer, B. (2004). Two insulin-responsive glucose transporter isoforms and the insulin receptor are developmentally expressed in rabbit preimplantation embryos. *Reproduction* 128, 503-516.

Nicholson, R. M., Kusne, Y., Nowak, L. A., LaFerla, F. M., Reiman, E. M., and Valla, J. (2010). Regional cerebral glucose uptake in the 3xTG model of Alzheimer's disease highlights common regional vulnerability across AD mouse models. *Brain research* 1347, 179-185.

Nualart, F., Godoy, A. and Reinicke, K. (1999). Expression of the hexose transporters GLUT1 and GLUT2 during the early development of the human brain. *Brain research* 824, 97-104.

O'Malley, D., Reimann, F., Simpson, A. K., and Gribble, F. M. (2006). Sodium-coupled glucose cotransporters contribute to hypothalamic glucose sensing. *Diabetes* 55, 3381-3386.

O'Neill, R. D., Fillenz, M., Sundstrom, L., and Rawlins, J. N. (1984). Voltammetrically monitored brain ascorbate as an index of excitatory amino acid release in the unrestrained rat. *Neuroscience letters* 52, 227-233.

Ogawa, M., Fukuyama, H., Ouchi, Y., Yamauchi, H., and Kimura, J. (1996). Altered energy metabolism in Alzheimer's disease. *Journal of the neurological sciences* 139, 78-82.

Olah, J., Klivenyi, P., Gardian, G., Vecsei, L., Orosz, F., Kovacs, G. G., Westerhoff, H. V., and Ovadi, J. (2008). Increased glucose metabolism and ATP level in brain tissue of Huntington's disease transgenic mice. *The FEBS journal* 275, 4740-4755.

Pardridge, W. M., Boado, R. J. and Farrell, C. R. (1990). Brain-type glucose transporter (GLUT-1) is selectively localized to the blood-brain barrier. Studies with quantitative western blotting and in situ hybridization. *The Journal of biological chemistry* 265, 18035-18040.

Pellerin, L. and Magistretti, P. J. (1994). Glutamate uptake into astrocytes stimulates aerobic glycolysis: a mechanism coupling neuronal activity to glucose utilization. *Proceedings of the National Academy of Sciences of the United States of America* 91, 10625-10629.

Pellerin, L. and Magistretti, P. J. (2012). Sweet sixteen for ANLS. *Journal of cerebral blood flow and metabolism: official journal of the International Society of Cerebral Blood Flow and Metabolism* 32, 1152-1166.

Perneczky, R., Drzezga, A., Boecker, H., Ceballos-Baumann, A. O., Granert, O., Forstl, H., Kurz, A., and Haussermann, P. (2008). Activities of daily living, cerebral glucose metabolism, and cognitive reserve in Lewy body and Parkinson's disease. *Dementia and geriatric cognitive disorders* 26, 475-481.

Pessin, J. E. and Bell, G. I. (1992). Mammalian facilitative glucose transporter family: structure and molecular regulation. *Annual review of physiology* 54, 911-930.

Petersen, R. C., Smith, G. E., Waring, S. C., Ivnik, R. J., Tangalos, E. G., and Kokmen, E. (1999). Mild cognitive impairment: clinical characterization and outcome. *Archives of neurology* 56, 303-308.

Phay, J. E., Hussain, H. B. and Moley, J. F. (2000). Strategy for identification of novel glucose transporter family members by using internet-based genomic databases. *Surgery* 128, 946-951.

Piper, R. C., Hess, L. J. and James, D. E. (1991). Differential sorting of two glucose transporters expressed in insulin-sensitive cells. *The American journal of physiology* 260, C570-580.

Plas, D. R., Talapatra, S., Edinger, A. L., Rathmell, J. C., and Thompson, C. B. (2001). Akt and Bcl-xL promote growth factor-independent survival through distinct effects on mitochondrial physiology. *The Journal of biological chemistry* 276, 12041-12048.

Podolsky, S., Leopold, N. A. and Sax, D. S. (1972). Increased frequency of diabetes mellitus in patients with Huntington's chorea. *Lancet* 1, 1356-1358.

Porras, O. H., Loaiza, A. and Barros, L. F. (2004). Glutamate mediates acute glucose transport inhibition in hippocampal neurons. *The Journal of neuroscience: the official journal of the Society for Neuroscience* 24, 9669-9673.

Pradat, P. F., Bruneteau, G., Gordon, P. H., Dupuis, L., Bonnefont-Rousselot, D., Simon, D., Salachas, F., Corcia, P., Frochot, V., Lacorte, J. M., et al. (2010). Impaired glucose tolerance in patients with amyotrophic lateral sclerosis. *Amyotrophic lateral sclerosis: official publication of the World Federation of Neurology Research Group on Motor Neuron Diseases* 11, 166-171.

Prapong, T., Buss, J., Hsu, W. H., Heine, P., West Greenlee, H., and Uemura, E. (2002). Amyloid beta-peptide decreases neuronal glucose uptake despite causing increase in GLUT3 mRNA transcription and GLUT3 translocation to the plasma membrane. *Experimental neurology* 174, 253-258.

Rajakumar, A., Thamotharan, S., Raychaudhuri, N., Menon, R. K., and Devaskar, S. U. (2004). Trans-activators regulating neuronal glucose transporter isoform-3 gene expression in mammalian neurons. *The Journal of biological chemistry* 279, 26768-26779.

Ramamurthy, S. and Ronnett, G. (2012). AMP-Activated Protein Kinase (AMPK) and Energy-Sensing in the Brain. *Experimental neurobiology* 21, 52-60.

Ramm, G., Larance, M., Guilhaus, M., and James, D. E. (2006). A role for 14-3-3 in insulin-stimulated GLUT4 translocation through its interaction with the RabGAP AS160. *The Journal of biological chemistry* 281, 29174-29180.

Rathmell, J. C., Elstrom, R. L., Cinalli, R. M., and Thompson, C. B. (2003). Activated Akt promotes increased resting T cell size, CD28-independent T cell growth, and development of autoimmunity and lymphoma. *European journal of immunology* 33, 2223-2232.

Rauch, M. C., Ocampo, M. E., Bohle, J., Amthauer, R., Yanez, A. J., Rodriguez-Gil, J. E., Slebe, J. C., Reyes, J. G., and Concha, I. I. (2006). Hexose transporters GLUT1 and GLUT3 are colocalized with hexokinase I in caveolae microdomains of rat spermatogenic cells. *Journal of cellular physiology* 207, 397-406.

Reagan, L. P., Rosell, D. R., Alves, S. E., Hoskin, E. K., McCall, A. L., Charron, M. J., and McEwen, B. S. (2002). GLUT8 glucose transporter is localized to excitatory and inhibitory neurons in the rat hippocampus. *Brain research* 932, 129-134.

Reiman, E. M., Chen, K., Alexander, G. E., Caselli, R. J., Bandy, D., Osborne, D., Saunders, A. M., and Hardy, J. (2004). Functional brain abnormalities in young adults at genetic risk for late-onset Alzheimer's dementia. *Proceedings of the National Academy of Sciences of the United States of America* 101, 284-289.

Ristow, M. (2004). Neurodegenerative disorders associated with diabetes mellitus. *Journal of molecular medicine* 82, 510-529.

Rogers, S., Macheda, M. L., Docherty, S. E., Carty, M. D., Henderson, M. A., Soeller, W. C., Gibbs, E. M., James, D. E., and Best, J. D. (2002). Identification of a novel glucose transporter-like protein-GLUT-12. *American journal of physiology Endocrinology and metabolism* 282, E733-738.

Rothman, D. L., Sibson, N. R., Hyder, F., Shen, J., Behar, K. L., and Shulman, R. G. (1999). In vivo nuclear magnetic resonance spectroscopy studies of the relationship between the glutamate-glutamine neurotransmitter cycle and functional neuroenergetics. *Philosophical transactions of the Royal Society of London Series B, Biological sciences* 354, 1165-1177.

Rowland, L. P. and Shneider, N. A. (2001). Amyotrophic lateral sclerosis. *The New England journal of medicine* 344, 1688-1700.

Rubinsztein, D. C. (2006). The roles of intracellular protein-degradation pathways in neurodegeneration. *Nature* 443, 780-786.

Sakyo, T. and Kitagawa, T. (2002). Differential localization of glucose transporter isoforms in non-polarized mammalian cells: distribution of GLUT1 but not GLUT3 to detergent-resistant membrane domains. *Biochimica et biophysica acta* 1567, 165-175.

Sakyo, T., Naraba, H., Teraoka, H., and Kitagawa, T. (2007). The intrinsic structure of glucose transporter isoforms Glut1 and Glut3 regulates their differential distribution to detergent-resistant membrane domains in nonpolarized mammalian cells. *The FEBS journal* 274, 2843-2853.

Salkovic-Petrisic, M., Tribl, F., Schmidt, M., Hoyer, S., and Riederer, P. (2006). Alzheimer-like changes in protein kinase B and glycogen synthase kinase-3 in rat frontal cortex and hippocampus after damage to the insulin signalling pathway. *Journal of neurochemistry* 96, 1005-1015.

Salminen, A., Kaarniranta, K., Haapasalo, A., Soininen, H., and Hiltunen, M. (2011). AMP-activated protein kinase: a potential player in Alzheimer's disease. *Journal of neurochemistry* 118, 460-474.

Saltiel, A. R. and Kahn, C. R. (2001). Insulin signalling and the regulation of glucose and lipid metabolism. *Nature* 414, 799-806.

Sandyk, R. (1993). The relationship between diabetes mellitus and Parkinson's disease. *The International journal of neuroscience* 69, 125-130.

Sano, H., Kane, S., Sano, E., Miinea, C. P., Asara, J. M., Lane, W. S., Garner, C. W., and Lienhard, G. E. (2003). Insulin-stimulated phosphorylation of a Rab GTPase-activating protein regulates GLUT4 translocation. *The Journal of biological chemistry* 278, 14599-14602.

Sasaki, S. and Iwata, M. (1999). Ultrastructural change of synapses of Betz cells in patients with amyotrophic lateral sclerosis. *Neuroscience letters* 268, 29-32.

Sathasivam, S., Ince, P. G. and Shaw, P. J. (2001). Apoptosis in amyotrophic lateral sclerosis: a review of the evidence. *Neuropathology and applied neurobiology* 27, 257-274.

Schapira, A. H. (2006). Mitochondrial disease. *Lancet* 368, 70-82.

Sibson, N. R., Dhankhar, A., Mason, G. F., Rothman, D. L., Behar, K. L., and Shulman, R. G. (1998). Stoichiometric coupling of brain glucose metabolism and glutamatergic neuronal activity. *Proceedings of the National Academy of Sciences of the United States of America* 95, 316-321.

Siesjo, B. K. (1978). Utilisation of substrates by brain tissues. In: *Brain energy metabolism* (New York: John Wiley and Sons), p. 308.

Silva-Alvarez, C., Carrasco, M., Balmaceda-Aguilera, C., Pastor, P., Garcia Mde, L., Reinicke, K., Aguayo, L., Molina, B., Cifuentes, M., Medina, R., et al. (2005). Ependymal cell differentiation and GLUT1 expression is a synchronous process in the ventricular wall. *Neurochemical research* 30, 1227-1236.

Silver, I. A. and Erecinska, M. (1994). Extracellular glucose concentration in mammalian brain: continuous monitoring of changes during increased neuronal activity and upon limitation in oxygen supply in normo-, hypo-, and hyperglycemic animals. *The Journal of neuroscience: the official journal of the Society for Neuroscience* 14, 5068-5076.

Simpson, I. A., Carruthers, A. and Vannucci, S. J. (2007). Supply and demand in cerebral energy metabolism: the role of nutrient transporters. *Journal of cerebral blood flow and metabolism: official journal of the International Society of Cerebral Blood Flow and Metabolism* 27, 1766-1791.

Simpson, I. A. and Davies, P. (1994). Reduced glucose transporter concentrations in brains of patients with Alzheimer's disease. *Annals of neurology* 36, 800-801.

Slomiany, M. G. and Rosenzweig, S. A. (2006). Hypoxia-inducible factor-1-dependent and -independent regulation of insulin-like growth factor-1-stimulated vascular endothelial growth factor secretion. *The Journal of pharmacology and experimental therapeutics* 318, 666-675.

Small, G. W., Ercoli, L. M., Silverman, D. H., Huang, S. C., Komo, S., Bookheimer, S. Y., Lavretsky, H., Miller, K., Siddarth, P., Rasgon, N. L., et al. (2000). Cerebral metabolic and cognitive decline in persons at genetic risk for Alzheimer's disease. *Proceedings of the National Academy of Sciences of the United States of America* 97, 6037-6042.

Sokoloff, L., Reivich, M., Kennedy, C., Des Rosiers, M. H., Patlak, C. S., Pettigrew, K. D., Sakurada, O., and Shinohara, M. (1977). The [14C]deoxyglucose method for the measurement of local cerebral glucose utilization: theory, procedure, and normal values in the conscious and anesthetized albino rat. *Journal of neurochemistry* 28, 897-916.

Stockli, J., Fazakerley, D. J. and James, D. E. (2011). GLUT4 exocytosis. *Journal of cell science* 124, 4147-4159.

Stuart, C. A., Howell, M. E., Zhang, Y., and Yin, D. (2009). Insulin-stimulated translocation of glucose transporter (GLUT) 12 parallels that of GLUT4 in normal muscle. *The Journal of clinical endocrinology and metabolism* 94, 3535-3542.

Thoidis, G., Kupriyanova, T., Cunningham, J. M., Chen, P., Cadel, S., Foulon, T., Cohen, P., Fine, R. E., and Kandror, K. V. (1999). Glucose transporter Glut3 is targeted to secretory vesicles in neurons and PC12 cells. *The Journal of biological chemistry* 274, 14062-14066.

Thorens, B., Sarkar, H. K., Kaback, H. R., and Lodish, H. F. (1988). Cloning and functional expression in bacteria of a novel glucose transporter present in liver, intestine, kidney, and beta-pancreatic islet cells. *Cell* 55, 281-290.

Treins, C., Giorgetti-Peraldi, S., Murdaca, J., Monthouel-Kartmann, M. N., and Van Obberghen, E. (2005). Regulation of hypoxia-inducible factor (HIF)-1 activity and expression of HIF hydroxylases in response to insulin-like growth factor I. *Molecular endocrinology* 19, 1304-1317.

Uemura, E. and Greenlee, H. W. (2001). Amyloid beta-peptide inhibits neuronal glucose uptake by preventing exocytosis. *Experimental neurology* 170, 270-276.

Uemura, E. and Greenlee, H. W. (2006). Insulin regulates neuronal glucose uptake by promoting translocation of glucose transporter GLUT3. *Experimental neurology* 198, 48-53.

Uldry, M., Ibberson, M., Horisberger, J. D., Chatton, J. Y., Riederer, B. M., and Thorens, B. (2001). Identification of a mammalian H(+)-myo-inositol symporter expressed predominantly in the brain. *The EMBO journal* 20, 4467-4477.

Vander Heiden, M. G., Plas, D. R., Rathmell, J. C., Fox, C. J., Harris, M. H., and Thompson, C. B. (2001). Growth factors can influence cell growth and survival through effects on glucose metabolism. *Molecular and cellular biology* 21, 5899-5912.

Vannucci, S. J., Reinhart, R., Maher, F., Bondy, C. A., Lee, W. H., Vannucci, R. C., and Simpson, I. A. (1998). Alterations in GLUT1 and GLUT3 glucose transporter gene expression following unilateral hypoxia-ischemia in the immature rat brain. *Brain research Developmental brain research* 107, 255-264.

Vannucci, S. J., Rutherford, T., Wilkie, M. B., Simpson, I. A., and Lauder, J. M. (2000). Prenatal expression of the GLUT4 glucose transporter in the mouse. *Developmental neuroscience* 22, 274-282.

Vemula, S., Roder, K. E., Yang, T., Bhat, G. J., Thekkumkara, T. J., and Abbruscato, T. J. (2009). A functional role for sodium-dependent glucose transport across the blood-brain barrier during oxygen glucose deprivation. *The Journal of pharmacology and experimental therapeutics* 328, 487-495.

Ward, M. W., Huber, H. J., Weisova, P., Dussmann, H., Nicholls, D. G., and Prehn, J. H. (2007). Mitochondrial and plasma membrane potential of cultured cerebellar neurons during glutamate-induced necrosis, apoptosis, and tolerance. *The Journal of neuroscience: the official journal of the Society for Neuroscience* 27, 8238-8249.

Weisova, P., Concannon, C. G., Devocelle, M., Prehn, J. H., and Ward, M. W. (2009). Regulation of glucose transporter 3 surface expression by the AMP-activated protein kinase mediates tolerance to glutamate excitation in neurons. *The Journal of neuroscience: the official journal of the Society for Neuroscience* 29, 2997-3008.

WHO [World Health Organisation] (2012). Dementia: A Public Health Priority (Geneva, Switzerland: WHO Press).

Wilson, C. M., Mitsumoto, Y., Maher, F., and Klip, A. (1995). Regulation of cell surface GLUT1, GLUT3, and GLUT4 by insulin and IGF-I in L6 myotubes. *FEBS letters* 368, 19-22.

Wright, E. M., Loo, D. D. and Hirayama, B. A. (2011). Biology of human sodium glucose transporters. *Physiological reviews* 91, 733-794.

Wright, E. M., Loo, D. D., Hirayama, B. A., and Turk, E. (2004). Surprising versatility of Na+-glucose cotransporters: SLC5. *Physiology* 19, 370-376.

Wright, E. M. and Turk, E. (2004). The sodium/glucose cotransport family SLC5. *Pflugers Archiv: European journal of physiology* 447, 510-518.

Wu, X. and Freeze, H. H. (2002). GLUT14, a duplicon of GLUT3, is specifically expressed in testis as alternative splice forms. *Genomics* 80, 553-557.

Wyss, M. T., Jolivet, R., Buck, A., Magistretti, P. J., and Weber, B. (2011). In vivo evidence for lactate as a neuronal energy source. *The Journal of neuroscience: the official journal of the Society for Neuroscience* 31, 7477-7485.

Yong, S. W., Yoon, J. K., An, Y. S., and Lee, P. H. (2007). A comparison of cerebral glucose metabolism in Parkinson's disease, Parkinson's disease dementia and dementia with Lewy bodies. *European journal of neurology: the official journal of the European Federation of Neurological Societies* 14, 1357-1362.

Yu, J., Li, J., Zhang, S., Xu, X., Zheng, M., Jiang, G., and Li, F. (2012). IGF-1 induces hypoxia-inducible factor 1alpha-mediated GLUT3 expression through PI3K/Akt/mTOR dependent pathways in PC12 cells. *Brain research* 1430, 18-24.

Yusa, T. (2001). Increased extracellular ascorbate release reflects glutamate re-uptake during the early stage of reperfusion after forebrain ischemia in rats. *Brain research* 897, 104-113.

Zambrano, A., Jara, E., Murgas, P., Jara, C., Castro, M. A., Angulo, C., and Concha, I. I. (2010a). Cytokine stimulation promotes increased glucose uptake via translocation at the plasma membrane of GLUT1 in HEK293 cells. *Journal of cellular biochemistry* 110, 1471-1480.

Zambrano, A., Otth, C., Maccioni, R. B., and Concha, I. I. (2010b). IL-3 controls tau modifications and protects cortical neurons from neurodegeneration. *Current Alzheimer research* 7, 615-624.

Zambrano, A., Otth, C., Mujica, L., Concha, I. I., and Maccioni, R. B. (2007). Interleukin-3 prevents neuronal death induced by amyloid peptide. *BMC neuroscience* 8, 82.

Zeller, K., Rahner-Welsch, S. and Kuschinsky, W. (1997). Distribution of Glut1 glucose transporters in different brain structures compared to glucose utilization and capillary density of adult rat brains. *Journal of cerebral blood flow and metabolism: official journal of the International Society of Cerebral Blood Flow and Metabolism* 17, 204-209.

Zhang, F., Wang, Y. Y., Liu, H., Lu, Y. F., Wu, Q., Liu, J., and Shi, J. S. (2012). Resveratrol Produces Neurotrophic Effects on Cultured Dopaminergic Neurons through Prompting Astroglial BDNF and GDNF Release. *Evidence-based complementary and alternative medicine: eCAM* 2012, 937605.

In: Glucose Uptake
Editors: Carter C. Johnson and Davis B. Williams
ISBN: 978-1-62618-670-5

Chapter 4

Role for the Microvasculature in Glucose Uptake in Skeletal Muscle

***Dino Premilovac*[1], *Huei L.H. Ng*[1], *Stephen M. Richards*[1], *Eloise A. Bradley*[1], *Renee M. Dwyer*[2], *Stephen Rattigan*[1] and *Michelle A. Keske*[1]**
[1]Menzies Research Institute Tasmania;
[2]School of Medicine, University of Tasmania, Hobart, Australia

Abstract

Skeletal muscle is an important site for insulin action and is responsible for ~80% of insulin-mediated glucose disposal following a meal. Insulin stimulates muscle glucose uptake in two ways: firstly insulin improves the delivery of glucose to muscle cells by increasing blood flow to the microvasculature and transport across the endothelium in contact with the myocyte, and secondly by increasing glucose uptake across the muscle cell membrane via activation of the insulin signalling pathway leading to the recruitment of a specific glucose transporter (GLUT4) to the cell membrane. During insulin resistance and type 2 diabetes, both of these responses to insulin can become impaired. Here we review the considerable body of literature showing that insulin-stimulated microvascular recruitment improves the delivery of insulin and glucose to the myocyte to augment insulin's metabolic action, and that vascular actions of insulin are lost during insulin resistance and type 2 diabetes. Furthermore we discuss more recent evidence to suggest that microvascular insulin responses are lost prior to and independently of the development of myocyte insulin resistance.

Insulin and Total Limb Blood Flow

Insulin-mediated vasodilation was first discovered in the 1930's when large doses of insulin were administered to humans [1]. These vascular effects of insulin were also reported in subsequent clinical studies by other investigators [2, 3]. In these early studies pharmacological doses of insulin were utilised without concomitant glucose infusion leading to severe hypoglycaemia. Thus, at the time it was unclear whether the increase in blood flow was due to insulin *per se*, or due to hypoglycaemia-induced epinephrine release. It was not until development of the hyperinsulinaemic euglycaemic clamp technique in 1975 by Norwich and colleagues that insulin's vascular actions could be dissociated from the effects of hypoglycaemia [4]. Using the clamp technique, Liang and colleagues demonstrated that insulin infusion (4 and 8 mU/kg/min) in dogs could significantly increase total limb blood flow independently of hypoglycaemia, establishing that insulin does indeed have direct actions on the vasculature [5].

Baron and colleagues were among the first to provide compelling evidence that insulin can stimulate limb blood flow in humans. These studies were performed in human subjects under hyperinsulinaemic euglycaemic conditions, excluding the possible contribution of hypoglycaemia-related epinephrine release [6, 7]. This research group demonstrated that insulin increases limb blood flow in a dose-dependent fashion, and that this increase was paralleled with enhancement of muscle glucose uptake. Thus, Baron and colleagues proposed that the insulin-mediated increase in total blood flow acts to augment insulin's metabolic actions in muscle by increasing delivery of both insulin and glucose to the myocyte [6, 7], thus facilitating increased insulin-mediated muscle glucose uptake. Since then a number of investigations have produced similar results that support these findings and have documented insulin's ability to increase total blood flow to skeletal muscle in both humans [8-13] and animal models [14-16].

For total blood flow to increase to skeletal muscle in response to insulin there must either be an increase in cardiac output or redistribution of blood flow between organs. Insulin infusion has been shown to increase cardiac output in healthy humans with little to no effects on mean arterial blood pressure [6, 12, 17, 18]. Therefore, peripheral vascular resistance (calculated as the mean arterial pressure divided by the cardiac output or organ blood flow) declines over the course of continued insulin infusion. Indeed, there is evidence that different vascular beds respond differently to elevations in circulating insulin concentrations. Using experimental animal models it has been reported that hindleg and renal vascular beds dilate, whereas the superior mesenteric vascular bed constricts in response to hyperinsulinaemia [19, 20]. In healthy humans, hyperinsulinaemia has been reported to induce a greater decrease in leg vascular resistance compared to the decrease in systemic vascular resistance [17]. The consensus view appears to be that insulin can differentially regulate vascular resistance between different organs. This conclusion led Baron and colleagues to propose that insulin preferentially vasodilates the vasculature in skeletal muscle [17].

The magnitude of the insulin-mediated increase in total blood flow to muscle has been quite variable between studies with some reports showing no significant increase in blood flow in response to insulin infusion [21-24]. A plausible explanation for these disparate findings arises from technical limitations/sensitivity of various techniques for assessing limb blood flow. Indeed, insulin's effects on total blood flow have been assessed with techniques ranging from thermodilution [6], plethysmography [8], dye dilution [25] and positron emission tomography (PET) coupled with $[^{15}O]H_2O$ [10]. Although each of these techniques can be used to measure total blood flow, they have differing degrees of sensitivity.

Other factors that appeared to be important in detecting the insulin-mediated increase in total blood flow were level of physical activity, muscularity and capillarization within muscles of the various subjects [12, 26, 27]. The dose and duration of insulin exposure used in the studies also impact on the increases in blood flow, which were more prominent at higher doses and following prolonged infusions of insulin [28, 29].

Although many of the studies that reported increases in total blood flow used supra-physiological/pharmacological concentrations of insulin [10, 30, 31], other reports demonstrated that physiological concentrations of insulin could increase blood flow [7, 8, 13]. However, the latter studies have been criticized since prolonged infusions of insulin (several hours) were required before changes in total blood flow could be detected. It should be noted that in healthy subjects insulin levels typically remain elevated for only 90-120 min following a meal [32, 33] and therefore whether the prolonged infusions of insulin during a clamp are physiologically relevant is questionable. Yki-Jarvinen and colleagues reported that insulin's metabolic effects, i.e. on glucose uptake in muscle, occur prior to the effects on total limb blood flow [34-36]. It should be noted, however, that this temporal discordance is due to the assumption that insulin's vascular action is restricted only to increases in total limb blood flow since in these studies this was the only vascular parameter assessed. Thus, there was considerable controversy over both the time-course of this vascular insulin action and whether physiological concentrations of insulin can stimulate an increase in total limb flow.

The discrepancy between total limb blood flow and muscle glucose uptake led our group, together with our collaborators, to hypothesise that the distribution of blood flow within the muscle microvasculature, and not only total flow, may be a more critical factor in determining muscle glucose uptake.

Insulin-Mediated Microvascular Recruitment

While the majority of studies have focused on insulin-mediated increases in total flow to skeletal muscle, what may be more physiologically relevant is the effect of insulin on the distribution of microvascular blood flow within muscle.

To understand this aspect of insulin action it is necessary to consider the anatomy of skeletal muscle vasculature.

Structure of Skeletal Muscle Microvasculature

The vasculature in skeletal muscle can be simply categorised into three types: arterioles, capillaries and venules (Figure 1). Blood enters the muscle through a feed artery which is generally classified as a 1st order arteriole. Subsequent arteriolar branches are numbered in increasing order to the level of the 5th order arterioles, which give rise to capillaries [37]. In muscle, 5th order arterioles have been reported to give rise to groups of approximately 15 capillaries, which are collectively termed a module [38].

A capillary network is the term used to describe a group of modules arising from one 4th order arteriole [38]. Total blood flow in skeletal muscle is initially controlled by constriction/dilation of the 1st-3rd order arterioles (resistance vessels) and flow distribution to the capillary networks in skeletal muscle is controlled by the 3rd-5th order arterioles (microvasculature) [38]. It has been proposed that at any one time not all capillaries are perfused [37, 39].

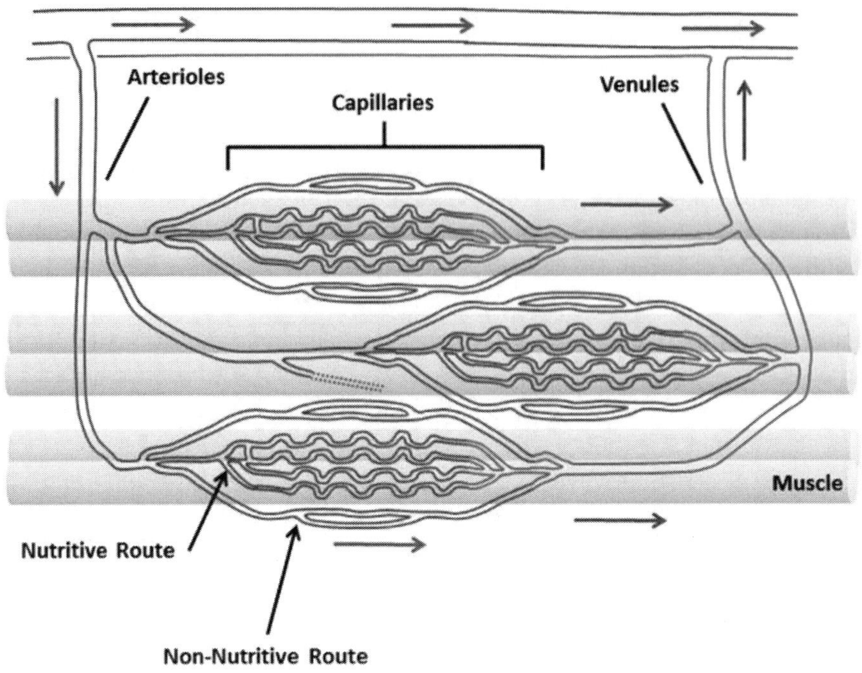

Figure 1. Schematic representation of the microvascular network in skeletal muscle. Schematic illustrating the two flow routes in skeletal muscle, termed nutritive and the non-nutritive. Blood flows in through the arteries and distributes between the nutritive (intimately associated with muscle cells) and non-nutritive routes (associated with connective tissue and adipocytes). During instances of low metabolic demand such as rest, blood flow is preferentially distributed to the non-nutritive route. Conversely, during times of increased metabolic demand (such as during exercise or following a meal) blood flow is preferentially distributed to the nutritive route to augment delivery of nutrients to the myocyte. The coloured arrows indicate the direction of blood flow.

Indeed, a number of studies have shown that, in both humans and animals, microvascular perfusion in resting skeletal muscle is neither continuous nor uniform, but intermittent and heterogeneous in distribution [36, 40-42].

Thus, it has been suggested that vessels with smooth muscle in muscle undergo vasomotion, that is, contraction and dilation at regular intervals to alternate blood flow through different capillary modules as required to maintain nutrient supply [43, 44]. Additionally, it has been proposed that the vascular system in muscle has a further level of complexity, involving modulation of blood flow between two distinct flow routes, termed the nutritive and non-nutritive flow routes [45].

Nutritive and Non-Nutritive Flow Routes

The first reports to suggest the existence of two flow routes in muscle were from Pappenheimer in the 1940's [46]. In these studies Pappenheimer reported that oxygen consumption across the hindleg could be increased or decreased using different hormonal substances and proposed that this difference was the result of flow redistribution within the hindleg muscle vasculature [46]. Since that time, our group, over the last 15+ years, has demonstrated in rats that pharmacologically shifting flow between these two flow routes results in different metabolic consequences for hindleg skeletal muscle [45, 47-49]. The non-nutritive flow route is likely composed of shorter capillaries of slightly larger diameter that consequently have lower resistance [50]. The non-nutritive flow route has been associated with less metabolically demanding tissues within skeletal muscle such as connective tissue (septa and tendon) [51] and adipocytes [52]. Conversely, the nutritive flow route is thought to consist of longer, more tortuous capillaries that wrap around muscle fibres to provide a larger surface area to allow for enhanced nutrient exchange between the blood and the interstitial space surrounding myocytes (See Figure 1) [45]. It has been proposed that the non-nutritive flow route acts as a functional flow reserve that, during increased metabolic demand (such as exercise or following a meal), from which blood flow can be rapidly redistributed to the nutritive route, thus allowing for increased delivery of nutrients to the myocyte [53]. This redistribution of flow within skeletal muscle is termed capillary- or microvascular recruitment.

Microvascular Recruitment

Microvascular recruitment is a term used to describe perfusion of previously underperfused capillaries that results in increased skeletal muscle capillary blood flow. This increased perfusion increases capillary surface area within muscle and leads to increased delivery of hormones and nutrients to the myocyte. Our laboratory has developed and adapted a number of independent methods that have allowed examination of insulin's action on the microcirculation in skeletal muscle. These include 1-methylxanthine (1-MX) metabolism [15] and contrast enhanced ultrasound (CEU) [54]. Using these techniques, we have reported that insulin increases microvascular recruitment in skeletal muscle.

A novel method for assessing the extent of microvascular recruitment during hyperinsulinaemia was developed by Rattigan and colleagues in 1997 [15]. The technique depends on metabolism of the exogenous substrate 1-MX by the endothelial bound enzyme xanthine oxidase (XO). 1-MX is not vasoactive and is converted to a single product, 1-methylurate (1-MU) by XO. 1-MU is not metabolised further in the body and the metabolism of 1-MX to 1-MU is stoichiometric. 1-MX and 1-MU can be readily separated from each other and quantified in deproteinised plasma using high performance liquid chromatography (HPLC) [48]. Previous studies have demonstrated that XO in bovine and human muscle is located primarily on capillary endothelial cells but not in endothelial cells of large arteries and veins, or in myocytes [55, 56]. However, it should be noted that in human muscle, XO has also been found in smooth muscle cells of small arterioles and venules [55]. Nevertheless, since XO is predominantly located in capillary endothelial cells, assessment of 1-MX metabolism by XO is a useful marker for the extent of microvascular recruitment in muscle.

The constant-flow pump-perfused rat hindleg preparation is an *ex vivo* technique that allows for examination of vascular perfusion and metabolism in muscle using an intact hindleg. We have previously demonstrated that infusion of serotonin increases vasoconstriction in this *ex vivo* preparation resulting in a shift in flow away from capillaries supplying myocytes to favour the non-nutritive flow route, culminating in marked attenuation of insulin-mediated metabolism in skeletal muscle [47, 49, 57-59]. Moreover, serotonin-mediated vasoconstriction in this setting significantly attenuated metabolism of 1-MX across the hindleg [48], thus implicating reduced nutritive blood flow in muscle with reduced metabolism of 1-MX. When we applied the 1-MX technique during a hyperinsulinaemic euglycaemic clamp (10mU/kg/min) in anaesthetised rats, a marked elevation in 1-MX metabolism was found when compared to saline infused rats [15], indicating an increase in the exposure of circulating 1-MX to XO during insulin infusion and thus an increase in microvascular recruitment within muscle. In this study, this increase in insulin-stimulated microvascular recruitment was accompanied by a parallel increase in femoral artery (total limb) blood flow [15]. To dissociate the effects of increased femoral artery blood flow from microvascular recruitment, we utilised an epinephrine infusion to produce a similar increase in femoral artery blood flow and showed that this did not affect microvascular recruitment (measured by 1-MX metabolism), indicating that these two processes can occur independently. This is a crucial observation, as it demonstrates that increases in total leg blood flow can occur without altering microvascular recruitment. Thus, it was proposed that insulin possesses a specific vascular action that redistributes blood flow in muscle, resulting in increased nutritive blood flow and ultimately in enhanced delivery of glucose and insulin to the myocyte. This ability of insulin to increase microvascular recruitment has been demonstrated at a variety of physiological concentrations of insulin [60]. Furthermore, we have also shown that low concentrations of insulin can increase microvascular recruitment without affecting femoral artery blood flow [60], suggesting that insulin can differentially affect vessels controlling total flow and microvascular flow distribution.

There are, however, some limitations associated with the 1-MX technique. Firstly, the technique is dependent on obtaining a constant arterial concentration of 1-MX. To achieve this, prolonged infusions of 1-MX are required. Secondly, relatively large arterial and venous blood samples (for experimental animals) are required for measurement of 1-MX metabolism across the hindlimb. For these reasons it is not feasible to obtain multiple time point measures of 1-MX metabolism during a hyperinsulinaemic euglycaemic clamp. Finally, the 1-MX technique is not suitable for human studies, since humans express a much lower level of XO compared to rats, thus making the arterio-venous 1-MX difference too small to be reliably measured [61]. For these reasons, other techniques (such as CEU) are preferred for assessment of microvascular recruitment in human subjects.

CEU is also a relatively new technique (last 15 years) that can measure skeletal muscle perfusion. Previous studies utilised CEU to quantify microvascular volume and perfusion in the myocardium [62-65], and more recently this has been adapted for use in skeletal muscle [16, 66, 67]. This technique involves infusion of albumin or phospholipid microbubbles filled with an inert gas (such as perflurocarbon) into the vasculature. The microbubbles are stable and possess a similar size and rheology to red blood cells, and importantly they remain exclusively intravascular [68]. When ultrasound is applied the microbubbles oscillate in size, and if the pulse is strong enough the bubbles can burst. The oscillation and/or destruction of microbubbles results in the generation of an acoustic signal that can subsequently be detected by the ultrasound transducer. The strength of the signal received is dependent on the number of microbubbles in the field of view of the ultrasound transducer. Since the microbubbles are confined to the vasculature, the signal received is itself a reflection of the extent of tissue perfusion under the ultrasound transducer. Following destruction of microbubbles using a high intensity ultrasound pulse, the rate of reappearance of microbubbles is a reflection of blood velocity, while the capillary blood volume can be measured by the extent of tissue opacification. The contribution of fast filling arteries, arterioles, veins and venules to the image can be subtracted along with the background tissue and thus the residual signal reflects only the microcirculation. Replenishment curves are fitted to the equation $y = A(1 - e^{-\beta t})$, where y is video intensity, t is the time from the destructive ultrasound pulse, A is plateau video intensity (an index of microvascular volume), and β is the rate constant, which provides a measure of flow velocity in the microvasculature.

Insulin-stimulated microvascular recruitment in skeletal muscle has been observed in both human and animal models using CEU [16, 67, 69-71]. Reports have demonstrated a strong correlation between the CEU and 1-MX methods for analysis of insulin-mediated microvascular recruitment in skeletal muscle [60, 71]. In these studies hyperinsulinaemic euglycaemic clamps, using various concentrations of insulin (1.5, 3.0 and 10mU/kg/min) revealed that both CEU and 1-MX methods showed similar increases in microvascular recruitment at each concentration of insulin [60, 71]. Furthermore, in both studies microvascular recruitment occurred at lower concentrations of insulin than changes in total flow, supporting the notion that changes in microvascular recruitment and total blood flow in response to insulin are independent of each other.

Mechanism of Microvascular Recruitment

Thus far, a number of studies have revealed that insulin can increase microvascular recruitment in skeletal muscle by redistributing flow to the capillary network associated with myocytes. The mechanisms responsible for this insulin-mediated microvascular recruitment have not been fully elucidated. While it appears that insulin's ability to increase total flow is dependent on dilation of 1st-3rd order arterioles, microvascular recruitment seems to be mediated by the 4th and 5th order arterioles [38, 72]. Insulin receptors are located on endothelial, vascular smooth muscle and skeletal muscle cells. Whether insulin acts on one or a combination of these cell types, and what the mechanisms and signalling molecules responsible for increases in total flow versus microvascular recruitment are, remain unclear.

Nitric oxide (NO) has been strongly implicated in insulin-mediated vascular activity [70, 73]. Studies in primary cultures of endothelial cells have elucidated a complete insulin signalling pathway from the insulin receptor, to activation of endothelial nitric oxide synthase (eNOS) and NO production by the IRS/PI3K/Akt pathway [74-76]. In turn NO can diffuse into adjacent vascular smooth muscle cells to activate soluble guanylyl cyclase, resulting in vasodilation. This mechanism is consistent with our studies in which insulin-mediated effects on total blood flow and microvascular recruitment were blocked by systemic infusion of the nitric oxide synthase inhibitor, Nω-nitro-L-arginine-methyl ester (L-NAME) [70, 77].

A second possibility for insulin's vascular actions exists whereby insulin can by-pass the endothelial cell and act directly on vascular smooth muscle cells to induce vasodilation. Trovatti and colleagues, using incubated human vascular smooth muscle cells, reported that insulin could stimulate a rapid increase in cyclic guanosine monophosphate (cGMP) and thus induce relaxation [78]. The authors reported that this was also mediated by a NO-dependent mechanism since the insulin-mediated increase in cGMP in incubated vascular smooth muscle cells was blocked by inhibiting nitric oxide synthase [78, 79].

Other vasoactive substances may also be involved in insulin-mediated microvascular recruitment. As well as mediating release of NO, insulin has also been reported to stimulate production of endothelin-1 (ET-1) via a MAPK-dependent pathway. This was evident because the vasodilator effects of insulin are potentiated by ET-1 receptor blockade [80, 81]. Anfossi and colleagues found that insulin stimulated ET-1 production in human vascular smooth muscle cells *in vitro* [82]. Furthermore, Misurski and colleagues reported that initial NO-mediated vasodilation can be overcome by generation of ET-1 when rat mesenteric vessels were exposed to high insulin concentrations [83]. This effect was blocked by addition of an ET-1 receptor antagonist [83]. Thus, it has been proposed that insulin stimulates generation of both NO and ET-1, and it is the balance between these two vasoactive agents that may control the extent of recruitment of nutritive capillary flow in skeletal muscle [80, 84].

It should be noted that the majority of investigations regarding insulin-mediated microvascular recruitment in muscle have involved systemic insulin infusion. Thus, there exists the possibility that a central insulin-mediated process may also influence muscle metabolism by controlling microvascular recruitment in muscle. Indeed, reports from Burcelin and associates have shown that centrally administered insulin can increase muscle glycogen synthesis (muscle metabolism) by approximately 100% [85] and also stimulate approximately a 40% increase in femoral artery blood flow [86]. However, whether this central effect of insulin to stimulate muscle metabolism is mediated by enhancement of microvascular recruitment in muscle was not assessed and other reports would argue against this concept. Coggins and associates demonstrated that local infusion of insulin (0.05mU/kg/min), which did not increase systemic insulin concentrations, increased insulin-mediated microvascular recruitment and muscle glucose uptake in the forearm of healthy human subjects [69]. The authors noted that the effects on microvascular recruitment were apparent in the absence of an increase in total forearm blood flow [69]. Additionally, we have shown that central (intracerebroventricular) administration of insulin (135mU/min/kg of brain) in healthy rats did not affect insulin's peripheral actions [87]. Indeed, during the central insulin infusion we found no effect on total femoral artery blood flow, microvascular recruitment or skeletal muscle glucose uptake. Together, these studies suggest that increased microvascular recruitment in muscle is likely to be mediated by local, rather than central effects of insulin.

In addition to the effects on NO release, direct actions on vascular smooth muscle cells, release of other vasoactive factors, changes in muscle metabolism and alterations in sympathetic nerve activity may contribute to insulin's vascular actions [88].

Thus, there may not be a single mechanism involved insulin-mediated vasodilation, but this action is likely to depend on a variety of factors, and modulation of these various processes could potentially alter overall insulin action and sensitivity.

Microvascular Recruitment and Muscle Glucose Uptake

Whilst it is generally accepted that insulin can increase blood flow to skeletal muscle, the physiological relevance of this in regulating glucose uptake and whether defects in this vascular action of insulin are involved in the development of insulin resistance remain somewhat controversial. Studies in both animal models and humans have suggested that insulin-mediated microvascular recruitment in skeletal muscle plays an integral role in regulating muscle glucose uptake [16, 33, 69, 77, 89]. The delivery of glucose and insulin to skeletal myocytes is dependent on blood flow, capillary surface area and permeability.

Capillaries in muscle are the site of nutrient exchange, and according to the Fick principle, glucose uptake can be calculated from the product of the A-V glucose difference multiplied by flow [90]. Thus, during states in which capillary surface area and permeability are small, an increase in total blood flow to muscle would have minimal effect on glucose uptake *per se*, since the rate limiting step in this equation would be glucose extraction [91].

However, during states in which microvascular perfusion in muscle is increased, such as following a glucose load or mixed meal, the resulting increase in both surface area and blood flow significantly augments insulin and glucose delivery to the myocyte [33, 92]. The end result is a significant increase in glucose disposal by muscle that correlates positively with increased microvascular recruitment.

Indeed, we have shown that this effect of insulin on the microvasculature is both rapid in onset and highly sensitive to elevations in circulating insulin concentrations [60, 77], consistent with its proposed importance in timely uptake and storage of glucose by skeletal myocytes. Physiological doses of insulin (~600 pmol/L) increase microvascular recruitment in muscle of anaesthetised rats within 10 min [77]. This increase in microvascular recruitment was found to precede activation of the insulin signalling pathway in muscle, stimulation of glucose disposal in muscle and insulin's effects on total blood flow [77].

In addition, infusion of relatively low doses of insulin (1.5 and 3 mU/kg/min) stimulated near-maximal microvascular recruitment in muscle without affecting total blood flow [60]. Together, these data suggest that insulin-stimulated microvascular recruitment in muscle is a highly sensitive, early event which likely plays an important role in facilitating enhancement of insulin-mediated muscle glucose uptake.

Insulin Resistance and Imparied Microvascular Responses

Insulin resistance is defined as a reduced biological response to a given dose of insulin and is strongly associated with the pathogenesis of many metabolic disorders including obesity, dyslipidaemia and type 2 diabetes [93, 94].

Together these diseases contribute significantly to increased risk of cardiovascular disease and type 2 diabetes. Given the evidence that insulin also possesses haemodynamic actions that contribute to glucose metabolism *in vivo*, it is probable that the development of insulin resistance also includes a haemodynamic origin.

Acute Models of Microvascular Insulin Resistance

The findings that insulin has robust and rapid actions on lower order arterioles suggest that microvascular recruitment by insulin is an important physiological response. The importance to muscle insulin sensitivity is evident when this microvascular effect of insulin is acutely attenuated. In a number of studies we have shown that acutely blocking insulin's vascular actions in muscle also impairs muscle glucose disposal. These include, but are not limited to, the effects produced by: elevated free fatty acids (FFAs), serotonin, tumour necrosis factor alpha (TNFα), glucosamine and ET-1 infusion.

Serotonin Infusion

Using the perfused hindleg preparation our group demonstrated that vasoconstriction induced by serotonin, which favours non-nutritive flow [49], reduced insulin-mediated muscle glucose uptake [57]. However, incubating muscles with serotonin had no effect on insulin-mediated muscle glucose uptake [57]. These data suggest that whilst serotonin does not directly inhibit insulin-mediated glucose uptake by the myocyte, it reduces insulin-mediated muscle glucose uptake by redistributing flow away from muscle, i.e. to the non-nutritive route. Using the anaesthetised rat preparation, acute serotonin (in the form of α-methyl-serotonin) infusion was found to significantly reduce insulin-mediated microvascular recruitment in muscle and this was associated with a ~50% reduction in insulin-mediated muscle glucose uptake [95]. Taken together, these data indicate that infusion of factors that inhibit the ability of insulin to increase microvascular recruitment in muscle also results in the development of acute muscle insulin resistance.

Tumour Necrosis Factor A

Obesity is associated with chronic, low grade inflammation that often co-exists with insulin resistance [96-98]. Increased production of proinflammatory cytokines from adipose tissue, such as TNFα, has been proposed as another factor that contributes to the development of insulin resistance [99]. TNFα has been reported to directly inhibit the insulin signalling pathway in various cell types [100-102], including vascular endothelial cells [103].

Acute *in vivo* infusion of TNFα in both humans and animals has been shown to reduce whole body insulin sensitivity [104, 105]. We have demonstrated that infusion of TNFα was associated with complete attenuation in insulin-mediated microvascular recruitment [106]. This was accompanied by a 50% reduction in insulin-mediated muscle glucose uptake [106]. However, it has been shown that acutely incubating muscles with TNFα does not significantly affect insulin-mediated muscle glucose uptake [107, 108]. Thus, as with serotonin, these data indicate that TNFα does not directly affect myocyte insulin sensitivity, but inhibits insulin-mediated microvascular recruitment and subsequently reduces insulin-mediated muscle glucose uptake *in vivo*.

Free Fatty Acids

Obesity, insulin resistance and type 2 diabetes have all been associated with increased circulating concentrations of FFAs [109]. Increased FFA concentrations have thus been proposed to contribute to the development of insulin resistance [109-111]. The increased FFA concentrations may induce insulin resistance by either alteration of the Randle cycle or through direct and indirect impairment of the insulin signalling pathway. The Randle cycle, or FFA cycle is a metabolic process involving competition of fatty acids and carbohydrates for cytosolic coenzymes [112]. Elevated circulating FFAs result in increased oxidation of fatty acids and thus increased mitochondrial acetyl-CoA content. This results in inhibition of pyruvate dehydrogenase activity [113] and leads to increased accumulation of citrate which in turn inhibits PFK-1 [114]. Inhibition of pyruvate dehydrogenase and PFK-1 results in an increase in the level of glucose-6-phosphate and thus inhibition of hexokinase, resulting in reduced glucose uptake. However, Boden and colleagues proposed that reduction in glucose oxidation was only responsible for approximately a third of fatty acid-induced inhibition of glucose uptake [115].

Indeed, other reports have suggested that FFA could induce insulin resistance by directly affecting insulin signalling in myocytes [116]. Increased circulating concentrations of FFAs lead to the accumulation of FFA metabolites fatty acyl-CoA, diacylglycerol and ceramides within non-adipose cells such as myocytes [117-119]. These metabolites have in turn been shown to impair insulin signalling at the PI3K/Akt level via activation of PKC and result in inhibitory serine phosphorylation of the insulin receptor substrate (IRS) [120, 121]. How much this direct inhibition of insulin signalling contributes to FFA induced insulin resistance is still unclear.

There is some evidence to suggest that FFA infusion may cause insulin resistance by impairing endothelial function and therefore delivery of insulin and glucose to the myocyte. Steinberg and colleagues reported that acutely raising FFAs in normal healthy subjects resulted in a 20% reduction in endothelium-dependent total leg blood flow in response to methacholine infusion [122].

In a subsequent study Steinberg and colleagues also found that long-term (8 hr) FFA infusion significantly attenuated limb NO production and reduced the insulin-stimulated increase in femoral artery blood flow in otherwise healthy subjects [123]. The authors showed that the reduction in insulin's effects on blood flow correlated strongly with reduced whole body glucose disposal [123]. These reports raised the possibility that increased FFA concentrations may result in the development of insulin resistance mediated by reduced insulin-mediated vasodilation and thus reduced glucose uptake in muscle. Other investigations have supported this hypothesis in both animal and human studies. We have reported that Intralipid and heparin infusion for 4 hrs attenuated insulin-mediated microvascular recruitment in muscle and this was associated with reduced insulin-mediated muscle glucose uptake [124]. These effects occurred even though the insulin-mediated changes in femoral artery blood flow were unaffected [124].

Similarly, Liu and colleagues demonstrated that infusion of Intralipid and heparin in humans also resulted in marked attenuation of insulin-mediated microvascular recruitment in muscle associated with reduced whole body glucose disposal [125]. Interestingly, Han and colleagues reported that insulin sensitivity in muscle of FFA infused rats could be restored by incubating excised muscles in a modified Krebs-Heneseleit buffer for 2 hrs, allowing for FFA washout [126].This finding suggests that acute FFA infusion (2-4 hrs), resulting in elevated circulating FFAs, does induce direct myocyte insulin resistance.

Together, these studies raise the possibility that increasing FFA concentrations alone can induce insulin resistance, and that this appears to be mediated not only by myocyte insulin resistance, but also by attenuation of insulin-mediated microvascular recruitment.

Glucosamine Infusion

Glucosamine, which increases flux through the hexosamine pathway, has also been shown to induce insulin resistance in a number of tissues including muscle, liver and adipose [127-130]. Whilst muscles incubated with glucosamine display reduced insulin-mediated glucose uptake, infusion of glucosamine has also been shown to induce muscle insulin resistance *in vivo* [128, 129, 131].

As well as inhibiting the metabolic effects of insulin, Holmang and colleagues demonstrated that glucosamine infusion also reduced insulin-stimulated total limb blood flow [132]. In a subsequent study we found that acute glucosamine infusion could block insulin's effects on both total blood flow and microvascular recruitment [131].

This glucosamine-induced vascular insulin resistance was associated with marked inhibition of insulin-mediated muscle glucose uptake *in vivo* [131]. Thus, these data suggest that as well as inducing myocyte insulin resistance, acute glucosamine infusion attenuates insulin-mediated microvascular recruitment and thus further supports a role for defects in microvascular insulin action as a contributor to insulin resistance.

Endothelin-1 Infusion

Increased insulin-stimulated ET-1 production in the endothelium has been proposed to contribute to the development of insulin resistance [133]. Some studies have reported that ET-1 decreases total blood flow to skeletal muscle, but these findings are not consistent [134-137]. We have found that acute infusion of ET-1 in the perfused rat hindleg preparation increased vasoconstriction in a dose-dependent manner [138]. At lower doses ET-1 increased oxygen consumption across the hindleg, however, at higher doses oxygen consumption was significantly reduced, suggesting a biphasic effect on blood flow distribution in muscle [138]. Furthermore, infusion of insulin was able to overcome and reverse this ET-1-mediated effect on the vasculature, at both low and high doses, and ET-1 did not affect insulin-mediated muscle glucose uptake in this *ex vivo* setting [138]. However, we have also shown that acute infusion of ET-1 in the anaesthetised rat preparation reduces insulin sensitivity. Indeed, we found that a 30 min systemic infusion of ET-1, prior to a hyperinsulinaemic euglycaemic clamp, attenuated insulin-mediated microvascular recruitment and reduced skeletal muscle glucose uptake [139]. Taken together, these studies suggest that ET-1 may not affect insulin-mediated muscle glucose uptake when assessed independently of insulin's haemodynamic effects. However, elevated levels of ET-1 *in vivo* may contribute to reduced insulin-mediated microvascular recruitment and thus may limit insulin and glucose delivery to the muscle.

Chronic Models of Microvascular Insulin Resistance

While acute treatment with factors such as TNFα and FFAs allow for examination of how individual factors inhibit insulin action directly, chronic models of insulin resistance have also exhibit defects in insulin's vascular actions.

Obese Zucker Rat

The obese Zucker (OZ) rat is a model of severe obesity that develops type 2 diabetes at a relatively early age. Obesity in these rats is the result of a recessive mutation, designated *fa*, that inactivates the leptin receptor [140]. The leptin receptor controls satiety and as a result the OZ rats cannot respond to leptin and thus hyperphagia ensues. In addition to excess food intake, OZ (*fa/fa*) rats have also been shown to exhibit decreased energy expenditure [141]. As a result the OZ rats weigh significantly more that their lean littermates (*Fa/?*) by 4-5 weeks of age [142]. The OZ rats also exhibit dyslipidaemia, hyperglycaemia, hyperinsulinaemia and insulin resistance at an early age and by 12 weeks develop severe glucose intolerance [143-145]. Furthermore, defects in muscle glucose uptake have been observed as early as 4 weeks of age in incubated muscle and *in vivo* [146, 147]. Some [148], but not all studies [149] have reported that these animals also develop hypertension.

Along with the metabolic dysfunction, OZ rats also develop large and small vessel vascular alterations resulting from endothelial dysfunction [150, 151]. Studies in isolated vessels have revealed that the ability of insulin to attenuate vasoconstriction is impaired in vessels from OZ rats compared to lean controls [152, 153]. During a hyperinsulinaemic euglycaemic clamp, OZ rats exhibit attenuation of both insulin-stimulation of total leg flow and insulin-mediated microvascular recruitment in muscle compared to lean control rats [149]. This vascular dysfunction was associated with marked attenuation of insulin-mediated whole body and skeletal muscle glucose disposal [149].

High Fat Fed Rat

For the last 50 years animal models have been used to examine the effects of increased caloric intake on the pathogenesis of obesity and insulin resistance [154-156]. Increasing dietary fat has been shown to induce whole body insulin resistance following just 4 weeks of intervention [157]. Kraegen and colleagues demonstrated that high fat fed rats exhibit increased body weight and especially fat mass, while only showing mild signs of hyperinsulinaemia and no apparent hyperglycaemia [157]. However, the animals exhibited significant attenuation of whole body insulin sensitivity, and specifically, reduced insulin-mediated glucose uptake by skeletal muscle [157]. The mechanisms of high fat diet-induced insulin resistance have been primarily attributed to one of two factors; (i) increased fatty acid metabolite build-up in non-adipose tissues, such as muscle, that inhibit insulin signalling [158] or (ii) increased inflammatory cytokine secretion (TNFα for example) by either adipocytes or infiltrating macrophages that can inhibit the various actions of insulin [97]. Others have proposed a role for the vasculature as another contributor to the development of insulin resistance following high fat feeding. Naderali and Williams reported that two days of high fat feeding was sufficient to reduce endothelium-dependent vasodilation in third order mesenteric arteries compared to arteries from control diet fed animals [159]. This defect was present without alterations in body weight, fat mass or plasma glucose and insulin levels [159]. Indeed, the only apparent effect of the high fat diet was an increase in the circulating FFA concentration, again implicating increased FFAs as a mediating factor in the development of endothelial dysfunction.

Other studies have revealed that long-term high fat feeding also results in significant attenuation of endothelium-dependent vasodilation in skeletal muscle arterioles [160, 161]. Erdei and colleagues demonstrated that arteries isolated from 10 week high fat diet-fed (36% fat wt./wt.) rats exhibited reduced responsiveness to acetylcholine-mediated vasodilation compared to arteries from control rats [161]. Our group recently reported that feeding rats a high fat diet (36% fat wt./wt.) for 4 weeks resulted in increased adiposity, hyperinsulinaemia and whole body insulin resistance compared to control rats [162]. High fat fed rats exhibited impairments in insulin-mediated effects on both total leg blood flow and microvascular recruitment in muscle [162]. This vascular insulin resistance was associated with marked reduction of insulin-mediated skeletal muscle glucose uptake compared to control fed animals [162]. Together, these data indicate that defects in insulin's vascular actions may in part be responsible for the insulin resistant state that develops following increased dietary fat intake.

Human Obesity and Vascualr Function

Obesity is becoming increasingly prevalent in society and is associated with the development of insulin resistance. It is now well recognised that obese individuals are at an increased risk of developing type 2 diabetes and/or cardiovascular disease. However, not all obese subjects develop insulin resistance, and the relationship between these two conditions appears to be complex and is currently poorly understood. One factor that has been proposed to link obesity and insulin resistance is impaired endothelial function. Indeed, Laakso and colleagues first reported that resistance arterioles of obese, insulin resistant subjects exhibited significantly decreased sensitivity to insulin-mediated vasodilation [6]. Subsequent studies have also reported a relationship between obesity, insulin resistance and attenuation of endothelial function [163-165]. While these previous studies were limited to assessment of large vessels and thus only the effects of insulin on total flow, more recent advances have identified that the microvascular response to insulin also appears to be reduced during obesity.

De Jongh and colleagues assessed insulin sensitivity and microvascular function in skin in healthy, lean (BMI < 25kg/m^2) and obese individuals (BMI > 35kg/m^2) during hyperinsulinaemic euglycaemic clamps [166]. The authors reported that obese, insulin resistant subjects exhibited reduced insulin-mediated microvascular recruitment in skin and that this correlated strongly with decreased whole body insulin resistance [166]. We took this concept further and assessed insulin-mediated microvascular recruitment in skeletal muscle in healthy (BMI < 25kg/m^2) and obese (BMI > 30kg/m^2) subjects using the hyperinsulinaemic euglycaemic clamp technique [67]. We showed that hyperinsulinaemia increased skeletal muscle microvascular recruitment associated with increased whole body glucose disposal in healthy, lean subjects [67]. In contrast, obese subjects displayed significant attenuation of insulin-mediated microvascular recruitment in muscle and approximately a 50% reduction in insulin-mediated whole body glucose disposal compared to lean subjects [67].

Together, these studies indicate that obesity is associated with impairment in insulin-mediated microvascular recruitment which may contribute to reduced muscle glucose disposal, thus worsening whole body insulin resistance.

While this microvascular action of insulin has been demonstrated during hyperinsulinaemic euglycaemic clamp studies, whether impairment of insulin-mediated microvascular recruitment in muscle has relevance in the development of insulin resistance in more physiological situations has not been clear until recently.

In healthy, lean subjects we have reported that microvascular recruitment in muscle was elevated up to 120 min after a mixed-meal [33, 167]. The increase in microvascular recruitment in skeletal muscle was associated with a concomitant increase in circulating insulin concentrations facilitating post-prandial glucose disposal [33, 167]. These data indicate that increased microvascular recruitment, in response to a meal, is a normal physiological response that occurs in healthy subjects and contributes to increased glucose disposal by muscle. In contrast, ingestion of the same mixed-meal in obese, insulin resistant subjects failed to stimulate any increase in skeletal muscle microvascular recruitment [167].

The absence of microvascular recruitment was apparent despite a 100% increase in post-prandial circulating insulin concentrations in the obese subjects (lean: ~230pmol/L vs. obese: ~450pmol/L), while the blood glucose concentrations were similar between lean and obese groups [167]. Together, these studies indicate that obese individuals exhibit impaired insulin-mediated microvascular recruitment in skeletal muscle and that this may contribute to the post-prandial hyperinsulinaemia, and subsequently worsen insulin resistance.

Knockout Models and Vascular Insulin Resistance

A variety of knockout mouse models have provided important links between insulin's vascular and metabolic actions. These include the eNOS, vascular insulin receptor, and endothelial IRS2 knockout mouse models.

The eNOS knockout mouse displays endothelial dysfunction, insulin resistance and hypertension [168]. These animals also have reduced capillary density and approximately a 40% reduction in insulin-mediated skeletal muscle glucose uptake [168, 169]. It has also been reported that animals with only partial eNOS deficiency are normotensive and insulin sensitive, and only develop hypertension and insulin resistance when challenged with a high fat diet [170]. Similarly, vascular endothelial insulin receptor knockout (VENRIKO) mice also develop insulin resistance, but only when the animals are maintained on either low or high salt diets, suggesting that insulin signalling in the vascular endothelium contributes to insulin-mediated glucose disposal in certain contexts [171].

Recently, Kubota and colleagues reported that impairment of insulin signalling in vascular endothelial cells reduces insulin-mediated muscle glucose uptake *in vivo* [172]. Endothelial cell IRS2 knockout mice exhibited normal levels of IRS2 in skeletal muscle, liver and white adipose tissue while endothelial cell IRS2 was virtually absent [172]. This reduction in IRS2 was associated with reduced expression of both Akt and eNOS in endothelial cells [172].

In hyperinsulinaemic euglycaemic clamp studies the IRS2 knockout mice exhibited a 50% reduction in whole body glucose disposal while liver glucose output was unaffected compared to control mice [172]. The authors showed that the knockout mice exhibited almost complete absence of insulin-mediated microvascular recruitment and approximately a 50% reduction in muscle glucose uptake during hyperinsulinaemia. These findings were associated with a marked reduction of insulin-stimulated phosphorylation of Akt within muscle cells [172]. Furthermore, when muscles from control and knockout animals were excised and incubated *in vitro* with insulin, no difference in insulin-mediated myocyte glucose uptake was detected [172].

These data suggest that impaired insulin signalling in endothelial cells, by way of reduced IRS2 and insulin-induced eNOS phosphorylation, reduces insulin-mediated muscle glucose uptake via, at least in part, decreased insulin-mediated microvascular recruitment in skeletal muscle.

In addition, these knockout animals seemingly simulate the pathogenesis of high fat diet-induced insulin resistance and implicate reduced insulin-mediated endothelial NO production (via reduced eNOS activation) as a contributor to reduced insulin-mediated microvascular recruitment and muscle glucose uptake.

While endothelial dysfunction is associated with development of muscle insulin resistance a recent report emerged illustrating that the extent of capillarization of skeletal muscle is also important.

Muscle specific vascular endothelial growth factor (VEGF) knockout animals had similar body weight, fat and muscle mass and as expected a significant reduction in capillary density in skeletal muscle when compared to controls [173]. Assessment of insulin sensitivity using the hyperinsulinaemic euglycaemic clamp revealed that the VEGF knockout animals had a significant reduction in insulin-mediated whole body glucose disposal, the majority of which was accounted for by a 50% decrease in insulin-stimulated muscle glucose uptake [173]. This reduction in muscle glucose uptake was apparent with a reduction in insulin- stimulated IRS-1 phosphorylation, while insulin-stimulated Ser^{473}-Akt phosphorylation was similar to that of control animals. Importantly, when the authors incubated the soleus and extensor digitorum longus muscles with 10 mU/mL insulin, no difference in insulin-mediated muscle glucose uptake was detected between the VEGF knockout and control animals [173]. Thus, this study illustrates that whilst myocyte insulin sensitivity may be normal in the VEGF knockout animals, a reduction in capillary density (and thus the ability of insulin to enhance microvascular recruitment in muscle) is sufficient to reduce muscle insulin sensitivity *in vivo* in otherwise lean, healthy animals. This finding is consistent with human data in which reduced capillary density is also associated with reduced insulin-mediated muscle glucose uptake [174, 175].

Myocyte Versus Microvascular Insulin Resistance

Although the mechanisms of muscle insulin resistance are multifactorial, there has been much focus on the impairment of insulin action on the myocyte as the primary defect. The development of myocyte insulin resistance certainly occurs when insulin resistance is fully established; however, the involvement of the microcirculation as the initial and independent event that requires consideration. The various models listed above demonstrate that the development of skeletal muscle insulin resistance is associated with the development of microvascular insulin resistance. Given that the majority of the chronic models mentioned above also exhibit direct myocyte insulin resistance, it has been difficult to dissociate between myocyte from microvascular insulin resistance. Very recently we have developed and characterised two new rat models in our laboratory that support the notion that microvascular insulin resistance develops before skeletal muscle insulin resistance.

The first of these animal models is the moderately elevated fat-fed rat. Rats fed a diet containing approximately a 2-fold increase in fat, from 5% to 9% (wt./wt.), for four weeks exhibit *in vivo* whole body and skeletal muscle insulin resistance (Figure 2A), but not liver insulin resistance, when assessed by hyperinsulinaemic euglycaemic clamp [176]. Whilst insulin stimulated an increase in femoral artery blood flow to a similar extent in both groups, a 50% reduction of insulin-stimulated microvascular recruitment in skeletal muscle was apparent in the 9% fat group.

When myocyte insulin sensitivity was assessed *ex vivo* using the constant-flow pump-perfused hindleg technique (in which insulin and glucose are delivered to the myocyte in the absence of insulin's vascular actions), no difference was detected between the 5% and 9% fat groups. Furthermore, in muscle samples from both the insulin clamp and perfused hindleg studies no difference in insulin-stimulated Ser^{473}-Akt phosphorylation between the 5% and 9% fat groups was detected, suggesting that muscle insulin signalling is intact in this model. Therefore, these data indicate that the 9% fat model does not have myocyte-derived insulin resistance; rather, the reduction in insulin-stimulated muscle glucose disposal *in vivo* is the result of attenuated microvascular insulin action. For the first time this study positions defects in microvascular insulin action as an early event that contributes significantly to fat-induced muscle insulin resistance.

The second model we have characterised is the high salt-fed rat model (unpublished). Rats fed a control (NS, 0.3% NaCl wt./wt.) or high salt (HS, 8.0% NaCl wt./wt.) diet for four weeks display *in vivo* whole body and skeletal muscle insulin resistance (Figure 2B) during a hyperinsulinaemic euglycaemic clamp, but interestingly exhibit enhanced liver insulin sensitivity. The HS animals displayed reduced basal femoral artery blood flow, however, the insulin-stimulated increase in femoral artery blood flow remained intact and was comparable to that of NS animals. In contrast, the HS animals exhibited complete attenuation of insulin-stimulated microvascular recruitment. Similarly to the 9% fat fed rat model described above, HS animals had normal myocyte insulin sensitivity when assessed *ex vivo* using the perfused hindleg preparation. Surprisingly, HS animals displayed an enhancement of insulin-stimulated Ser^{473}-Akt phosphorylation in muscle compared to NS animals during hyperinsulinaemic euglycaemic clamps. Together these data imply that myocyte insulin sensitivity is intact (if not enhanced) in HS animals following four weeks of intervention and that impairment of microvascular recruitment by insulin is sufficient to induce muscle insulin resistance *in vivo*. Importantly, these effects on insulin sensitivity were apparent in the absence of hypertension, demonstrating that HS-induced muscle insulin resistance can develop independently of vascular defects associated with hypertension.

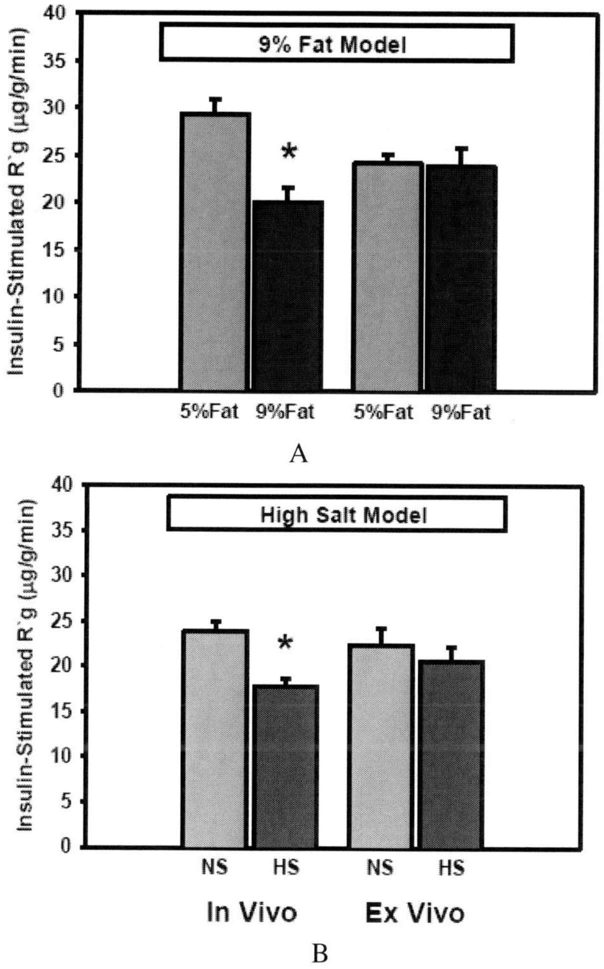

Figure 2. Insulin-stimulated glucose uptake in skeletal muscle. The rate of muscle specific glucose uptake (R`g) was determined using the radioactive glucose tracer 14C-2-deoxyglucose in the 9% fat (Panel A) and HS (Panel B) models during either hyperinsulinaemic euglycaemic clamp (In Vivo; 10mU/min/kg insulin) or constant-flow pump-perfused hindleg preparation (Ex Vivo; 1.5nM insulin). * Significantly different ($p<0.05$) versus respective control group.

It is of note that while both the 9% fat and HS models exhibited microvascular and skeletal muscle insulin resistance *in vivo*, the insulin-mediated increase in femoral artery blood flow remained intact. These data add further support to the hypothesis that it is redistribution of blood flow within skeletal muscle microvasculature and not insulin's effects on resistance arteries (such as the femoral artery) that determines the extent of insulin-mediated muscle glucose uptake.

This is further illustrated when the rate of insulin-stimulated muscle glucose uptake (R`g) is plotted against either the insulin-stimulated change in femoral artery blood flow (Figure 3A) or insulin-stimulated 1-MX metabolism (Figure 3B) *in vivo*. As can be seen, there is no evident relationship between insulin's effects on R`g and femoral artery blood flow. In contrast, there is a strong, positive relationship between insulin-stimulated R`g and 1-MX metabolism (microvascular recruitment). Therefore, these data indicate that it is microvascular, and not large blood vessel responses that are highly sensitive to insulin (and indeed resistance to insulin), positioning microvascular insulin resistance as an early defect that can directly contribute to skeletal muscle insulin resistance.

There has been much debate as to whether insulin or glucose delivery to the myocyte is rate-limiting for glucose uptake. Recently it has been proposed that insulin delivery is rate-limiting for insulin-stimulated glucose uptake by skeletal muscle (reviewed by Barrett and colleagues [177, 178]).

With the 9% fat and HS studies, we provide two pieces of evidence that implicate glucose and not insulin delivery as rate-limiting. The first piece of evidence comes from the absence of a reduction in insulin-stimulated Ser^{473}-Akt phosphorylation in the hyperinsulinaemic clamp studies despite significant attenuation of muscle glucose uptake in both the 9% fat and HS groups.

One would expect that if insulin delivery was reduced that this would be reflected in reduced insulin signalling within the myocyte. This was not the case. The second piece of evidence comes from the constant-flow pump-perfused hindleg studies in which delivery of both insulin and glucose to the myocyte is via an intact vascular network. In this setting, the 9% fat model displayed normal glucose uptake and had normal myocyte insulin signalling, thus indicating that when glucose delivery is restored, insulin-stimulated glucose uptake by skeletal myocytes is also restored.

While the data for the 9% fat and HS models implicate glucose delivery as the rate-limiting factor in determining insulin-stimulated glucose uptake by skeletal muscle, whether this is also the case in humans is currently unclear.

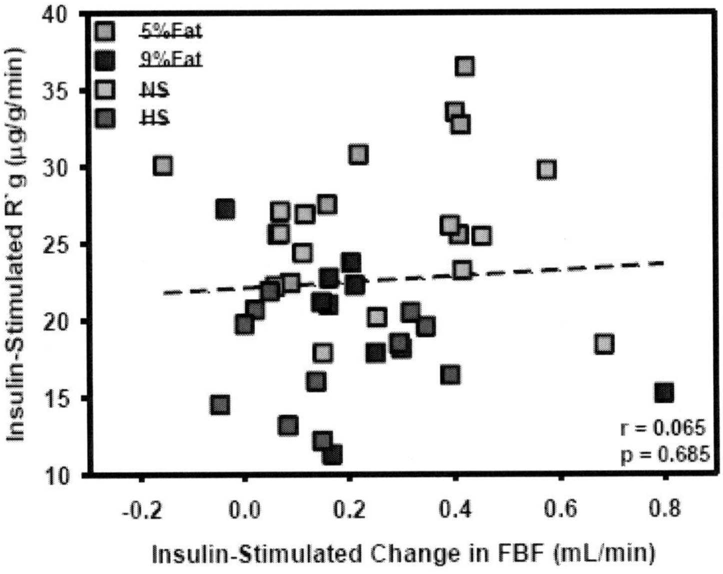

A

Figure 3. (Continued)

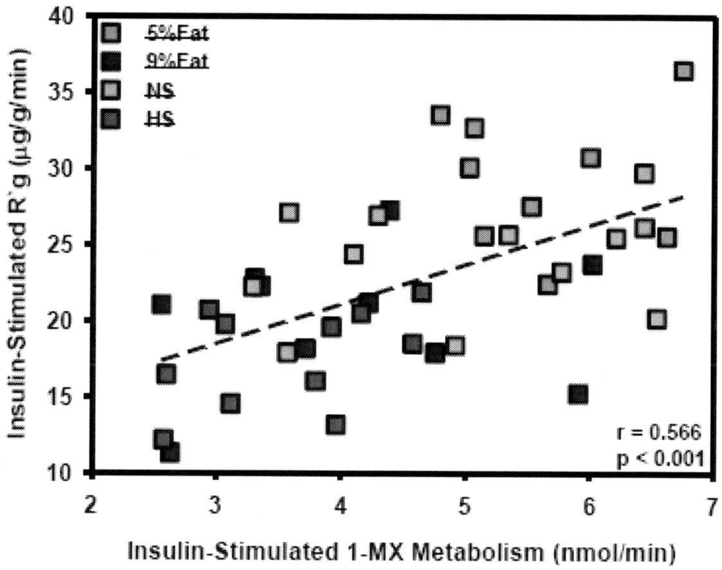

B

Figure 3. Relationship between insulin's vascular and metabolic actions in skeletal muscle *in vivo*. Insulin-stimulated change in muscle specific glucose uptake (R`g) was plotted against either the insulin-stimulated change in femoral artery blood flow (Panel A; FBF) from baseline or against insulin-stimulated 1-MX metabolism (Panel B). Data points represent individual values for 5% fat, 9% fat, NS and HS animals at the conclusion of the hyperinsulinaemic euglycaemic clamp studies. Linear regression was utilised to determine the relationship between these data.

Conclusion

Insulin-mediated microvascular recruitment in skeletal muscle is important for glucose delivery and uptake by muscle for prompt storage as glycogen following a meal. When insulin-mediated microvascular recruitment is absent, for example due to acute inhibition, there is a 50% reduction in muscle glucose uptake, suggesting that the loss of microvascular recruitment in insulin resistant states contributes significantly to skeletal muscle insulin resistance. In the current review we describe two novel models in which loss of microvascular insulin sensitivity occurs in the absence of skeletal myocyte insulin resistance. We have shown that despite impairment of insulin's actions within the microvasculature, insulin's effects on the larger resistance arteries in muscle remain intact in both the 9% fat and HS models.

These two pieces of evidence indicate that the microvasculature in skeletal muscle can become resistant to insulin independently of and prior to either the skeletal myocytes or larger resistance arteries. Moreover, the sensitivity of the microvasculature to even moderate increases in dietary fat positions "microvascular insulin resistance" as one of the earliest events in the aetiology of obesity-induced insulin resistance. Thus, while it is important to recognise that amelioration of insulin sensitivity in humans requires addressing defects in skeletal myocyte insulin signalling, to do so without correcting impaired microvascular responsiveness may limit potential therapeutic benefits.

References

[1] Abramson DI, Schkloven N, Margolis MN, Mirsky IA. Influence of massive doses of insulin on peripheral blood flow in man. *American Journal of Physiology.* 1939;128:124-32.

[2] Allwood MJ, Hensel H, Papenberg J. Muscle and skin blood flow in the human forearm during insulin hypoglycaemia. *Journal of Physiology.* 1959;147:269-73.

[3] Dosekun FO, Grayson J, Mendel D. The measurement of metabolic and vascular responses in liver and muscle with observations on their responses to insulin and glucose. *Journal of Physiology*. 1960;150:581-606.

[4] Norwich KH, Fluker G, Anthony J, Popescu I, Pagurek B, Hetenyi G, Jr. The development of a glucose clamp. *Metabolism.* 1975 Nov;24(11):1221-30.

[5] Liang C, Doherty JU, Faillace R, Maekawa K, Arnold S, Gavras H, et al. Insulin infusion in conscious dogs. Effects on systemic and coronary hemodynamics, regional blood flows, and plasma catecholamines. *Journal of Clinical Investigation.* 1982;69(6):1321-36.

[6] Laakso M, Edelman SV, Brechtel G, Baron AD. Decreased effect of insulin to stimulate skeletal muscle blood flow in obese man. *Journal of Clinical Investigation. 1990*;85:1844-52.

[7] Laakso M, Edelman SV, Olefsky JM, Brechtel G, Wallace P, Baron AD. Kinetics of in vivo muscle insulin-mediated glucose uptake in human obesity. *Diabetes.* 1990;39:965-74.

[8] Anderson EA, Hoffman RP, Balon TW, Sinkey CA, Mark AL. Hyperinsulinemia produces both sympathetic neural activation and vasodilation in normal humans. *Journal of Clinical Investigation.* 1991;87:2246-52.

[9] Dela F, Larssen JJ, Mikines KJ, Galbo H. Normal effect of insulin to stimulate leg blood flow in NIDDM. *Diabetes.* 1995;44:221-6.

[10] Raitakari M, Nuutila P, Ruotsalainen U, Laine H, Teras M, Iida H, et al. Evidence for dissociation of insulin stimulation of blood flow and glucose uptake in human skeletal muscle - Studies using [15 O]H 2 O, [18 F]fluoro-2-deoxy-D-glucose, and positron emission tomography. *Diabetes.* 1996;45:1471-7.

[11] Tack CJ, Lutterman JA, Vervoort G, Thien T, Smits P. Activation of the sodium-potassium pump contributes to insulin-induced vasodilation in humans. *Hypertension.* 1996;28(3):426-32.

[12] Utriainen T, Malmstrom R, Makimattila S, Yki-Jarvinen H. Methodological aspects, dose-response characteristics and causes of interindividual variation in insulin stimulation of limb blood flow in normal subjects. *Diabetologia.* 1995;38(5):555-64.

[13] Vollenweider P, Tappy L, Randin D, Schneiter P, Jequier E, Nicod P, et al. Differential effects of hyperinsulinemia and carbohydrate metabolism on sympathetic nerve activity and muscle blood flow in humans. *Journal of Clinical Investigation*. 1993;92:147-54.

[14] Renaudin C, Michoud E, Rapin JR, Lagarde M, Wiernsperger N. Hyperglycaemia modifies the reaction of microvessels to insulin in rat skeletal muscle. *Diabetologia.* 1998;41(1):26-33.

[15] Rattigan S, Clark MG, Barrett EJ. Hemodynamic actions of insulin in rat skeletal muscle: evidence for capillary recruitment. *Diabetes.* 1997;46(9):1381-8.

[16] Vincent MA, Dawson D, Clark AD, Lindner JR, Rattigan S, Clark MG, et al. Skeletal muscle microvascular recruitment by physiological hyperinsulinemia precedes increases in total blood flow. *Diabetes.* 2002;51(1):42-8.

[17] Baron AD, Brechtel G. Insulin differentially regulates systemic and skeletal muscle vascular resistance. *American Journal of Physiology.* 1993;265:E61-E7.

[18] Natali A, Taddei S, Galvan AQ, Camastra S, Baldi S, Frascerra S, et al. Insulin sensitivity, vascular reactivity, and clamp-induced vasodilatation in essential hypertension. *Circulation.* 1997;96(3):849-55.

[19] Gaudreault N, Santure M, Pitre M, Nadeau A, Marette A, Bachelard H. Effects of insulin on regional blood flow and glucose uptake in Wistar and Sprague-Dawley rats. Metabolism: *Clinical and Experimental.* 2001;50(1):65-73.

[20] Pitre M, Nadeau A, Bachelard H. Insulin sensitivity and hemodynamic responses to insulin in Wistar-Kyoto and spontaneously hypertensive rats. *American Journal of Physiology.* 1996;271(4 Pt 1):E658-E68.

[21] Bonadonna RC, Del Prato S, Saccomani MP, Bonora E, Gulli G, Bier D, et al. Transmembrane glucose transport in skeletal muscle of patients with non-insulin-dependent diabetes. *Journal of Clinical Investigation.* 1993;92:486-94.

[22] Ebeling P, Bourey R, Koranyi L, Tuominen JA, Groop LC, Henriksson J, et al. Mechanism of enhanced insulin sensitivity in athletes. Increased blood flow, muscle glucose transport protein (GLUT-4) concentration, and glycogen synthase activity. *Journal of Clinical Investigation.* 1993;92:1623-31.

[23] Jackson RA, Hamling JB, Blix PM, Sim BM, Hawa MI, Jaspan JB, et al. The influence of graded hyperglycemia with and without physiological hyperinsulinemia on forearm glucose uptake and other metabolic responses in man. *Journal of Clinical Endocrinology and Metabolism.* 1986 Sep;63(3):594-604.

[24] Yki-Jarvinen H, Young AA, Lamkin C, Foley JE. Kinetics of glucose disposal in whole body and across the forearm in man. *Journal of Clinical Investigation.* 1987;79(6):1713-9.

[25] Gelfand RA, Barrett EJ. Effects of physiologic hyperinsulinemia on skeletal muscle protein synthesis and breakdown in man. *Journal of Clinical Investigation.* 1987;80:1-6.

[26] Utriainen T, Holmang A, Bj”rntorp P, M„kimattila S, Sovij„rvi A, Lindholm H, et al. Physical fitness, muscle morphology, and insulin-stimulated limb blood flow in normal subjects. *American Journal of Physiology.* 1996;270:E905-E11.

[27] Utriainen T, Makimattila S, Virkamaki A, Lindholm H, Sovijarvi A, Yki-Jarvinen H. Physical fitness and endothelial function (nitric oxide synthesis) are independent determinants of insulin-stimulated blood flow in normal subjects. *Journal of Clinical Endocrinology and Metabolism.* 1996;81(12):4258-63.

[28] Yki-Jarvinen H, Utriainen T. Insulin-induced vasodilatation: physiology or pharmacology? *Diabetologia.* 1998;41(4):369-79.

[29] Steinberg HO, Baron AD. Insulin-mediated vasodilation: why one's physiology could be the other's pharmacology [letter; comment] [see comments]. *Diabetologia.* 1999;42(4):493-5.

[30] Baron AD, Brechtel-Hook G, Johnson A, Hardin D. Skeletal muscle blood flow. A possible link between insulin resistance and blood pressure. *Hypertension.* 1993;21:129-35.

[31] Bonadonna RC, Saccomani MP, Del Prato S, Bonora E, Defronzo RA, Cobelli C. Role of tissue-specific blood flow and tissue recruitment in insulin-mediated glucose uptake of human skeletal muscle. *Circulation.* 1998;98(3):234-41.

[32] Fugmann A, Sarabi M, Karlstrom B, Berne C, Lithell H, Lind L. Blood flow is an important determinant of forearm glucose uptake following a mixed meal. *Acta Diabetologica.* 2003;40(3):113-7.

[33] Vincent MA, Clerk LH, Lindner JR, Price WJ, Jahn LA, Leong-Poi H, et al. Mixed meal and light exercise each recruit muscle capillaries in healthy humans. *American Journal of Physiology Endocrinology and Metabolism.* 2006;290(6):E1191-E7.

[34] Nuutila P, Raitakari M, Laine H, Kirvela O, Takala T, Utriainen T, et al. Role of blood flow in regulating insulin-stimulated glucose uptake in humans. Studies using bradykinin, [15O]water, and [18F]fluoro-deoxy- glucose and positron emission tomography. *Journal of Clinical Investigation. 1996*;97(7):1741-7.

[35] Vicini P, Bonadonna RC, Utriainen T, Nuutila P, Raitakari M, Yki-Jarvinen H, et al. Estimation of blood flow heterogeneity distribution in human skeletal muscle from positron emission tomography data. *Annals of Biomedical Engineering.* 1997;25(5):906-10.

[36] Utriainen T, Nuutila P, Takala T, Vicini P, Ruotsalainen U, Ronnemaa T, et al. Intact insulin stimulation of skeletal muscle blood flow, its heterogeneity and redistribution, but not of glucose uptake in non-insulin-dependent diabetes mellitus. *Journal of Clinical Investigation.* 1997;100(4):777-85.

[37] Sweeney TE, Sarelius IH. Arteriolar control of capillary cell flow in striated muscle. *Circulation Research.* 1989;64:112-20.

[38] Murrant CL, Sarelius IH. Coupling of muscle metabolism and muscle blood flow in capillary units during contraction. *Acta Physiologica Scandinavica.* 2000;168(4):531-41.

[39] Honig CR, Odoroff CL, Frierson JL. Active and passive capillary control in red muscle at rest and in exercise. *American Journal of Physiology.* 1982;243:H196-H206.

[40] Iversen PO, Nicolaysen G. High correlation of fractals for regional blood flows among resting and exercising skeletal muscles. *American Journal of Physiology.* 1995;269:H7-H13.

[41] Lindbom L, Arfors KE. Non-homogeneous blood flow distribution in the rabbit tenuissimus muscle Differential control of total blood flow and capillary perfusion. *Acta Physiologica Scandinavica.* 1984;122:225-33.

[42] Piiper J, Marconi C, Heisler N, Meyer M, Weitz H, Pendergast DR, et al. Spatial and temporal variability of blood flow in stimulated dog gastrocnemius muscle. *Advances in Experimental Medicine and Biology.* 1989;248:719-43.

[43] Intaglietta M. Vasomotor activity, time-dependent fluid exchange and tissue pressure. *Microvasc Res.* 1981 Mar;21(2):153-64.

[44] Intaglietta M, Gross JF. Vasomotion, tissue fluid flow and the formation of lymph. International Journal of Microcirculation: *Clinical and Experimental.* 1982;1(1):55-65.

[45] Clark MG, Colquhoun EQ, Rattigan S, Dora KA, Eldershaw TP, Hall JL, et al. Vascular and endocrine control of muscle metabolism. *American Journal of Physiology.* 1995;268(5 Pt 1):E797-E812.

[46] Pappenheimer JR. Vasoconstrictor nerves and oxygen consumption in the isolated perfused hindlimb muscles of the dog. *Journal of Physiology.* 1941;99:182-200.

[47] Dora KA, Richards SM, Rattigan S, Colquhoun EQ, Clark MG. Serotonin and norepinephrine vasoconstriction in rat hindlimb have different oxygen requirements. *American Journal of Physiology.* 1992;262:H698-H703.

[48] Rattigan S, Appleby GJ, Miller KA, Steen JT, Dora KA, Colquhoun EQ, et al. Serotonin inhibition of 1-methylxanthine metabolism parallels its vasoconstrictor activity and inhibition of oxygen uptake in perfused rat hindlimb. *Acta Physiologica Scandinavica.* 1997;161(2):161-9.

[49] Newman JM, Dora KA, Rattigan S, Edwards SJ, Colquhoun EQ, Clark MG. Norepinephrine and serotonin vasoconstriction in rat hindlimb control different vascular flow routes. *American Journal of Physiology.* 1996;270(4 Pt 1):E689-E99.

[50] Vincent MA, Rattigan S, Clark MG, Bernard SL, Glenny RW. Spatial distribution of nutritive and nonnutritive vascular routes in perfused rat hindlimb muscle using microspheres. *Microvascular Research.* 2001;61(1):111-21.

[51] Newman JM, Steen JT, Clark MG. Vessels supplying septa and tendons as functional shunts in perfused rat hindlimb. *Microvascular Research.* 1997;54(1):49-57.

[52] Clerk LH, Smith ME, Rattigan S, Clark MG. Increased chylomicron triglyceride hydrolysis by connective tissue flow in perfused rat hindlimb. Implications for lipid storage. *Journal of Lipid Research.* 2000;41(3):329-35.

[53] Clark MG, Rattigan S, Clerk LH, Vincent MA, Clark AD, Youd JM, et al. Nutritive and non-nutritive blood flow: rest and exercise. *Acta Physiologica Scandinavica.* 2000;168(4):519-30.

[54] Wei K, Le E, Bin JP, Coggins M, Thorpe J, Kaul S. Quantification of renal blood flow with contrast-enhanced ultrasound. *Journal of the American College of Cardiology.* 2001;37(4):1135-40.

[55] Hellsten-Westing Y. Immunohistochemical localization of xanthine oxidase in human cardiac and skeletal muscle. *Histochemistry.* 1993;100:215-22.

[56] Jarasch ED, Bruder G, Heid HW. Significance of xanthine oxidase in capillary endothelial cells. *Acta Physiologica Scandinavica Supplementum.* 1986;548:39-46.

[57] Rattigan S, Dora KA, Colquhoun EQ, Clark MG. Serotonin-mediated acute insulin resistance in the perfused rat hindlimb but not in incubated muscle: a role for the vascular system. *Life Sciences.* 1993;53(20):1545-55.

[58] Dora KA, Colquhoun EQ, Hettiarachchi M, Rattigan S, Clark MG. The apparent absence of serotonin-mediated vascular thermogenesis in perfused rat hindlimb may result from vascular shunting. *Life Sciences.* 1991;48:1555-64.

[59] Zhang L, Newman JM, Richards SM, Rattigan S, Clark MG. Microvascular flow routes in muscle controlled by vasoconstrictors. *Microvascular Research.* 2005;70(1-2):7-16.

[60] Zhang L, Vincent MA, Richards SM, Clerk LH, Rattigan S, Clark MG, et al. Insulin sensitivity of muscle capillary recruitment in vivo. *Diabetes.* 2004;53(2):447-53.

[61] Rattigan S, Wheatley C, Richards SM, Barrett EJ, Clark MG. Exercise and insulin-mediated capillary recruitment in muscle. *Exercise and Sport Sciences Reviews.* 2005;33(1):43-8.

[62] Lindner JR, Villanueva FS, Dent JM, Wei K, Sklenar J, Kaul S. Assessment of resting perfusion with myocardial contrast echocardiography: theoretical and practical considerations. *Am. Heart J.* 2000 Feb;139(2 Pt 1):231-40.

[63] Pelberg RA, Wei K, Kamiyama N, Sklenar J, Bin J, Kaul S. Potential advantage of flash echocardiography for digital subtraction of B-mode images acquired during

myocardial contrast echocardiography. *J. Am Soc. Echocardiogr.* 1999 Feb;12(2):85-93.

[64] Wei K, Jayaweera AR, Firoozan S, Linka A, Skyba DM, Kaul S. Quantification of myocardial blood flow with ultrasound-induced destruction of microbubbles administered as a constant venous infusion. *Circulation.* 1998;97(5):473-83.

[65] Wei K, Ragosta M, Thorpe J, Coggins M, Moos S, Kaul S. Noninvasive quantification of coronary blood flow reserve in humans using myocardial contrast echocardiography. *Circulation.* 2001;103(21):2560-5.

[66] Vincent MA, Dawson D, Clark AD, Leong-Poi H, Clark MG, Rattigan S, et al. Physiologic hyperinsulinemia mimics the capillary recruitment induced by exercise in skeletal muscle in vivo. Diabetes. 2000:#451778.

[67] Clerk LH, Vincent MA, Jahn LA, Liu Z, Lindner JR, Barrett EJ. Obesity blunts insulin-mediated microvascular recruitment in human forearm muscle. *Diabetes.* 2006;55(5):1436-42.

[68] Jayaweera AR, Edwards N, Glasheen WP, Villanueva FS, Abbott RD, Kaul S. In vivo myocardial kinetics of air-filled albumin microbubbles during myocardial contrast echocardiography. Comparison with radiolabeled red blood cells. *Circulation Research.* 1994;74(6):1157-65.

[69] Coggins M, Lindner J, Rattigan S, Jahn L, Fasy E, Kaul S, et al. Physiologic hyperinsulinemia enhances human skeletal muscle perfusion by capillary recruitment. *Diabetes.* 2001;50(12):2682-90.

[70] Vincent MA, Barrett EJ, Lindner JR, Clark MG, Rattigan S. Inhibiting NOS blocks microvascular recruitment and blunts muscle glucose uptake in response to insulin. *American Journal of Physiology Endocrinology and Metabolism.* 2003;285(1):E123-E9.

[71] Dawson D, Vincent MA, Barrett EJ, Kaul S, Clark A, Leong-Poi H, et al. Vascular recruitment in skeletal muscle during exercise and hyperinsulinemia assessed by contrast ultrasound. *American Journal of Physiology Endocrinology and Metabolism.* 2002;282(3):E714-E20.

[72] Sarelius IH, Cohen KD, Murrant CL. Role for capillaries in coupling blood flow with metabolism. *Clinical and Experimental Pharmacology and Physiology.* 2000;27(10):826-9.

[73] Steinberg HO, Brechtel G, Johnson A, Fineberg N, Baron AD. Insulin-mediated skeletal muscle vasodilation is nitric oxide dependent. A novel action of insulin to increase nitric oxide release. *Journal of Clinical Investigation.* 1994;94:1172-9.

[74] Montagnani M, Ravichandran LV, Chen H, Esposito DL, Quon MJ. Insulin receptor substrate-1 and phosphoinositide-dependent kinase-1 are required for insulin-stimulated production of nitric oxide in endothelial cells. *Mol. Endocrinol.* 2002 Aug;16(8):1931-42.

[75] Zeng GY, Quon MJ. Insulin-stimulated production of nitric oxide is inhibited by wortmannin - Direct measurement in vascular endothelial cells. *Journal of Clinical Investigation.* 1996 Aug 15;98(4):894-8.

[76] Zeng G, Nystrom FH, Ravichandran LV, Cong LN, Kirby M, Mostowski H, et al. Roles for insulin receptor, PI3-kinase, and Akt in insulin-signaling pathways related to production of nitric oxide in human vascular endothelial cells. *Circulation.* 2000;101(13):1539-45.

[77] Vincent MA, Clerk LH, Lindner JR, Klibanov AL, Clark MG, Rattigan S, et al. Microvascular recruitment is an early insulin effect that regulates skeletal muscle glucose uptake in vivo. *Diabetes.* 2004;53(6):1418-23.

[78] Trovati M, Massucco P, Mattiello L, Cavalot F, Mularoni E, Hahn A, et al. Insulin increases cyclic nucleotide content in human vascular smooth muscle cells: a mechanism potentially involved in insulin-induced modulation of vascular tone. *Diabetologia.* 1995 Aug;38(8):936-41.

[79] Trovati M, Massucco P, Mattiello L, Costamagna C, Aldieri E, Cavalot F, et al. Human vascular smooth muscle cells express a constitutive nitric oxide synthase that insulin rapidly activates, thus increasing guanosine 3':5'-cyclic monophosphate and adenosine 3':5'-cyclic monophosphate concentrations. *Diabetologia.* 1999;42(7):831-9.

[80] Cardillo C, Nambi SS, Kilcoyne CM, Choucair WK, Katz A, Quon MJ, et al. Insulin stimulates both endothelin and nitric oxide activity in the human forearm. Circulation. 1999;100(8):820-5.

[81] Cardillo C, Campia U, Bryant MB, Panza JA. Increased activity of endogenous endothelin in patients with Type II diabetes mellitus. *Circulation.* 2002;106(14):1783-7.

[82] Anfossi G, Cavalot F, Massucco P, Mattiello L, Mularoni EM, Burzacca S, et al. Insulin increases endothelin-1 production by vascular smooth muscle cells derived from human microvessels. *Frontiers in Diabetes.* 1993;12:272-4.

[83] Misurski DA, Wu SQ, McNeill JR, Wilson TW, Gopalakrishnan V. Insulin-induced biphasic responses in rat mesenteric vascular bed: role of endothelin. *Hypertensio*n. 2001;37(5):1298-302.

[84] Eringa EC, Stehouwer CD, Walburg K, Clark AD, Nieuw Amerongen GP, Westerhof N, et al. Physiological concentrations of insulin induce endothelin-dependent vasoconstriction of skeletal muscle resistance arteries in the presence of tumor necrosis factor-{alpha} dependence on c-Jun N-terminal kinase. *Arteriosclerosis, Thrombosis, and Vascular Biology.* 2005;26(2):274-80.

[85] Perrin C, Knauf C, Burcelin R. Intracerebroventricular infusion of glucose, insulin, and the adenosine monophosphate-activated kinase activator, 5-aminoimidazole-4-carboxamide-1-beta-D-ribofuranoside, controls muscle glycogen synthesis. *Endocrinology.* 2004;145(9):4025-33.

[86] Cabou C, Cani PD, Campistron G, Knauf C, Mathieu C, Sartori C, et al. Central insulin regulates heart rate and arterial blood flow: an endothelial nitric oxide synthase-dependent mechanism altered during diabetes. *Diabetes.* 2007;56:2872-7.

[87] Bradley EA, Willson KJ, Choi-Lundberg D, Clark MG, Rattigan S. Effects of central administration of insulin or l-NMMA on rat skeletal muscle microvascular perfusion. *Diabetes Obes Metab.* 2010 Oct;12(10):900-8.

[88] Muniyappa R, Montagnani M, Koh KK, Quon MJ. Cardiovascular Actions of Insulin. *Endocrine Reviews.* 2007;28(5):463-91.

[89] Rattigan S, Bradley EA, Richards SM, Clark MG. Muscle metabolism and control of capillary blood flow: insulin and exercise. *Essays in Biochemistry.* 2006;42:133-44.

[90] Zierler KL. Theory of the use of arteriovenous concentration differences for measuring metabolism in steady and non-steady states. *J. Clin. Invest.* 1961 Dec;40(12):2111-25.

[91] Renkin EM, Renkin EM, Michel CC, Geiger SR. Control of microcirculation and blood-tissue exchange. Handbook of Physiology - The Cardiovascular System IV: *American Physiological Society Bethesda*; 1984. p. 627-87.

[92] Gudbjornsdottir S, Sjostrand M, Strindberg L, Wahren J, Lonnroth P. Direct measurements of the permeability surface area for insulin and glucose in human skeletal muscle. *Journal of Clinical Endocrinology and Metabolism.* 2003;88(10):4559-64.
[93] Defronzo RA, Ferrannini E. Insulin resistance. A multifaceted syndrome responsible for NIDDM, obesity, hypertension, dyslipidemia, and atherosclerotic cardiovascular disease. *Diabetes Care.* 1991;14:173-94.
[94] Reaven GM. Insulin resistance and compensatory hyperinsulinemia: role in hypertension, dyslipidemia, and coronary heart disease. *American Heart Journal.* 1991;121:1283-8.
[95] Rattigan S, Clark MG, Barrett EJ. Acute vasoconstriction-induced insulin resistance in rat muscle in vivo. *Diabetes.* 1999;48(3):564-9.
[96] Hotamisligil GS. Inflammatory pathways and insulin action. International *Journal of Obesity and Related Metabolic Disorders.* 2003;27(Suppl 3):S53-S5.
[97] Hotamisligil GS. Inflammation and metabolic disorders. *Nature.* 2006;444(7121):860-7.
[98] Yudkin JS. Adipose tissue, insulin action and vascular disease: inflammatory signals. *International Journal of Obesity and Related Metabolic Disorders.* 2003;27(Suppl 3):S25-S8.
[99] Yudkin JS. Inflammation, obesity, and the metabolic syndrome. *Hormone and Metabolic Research.* 2007;39(10):707-9.
[100] Hotamisligil GS, Shargill NS, Spiegelman BM. Adipose expression of tumor necrosis factor-alpha: Direct role in obesity-linked insulin resistance. *Science.* 1993;259:87-91.
[101] Hotamisligil GS, Murray DL, Choy LN, Spiegelman BM. Tumor necrosis factor alpha inhibits signaling from the insulin receptor. *Proceedings of the National Academy of Sciences of the United States of America.* 1994;91(11):4854-8.
[102] Hotamisligil GS. Mechanisms of TNF-alpha-induced insulin resistance. *Experimental and Clinical Endocrinology and Diabetes.* 1999;107(2):119-25.
[103] Li G, Barrett EJ, Barrett MO, Cao W, Liu Z. Tumor necrosis factor-alpha induces insulin resistance in endothelial cells via a p38 mitogen-activated protein kinase-dependent pathway. *Endocrinology.* 2007;148(7):3356-63.
[104] Zhang L, Wheatley CM, Richards SM, Barrett EJ, Clark MG, Rattigan S. TNF-alpha acutely inhibits vascular effects of physiological but not high insulin or contraction. *American Journal of Physiology Endocrinology and Metabolism.* 2003;285(3):E654-E60.
[105] Plomgaard P, Bouzakri K, Krogh-Madsen R, Mittendorfer B, Zierath JR, Pedersen BK. Tumor necrosis factor-alpha induces skeletal muscle insulin resistance in healthy human subjects via inhibition of Akt substrate 160 phosphorylation. *Diabetes.* 2005;54(10):2939-45.
[106] Youd JM, Rattigan S, Clark MG. Acute impairment of insulin-mediated capillary recruitment and glucose uptake in rat skeletal muscle in vivo by TNFa. *Diabetes.* 2000;49(11):1904-9.
[107] Fürnsinn C, Neschen S, Wagner O, Roden M, Bisschop M, Waldhausl W. Acute and chronic exposure to tumor necrosis factor-alpha fails to affect insulin-stimulated glucose metabolism of isolated rat soleus muscle. *Endocrinology.* 1997;138(7):2674-9.

[108] Nolte LA, Hansen PA, Chen MM, Schluter JM, Gulve EA, Holloszy JO. Short-term exposure to tumor necrosis factor-alpha does not affect insulin-stimulated glucose uptake in skeletal muscle. *Diabetes.* 1998;47(5):721-6.

[109] Reaven GM. Role of insulin resistance in human disease. *Diabetes.* 1988;37:1595-607.

[110] Boden G. Free fatty acids-the link between obesity and insulin resistance. *Endocr. Pract.* 2001 Jan-Feb;7(1):44-51.

[111] Boden G. Obesity and free fatty acids. *Endocrinol. Metab. Clin. North Am.* 2008 Sep;37(3):635-46, viii-ix.

[112] Randle PJ, Garland PB, Hales CN, Newsholme EA. The glucose fatty-acid cycle - its role in insulin sensitivity and the metabolic disturbances of diabetes mellitus. *Lancet.* 1963:785-9.

[113] Sugden MC, Orfali KA, Holness MJ. The pyruvate dehydrogenase complex: Nutrient control and the pathogenesis of insulin resistance. *Journal of Nutrition.* 1995;125 Suppl.:1746S-52S.

[114] Garland PB, Randle PJ, Newsholme EA. Citrate as an intermediary in the inhibition of phosphofructokinase in rat heart muscle by fatty acids, ketone bodies, pyruvate, diabetes and starvation. *Nature.* 1963;200:169-70.

[115] Boden G, Chen X, Ruiz J, White JV, Rossetti L. Mechanisms of fatty acid-induced inhibition of glucose uptake. *J. Clin. Invest.* 1994 Jun;93(6):2438-46.

[116] Shulman GI. Cellular mechanisms of insulin resistance. *Journal of Clinical Investigation.* 2000;106(2):171-6.

[117] Brechtel K, Dahl DB, Machann J, Bachmann OP, Wenzel I, Maier T, et al. Fast elevation of the intramyocellular lipid content in the presence of circulating free fatty acids and hyperinsulinemia: a dynamic 1H-MRS study. *Magnetic Resonance in Medicine.* 2001;45(2):179-83.

[118] Machann J, Haring H, Schick F, Stumvoll M. Intramyocellular lipids and insulin resistance. *Diabetes, Obesity and Metabolism.* 2004;6(4):239-48.

[119] Summers SA. Ceramides in insulin resistance and lipotoxicity. *Progress in Lipid Research.* 2006;45(1):42-72.

[120] Yu C, Chen Y, Cline GW, Zhang D, Zong H, Wang Y, et al. Mechanism by which fatty acids inhibit insulin activation of insulin receptor substrate-1 (IRS-1)-associated phosphatidylinositol 3-kinase activity in muscle. *Journal of Biological Chemistry.* 2002;277(52):50230-6.

[121] Dresner A, Laurent D, Marcucci M, Griffin ME, Dufour S, Cline GW, et al. Effects of free fatty acids on glucose transport and IRS-1-associated phosphatidylinositol 3-kinase activity. *J Clin Invest.* 1999 Jan;103(2):253-9.

[122] Steinberg HO, Tarshoby M, Monestel R, Hook G, Cronin J, Johnson A, et al. Elevated circulating free fatty acid levels impair endothelium-dependent vasodilation. *Journal of Clinical Investigation.* 1997;100(5):1230-9.

[123] Steinberg HO PG, Hook G, Crowder K, Cronin J, Baron AD. Free fatty acid elevation impairs insulin-mediated vasodilation and nitric oxide production. *Diabetes.* 2000;49(July):1231-8.

[124] Clerk LH, Rattigan S, Clark MG. Lipid infusion impairs physiologic insulin-mediated capillary recruitment and muscle glucose uptake in vivo. *Diabetes.* 2002;51(4):1138-45.

[125] Liu Z, Liu J, Jahn LA, Fowler DE, Barrett EJ. Infusing Lipid Raises Plasma Free Fatty Acids and Induces Insulin Resistance in Muscle Microvasculature. *Journal of Clinical Endocrinology and Metabolism*. 2009;94(9):3543-9.

[126] Han DH, Hancock C, Jung SR, Holloszy JO. Is "fat-induced" muscle insulin resistance rapidly reversible? *Am. J. Physiol. Endocrinol. Metab.* 2009 Jul;297(1):E236-41.

[127] Marshall S, Bacote V, Traxinger RR. Discovery of a metabolic pathway mediating glucose-induced desensitization of the glucose transport system. *Journal of Biological Chemistry*. 1991;266:4706-12.

[128] Baron AD, Zhu JS, Weldon J, Maianu L, Garvey WT. Glucosamine induces insulin resistance in vivo by affecting GLUT 4 translocation in skeletal muscle - Implications for glucose toxicity. *Journal of Clinical Investigation*. 1995;96:2792-801.

[129] Fürnsinn C, Sanderson AL, Radda GK, Leighton B. Effects of glucosamine on insulin-stimulated glucose metabolism in rat soleus muscle. International *Journal of Biochemistry and Cell Biology*. 1995;27:805-14.

[130] Barzilai N, Hawkins M, Angelov I, Hu M, Rossetti L. Glucosamine-induced inhibition of liver glucokinase impairs the ability of hyperglycemia to suppress endogenous glucose production. *Diabetes*. 1996;45(10):1329-35.

[131] Wallis MG, Smith ME, Kolka CM, Zhang L, Richards SM, Rattigan S, et al. Acute glucosamine-induced insulin resistance in muscle in vivo is associated with impaired capillary recruitment. *Diabetologia*. 2005;48:2131-9.

[132] Holmang A, Nilsson C, Niklasson M, Larsson BM, Lonroth P. Induction of insulin resistance by glucosamine reduces blood flow but not interstitial levels of either glucose or insulin. *Diabetes*. 1999;48(1):106-11.

[133] Mather K, Anderson TJ, Verma S. Insulin action in the vasculature: physiology and pathophysiology. *Journal of Vascular Research*. 2001;38(5):415-22.

[134] Mitchell D, Bihari A, Sandig M, Tyml K. Endothelin-a receptor in rat skeletal muscle microvasculature. *Microvascular Research*. 2002;64(1):179-85.

[135] Mather KJ, Mirzamohammadi B, Lteif A, Steinberg HO, Baron AD. Endothelin contributes to basal vascular tone and endothelial dysfunction in human obesity and type 2 diabetes. *Diabetes*. 2002;51(12):3517-23.

[136] Lteif AA, Fulford AD, Considine RV, Gelfand I, Baron AD, Mather KJ. Hyperinsulinemia fails to augment ET-1 action in the skeletal muscle vascular bed in vivo in humans. *American Journal of Physiology Endocrinology and Metabolism*. 2008;295(6):E1510-E7.

[137] Lteif A, Vaishnava P, Baron AD, Mather KJ. Endothelin limits insulin action in obese/insulin-resistant humans. *Diabetes*. 2007;56(3):728-34.

[138] Kolka CM, Rattigan S, Richards S, Clark MG. Metabolic and vascular actions of endothelin-1 are inhibited by insulin-mediated vasodilation in perfused rat hindlimb muscle. *British Journal of Pharmacology*. 2005;145(7):992-1000.

[139] Ross RM, Kolka CM, Rattigan S, Clark MG. Acute blockade by endothelin-1 of haemodynamic insulin action in rats. *Diabetologia*. 2007;50(2):443-51.

[140] Phillips MS, Liu Q, Hammond HA, Dugan V, Hey PJ, Caskey CJ, et al. Leptin receptor missense mutation in the fatty Zucker rat. *Nat Genet.* 1996 May;13(1):18-9.

[141] Moore BJ, Armbruster SJ, Horwitz BA, Stern JS. Energy expenditure is reduced in preobese 2-day Zucker fa/fa rats. *Am. J Physiol.* 1985 Aug;249(2 Pt 2):R262-5.

[142] Zucker LM. Hereditary obesity in the rat associated with hyperlipemia. *Ann. N Y Acad. Sci.* 1965 Oct 8;131(1):447-58.

[143] Stern J, Johnson PR, Greenwood MR, Zucker LM, Hirsch J. Insulin resistance and pancreatic insulin release in the genetically obese Zucker rat. *Proc. Soc. Exp. Biol. Med.* 1972 Jan;139(1):66-9.

[144] Kemmer FW, Berger M, Herberg L, Gries FA, Wirdeier A, Becker K. Glucose metabolism in perfused skeletal muscle. Demonstration of insulin resistance in the obese Zucker rat. *Biochemical Journal.* 1979;178(3):733-41.

[145] Apweiler R, Freund P. Development of glucose intolerance in obese (fa/fa) Zucker rats. *Hormone and Metabolic Research.* 1993;25:521-4.

[146] Crettaz M, Prentki M, Zaninetti D, Jeanrenaud B. Insulin resistance in soleus muscle from obese Zucker rats. *Biochemical Journal.* 1980;186:525-34.

[147] Penicaud L, Ferre P, Terretaz J, Kinebanyan MF, Leturque A, Dore E, et al. Development of obesity in Zucker rats. Early insulin resistance in muscles but normal sensitivity in white adipose tissue. *Diabetes.* 1987 May;36(5):626-31.

[148] Alonso-Galicia M, Brands MW, Zappe DH, Hall JE. Hypertension in obese Zucker rats. Role of angiotensin II and adrenergic activity. *Hypertension.* 1996;28(6):1047-54.

[149] Wallis MG, Wheatley CM, Rattigan S, Barrett EJ, Clark AD, Clark MG. Insulin-mediated hemodynamic changes are impaired in muscle of zucker obese rats. *Diabetes.* 2002;51(12):3492-8.

[150] Nishimatsu H, Suzuki E, Satonaka H, Takeda R, Omata M, Fujita T, et al. Endothelial dysfunction and hypercontractility of vascular myocytes are ameliorated by fluvastatin in obese Zucker rats. *American Journal of Physiology Heart and Circulatory Physiology.* 2005;288(4):H1770-H6.

[151] Oltman CL, Richou LL, Davidson EP, Coppey LJ, Lund DD, Yorek MA. Progression of coronary and mesenteric vascular dysfunction in Zucker obese and Zucker diabetic fatty rats. *American Journal of Physiology Heart and Circulatory Physiology.* 2006;291(4):H1780-H7.

[152] Zemel MB, Reddy S, Sowers JR. Insulin attenuation of vasoconstrictor responses to phenylephrine in Zucker lean and obese rats. *American Journal of Hypertension.* 1991;4:537-9.

[153] Walker AB, Dores J, Buckingham RE, Savage MW, Williams G. Impaired insulin-induced attenuation of noradrenaline-mediated vasoconstriction in insulin-resistant obese Zucker rats. *Clinical Science.* 1997;93(3):235-41.

[154] Masek J, Fabry P. High-fat diet and the development of obesity in albino rats. *Experientia.* 1959 Nov 15;15:444-5.

[155] Storlien LH, James DE, Burleigh KM, Chisholm DJ, Kraegen EW. Fat feeding causes widespread in vivo insulin resistance, decreased energy expenditure, and obesity in rats. *American Journal of Physiology.* 1986;251:E576-E83.

[156] Buettner R, Scholmerich J, Bollheimer LC. High-fat diets: modeling the metabolic disorders of human obesity in rodents. *Obesity* (Silver Spring). 2007 Apr;15(4):798-808.

[157] Kraegen EW, James DE, Storlien LH, Burleigh KM, Chisholm DJ. In vivo insulin resistance in individual peripheral tissues of the high fat fed rat: assessment by euglycaemic clamp plus deoxyglucose administration. *Diabetologia.* 1986;29:192-8.

[158] Kraegen EW, Cooney GJ. Free fatty acids and skeletal muscle insulin resistance. *Curr. Opin. Lipidol.* 2008 Jun;19(3):235-41.
[159] Naderali EK, Williams G. Effects of short-term feeding of a highly palatable diet on vascular reactivity in rats. *Eur. J. Clin. Invest.* 2001 Dec;31(12):1024-8.
[160] Song GY, Gao Y, Di YW, Pan LL, Zhou Y, Ye JM. High-fat feeding reduces endothelium-dependent vasodilation in rats: differential mechanisms for saturated and unsaturated fatty acids? *Clin. Exp. Pharmacol. Physiol.* 2006 Aug;33(8):708-13.
[161] Erdei N, Toth A, Pasztor ET, Papp Z, Edes I, Koller A, et al. High-fat diet-induced reduction in nitric oxide-dependent arteriolar dilation in rats: role of xanthine oxidase-derived superoxide anion. American *Journal of Physiology Heart and Circulatory Physiology.* 2006;291(5):H2107-H15.
[162] St-Pierre P, Genders AJ, Keske MA, Richards SM, Rattigan S. Loss of insulin-mediated microvascular perfusion in skeletal muscle is associated with the development of insulin resistance. *Diabetes, Obesity and Metabolism.* 2010;12(9):798-805.
[163] Steinberg HO, Chaker H, Leaming R, Johnson A, Brechtel G, Baron AD. Obesity/insulin resistance is associated with endothelial dysfunction. Implications for the syndrome of insulin resistance. *Journal of Clinical Investigation.* 1996;97(11):2601-10.
[164] Hashimoto M, Akishita M, Eto M, Kozaki K, Ako J, Sugimoto N, et al. The impairment of flow-mediated vasodilatation in obese men with visceral fat accumulation. *International Journal of Obesity and Related Metabolic Disorders.* 1998;22(5):477-84.
[165] Westerbacka J, Vehkavaara S, Bergholm R, Wilkinson I, Cockcroft J, Yki-Jarvinen H. Marked resistance of the ability of insulin to decrease arterial stiffness characterizes human obesity. *Diabetes.* 1999 Apr;48(4):821-7.
[166] de Jongh RT, Serne EH, Ijzerman RG, De Vries G, Stehouwer CD. Impaired microvascular function in obesity: implications for obesity-associated microangiopathy, hypertension, and insulin resistance. *Circulation.* 2004;109(21):2529-35.
[167] Keske MA, Clerk LH, Price WJ, Jahn LA, Barrett EJ. Obesity blunts microvascular recruitment in human forearm muscle after a mixed meal. *Diabetes Care.* 2009 Sep;32(9):1672-7.
[168] Duplain H, Burcelin R, Sartori C, Cook S, Egli M, Lepori M, et al. Insulin resistance, hyperlipidemia, and hypertension in mice lacking endothelial nitric oxide synthase. *Circulation.* 2001;104(3):342-5.
[169] Kubis N, Richer C, Domergue V, Giudicelli JF, Levy BI. Role of microvascular rarefaction in the increased arterial pressure in mice lacking for the endothelial nitric oxide synthase gene (eNOS3pt-/-). *Journal of Hypertension.* 2002;20(8):1581-7.
[170] Cook S, Hugli O, Egli M, Menard B, Thalmann S, Sartori C, et al. Partial gene deletion of endothelial nitric oxide synthase predisposes to exaggerated high-fat diet-induced insulin resistance and arterial hypertension. *Diabetes.* 2004;53(8):2067-72.
[171] Vicent D, Ilany J, Kondo T, Naruse K, Fisher SJ, Kisanuki YY, et al. The role of endothelial insulin signaling in the regulation of vascular tone and insulin resistance. *Journal of Clinical Investigation.* 2003;111(9):1373-80.
[172] Kubota T, Kubota N, Kumagai H, Yamaguchi S, Kozono H, Takahashi T, et al. Impaired insulin signaling in endothelial cells reduces insulin-induced glucose uptake by skeletal muscle. *Cell Metab.* 2011 Mar 2;13(3):294-307.

[173] Bonner JS, Lantier L, Hasenour CM, James FD, Bracy DP, Wasserman DH. Muscle-Specific Vascular Endothelial Growth Factor Deletion Induces Muscle Capillary Rarefaction Creating Muscle Insulin Resistance. *Diabetes.* 2012.

[174] Gavin TP, Stallings HW, III, Zwetsloot KA, Westerkamp LM, Ryan NA, Moore RA, et al. Lower capillary density but no difference in VEGF expression in obese vs. lean young skeletal muscle in humans. *Journal of Applied Physiology.* 2005;98(1):315-21.

[175] Gudbjornsdottir S, Sjostrand M, Strindberg L, Lonnroth P. Decreased muscle capillary permeability surface area in type 2 diabetic subjects. *Journal of Clinical Endocrinology and Metabolism.* 2004;90(2):1078-82.

[176] Premilovac D, Bradley EA, Ng HL, Richards SM, Rattigan S, Keske MA. Muscle insulin resistance resulting from impaired microvascular insulin sensitivity in Sprague Dawley rats. *Cardiovasc Res.* 2013 Feb 17.

[177] Barrett EJ, Eggleston EM, Inyard AC, Wang H, Li G, Chai W, et al. The vascular actions of insulin control its delivery to muscle and regulate the rate-limiting step in skeletal muscle insulin action. *Diabetologia.* 2009;52(5):752-64.

[178] Barrett EJ, Liu Z. The endothelial cell: An "early responder" in the development of insulin resistance. *Rev. Endocr. Metab. Disord.* 2013 Jan 10.

In: Glucose Uptake
Editors: Carter C. Johnson and Davis B. Williams
ISBN: 978-1-62618-670-5

Chapter 5

Implications of Diabetes on Sperm Glucose Uptake and Metabolism

*T. R. Dias, M. G. Alves, A. Neuhaus-Oliveira, S. Socorro, B. M. Silva and P. F. Oliveira**
CICS – UBI – Health Sciences Research Centre,
University of Beira Interior, Covilhã, Portugal

Abstract

Diabetes mellitus (DM) is described as a metabolic disorder characterized by hyperglycaemia resulting from defective insulin secretion, resistance to insulin, or both, and represents one of the greatest threats to modern global health as its incidence is rapidly rising worldwide. Type 1 DM results from an absolute deficiency of insulin due to an autoimmune destruction of the pancreatic beta cells while type 2 DM is characterized by impaired insulin secretion and increased insulin resistance.

It is well known that glucose regulation is crucial for normal spermatogenesis and fertility. In this process, insulin plays a crucial role since its dysfunction is connected with decreased cellular glucose transport. Both clinical and experimental reports suggest that fertility is highly decreased in patients or animals with DM. Both DM types impair male fertility and numerous studies in male diabetic individuals have demonstrated a marked reduction in fecundity, as well as impairment of sperm quality and higher percentage of spermatozoa with nuclear DNA damage. All these effects are known to be related with metabolic signaling pathways in testis that results in defective sperm metabolism. The different regions and structures of the sperm flagellum are of great importance because the metabolic pathways are compartmentalized in them. Spermatozoa metabolize several substrates as energy sources, such as hexoses (glucose, mannose, and fructose) or other metabolites (lactate and citrate). Any alteration in the ability of the spermatozoa to utilize substrates involved in ATP production is expected to compromise motility and subsequently fertility. For glycolysis to occur in sperm, these cells need

* Corresponding author: Pedro Fontes Oliveira, Health Sciences Research Centre, Faculty of Health Sciences, University of Beira Interior, Av. Infante D. Henrique, 6201-506 Covilhã, Portugal, Email:poliveira@fcsaude.ubi.pt.

specific carriers to transport energy sources through the cellular membrane, namely glucose transporters (GLUTs). DM is known to modulate spermatozoa substrate consumption and/or production due to altered glycolysis. The transport of hexoses via GLUTs is also known to be highly dysregulated in diabetic male individuals.

Throughout this chapter we will discuss the effects of DM in sperm glucose uptake and metabolism. Understanding the functioning and regulation of these processes is crucial to identify the key mechanisms associated with male (in)fertility in order to develop possible therapeutics.

Introduction

Diabetes mellitus (DM) is the most common metabolic disease in the world and has become a serious problem of modern society due to the severe long-term health associated complications [1]. It is a potentially devastating, expensive, treatable but incurable lifelong disease [2]. According to a widely accepted estimation, in the year 2000, 171 million people had DM, and this is expected to double by year 2030 [3, 4]. The proportion of individuals with DM is increasing due to population growth, aging, urbanization, and increasing prevalence of obesity and physical inactivity [5]. DM is known to cause many systemic complications such as cardiovascular diseases and hypertension [6]. Moreover, male reproductive alterations have also been widely reported in individuals with DM [7]. Numerous studies in male diabetic individuals have demonstrated a marked reduction in fecundity [7-11], as well as impairment of sperm quality [11, 12] and higher percentage of spermatozoa with nuclear DNA damage [3]. DM-induced effects on testicular function have been attributed to the lack of insulin [7], which is the leading hormone responsible for glucose homeostasis regulation [13]. Maintenance of spermatogenesis *in vivo* and the fertility capacity of male sperm depends on glucose metabolism [14, 15]. Sperm cells can effectively use several simple sugars such as glucose, fructose and mannose [16] and the spermatozoa energy production requires catabolism of glucose to pyruvate and lactate by the glycolytic pathway enzymes. Lactate production is made by Sertoli cells (SCs) to maintain germ cells survival and this process has been shown to be predominantly under the control of the endocrine system, primarily by sex steroid hormones [17-19], follicle-stimulating hormone (FSH) and insulin [20]. An alteration in spermatozoa ability to utilize the substrates involved in ATP production would be expected to compromise sperm motility and subsequently fertility [21]. Spermatozoa need specific carriers, known as glucose transportes (GLUTs) to mediate the glucose uptake from the surrounding medium into the cell [22]. DM has been shown to be associated with a depletion of GLUTs [23]. Therefore, diabetic individuals are known to possess an inability to transport glucose, which supports an association of this disease with disruptions in sperm metabolism and consequently subfertility or even infertility. Thus, there is a growing interest in developing an efficient treatment to prevent DM or at least decrease its associated problems. Many of the already existing drugs to manage DM fail as a curative agent for diabetic complications and also have a number of serious adverse effects that can discourage patient compliance [24]. Over the years, the use of medicinal plants has become a feasible alternative for the treatment of DM or to reinforce the currently used treatments [1, 25]. Plants often contain substantial amounts of antioxidant phytochemicals and its antioxidant action is thought to be associated with a hypoglycaemic effect on DM [26, 27].

More investigations must be carried out to evaluate the exact molecular mechanisms of edible and medicinal plants action with antidiabetic and insulinomimetic activity. As DM is rapidly rising worldwide and its prevalence is very high in men in reproductive years, it is expected that subfertility or infertility associated with DM will also dramatically rise in the upcoming years. So, there is an endless search for a new therapy to counteract these alarming consequences of DM.

Diabetes Mellitus at a Glance

Diabetes Mellitus (DM) is a metabolic disorder of multiple aetiologies, characterized by hyperglycaemia and abnormalities in carbohydrate, fat, and protein metabolisms emanating from deficiencies or disruptions in insulin secretion and/or insulin action [6, 28], defects in reactive oxygen species (ROS) scavenging enzymes [29], and high oxidative stress (OS) impairing pancreatic beta cells [30, 31]. The chronic hyperglycaemia in diabetic individuals is associated with long-term damage, dysfunction, and failure of different organs, especially the eyes, kidneys, nerves, heart, and blood vessels [6].

There are two major forms of DM, referred as type 1 or insulin-dependent diabetes mellitus (T1DM) and type 2 or non-insulin-dependent diabetes mellitus (T2DM) [6]. T1DM, which generally develops before 30 years [3], is characterized by a cellular-mediated autoimmune destruction of insulin-producing pancreatic beta cells in genetically susceptible individuals [32, 33]. This destruction of the insulin secreting beta cells is progressive, leading to an absolute insulin deficiency and the need of exogenous insulin treatment for survival. The pathogenic factors that lead to T1DM are not yet fully elucidated, but there is clear evidence that it appears due to an alteration of the immune regulation [34]. T2DM, which is responsible for approximately 90–95% of DM cases [6], results from an imbalance between insulin sensitivity and insulin secretion [35]. Insulin is the leading hormone responsible for regulating glucose homeostasis, primarily through the suppression of hepatic glucose production and stimulation of glucose uptake by muscle and adipose tissues from the circulatory system [13]. The profound dysregulations of these processes lead to insulin resistance (IR) [6], which is described as the inability of cells (liver, muscle and fat cells) to respond to normal levels of circulating insulin [36], leading to the development of T2DM. Because the pancreas is able to appropriately augment its secretion of insulin to offset IR, glucose tolerance remains normal. However, with time, pancreatic beta cells fail to maintain its high rate of insulin secretion to compensate for the IR, leading to the development of impaired glucose tolerance (IGT) [35]. IGT is an intermediate category between normal glucose tolerance and overt T2DM, referred as prediabetes [37], and it can be identified by an oral glucose-tolerance test [38]. The progression from prediabetes to T2DM occurs over many years and there is strong evidence to support interventions to prevent the disease progression to DM [39, 40]. Despite genetic predisposition, the risk of developing T2DM in humans increases with age, obesity, cardiovascular diseases and the lack of physical activity [41, 42]. At least initially, and often throughout their lifetime, these individuals do not need insulin treatment to survive [6].

Patients with DM experience significant morbidity and mortality due to the development of microvascular (retinopathy, neuropathy, and nephropathy) and macrovascular (heart attack, stroke and peripheral vascular disease) long-term complications [43]. Hypertension, sexual dysfunction and abnormalities of lipoprotein metabolism are often found in people with DM [6]. These dramatic consequences highly increase the complexity of the disease and make DM a preferential field of investigation for researchers all over the world.

Diabetes Mellitus and Male Fertility

Male fertility is compromised by the hormonal and metabolic changes associated with both types of DM [4, 44], obesity [45] and the metabolic syndrome, the latter sharing essential pathological features with DM [46]. According to the World Health Organization (WHO), infertility is defined as the inability to conceive after 1 year of unprotected intercourse [47] and affects about 13%–18% of couples [48]. In about half of them, male factor is the sole cause for the infertility problem [49]. A large number of the male infertility cases are also associated to suboptimal sperm quality due to abnormal parameters - motility, morphology, concentration and DNA fragmentation [50, 51]. The deleterious influence of DM and obesity on fertility is receiving increasing attention since their prevalence and incidence is escalating worldwide, while the age of first diagnosis for both diseases is in continuous decline [52, 53]. Consequently, the fertility of a growing number of individuals is affected prior to and during their reproductive years [54, 55].

DM may affect male reproductive function at multiple levels as a result of its effects on the endocrine control of spermatogenesis, spermatogenesis itself, or by impairing penile erection and ejaculation [56]. It is well-recognized that DM is a cause of male sexual dysfunction, which by itself may contribute to subfertility or even infertility [3]. The prevalence of subfertility in diabetic male individuals has been reported to account 50% of the cases [57]. Data from animal models demonstrated a marked reduction in fecundity when male animals are diabetic [7-11]. In particular, decreased sperm concentration and motility, abnormal sperm morphology and increased seminal plasma abnormalities were detected [58]. Furthermore, DM is associated with increased OS which damages sperm nuclear and mitochondrial DNA [46, 59]. Other disturbances such as retrograde ejaculation, premature ejaculation, decreased libido, delayed sexual maturation and impotence are also known to occur in diabetic males [60, 61]. DM-related effects on testicular function have been attributed to the lack of insulin [7], which is a direct consequence of a severe disruption in the unique glucose metabolism that testicular cells present. Insulin expression in the testes also seems to be affected by DM and for instance, streptozotocin (STZ)-induced diabetic rats express less than half of the insulin protein compared with nondiabetic controls [62]. In addition, observations of a direct insulin effect on both Leydig cells [63, 64] and SCs [65, 66] have been reported. Nonetheless, the data are confusing, and the exact role for insulin in the regulation of the male reproductive function still unclear [7].

The link between T1DM and fertility has long been established. Accounts dating as far back as the 11th century have described the disease as "a collapse of sexual functions" highlighting the importance of insulin in the reproductive system. It remains unclear whether the damage to sperm is attributable to local effects from hyperglycaemia or to alterations in hormone levels that disrupt the hypothalamic-pituitary-gonadal (HPG) axis [67]. However, it is thought that spermatogenesis disruption and germ cell apoptosis in T1DM may be related to a local autoimmune damage [60].

A recent study reported that the prevalence of infertility in T2DM men was 35.1% [60]. Also, patients with T2DM, IR, obesity and other related co-morbidities present impaired sperm parameters and decreased testosterone serum levels [60]. Low testosterone levels have also been found to predict IR and the future development of T2DM from a prediabetic state [68].

As male infertility/subfertility problems may become more widespread while DM rates increase, the enlightenment of the key regulatory mechanisms by which sperm production is affected in this disease is critical in order to highlight new therapeutic approaches.

Sperm Metabolism and Glucose Uptake

Spermatogenesis is a metabolically active process that is under strict hormonal control and depends upon the metabolic cooperation between different testicular cells [69]. Glucose is one of the most important energetic substrates for mammalian cells and spermatogenesis maintenance *in vivo* relies on glucose metabolism. Nonetheless, glucose is present at low levels within the tubular fluid and the blood-to-germ cells glucose transport is highly controlled. Spermatozoa production takes several days, occurring within the seminiferous tubules under endocrine and paracrine control through the SCs [56] which are known as "nurse-cells" and are responsible for glucose conversion to lactate [70]. This is a very important event in spermatogenesis since developing germ cells do not metabolize glucose but rather lactate [71]. After sperm cells are finally produced, they present specific metabolic needs. Spermatozoa can use not only external hexoses such as fructose, mannose and glucose but also lactate, citrate, lipids and aminoacids [72-76]. As a result they can use glycolysis and/or oxidative phosphorylation. Besides, the spermatozoa metabolic organization is very complex. There is a mitochondria sheath in the midpiece (Figure 1), which indicates that the oxidative processes occur at this specific place while the most important glycolytic enzymes are located in the principal piece of the tail [77-80]. Interestingly, sperm hyperactivation and/or capacitation are highly dependent on glucose. Nevertheless, the exact role of this sugar as a direct energy source or as a precursor for other metabolic substrates has not been fully disclosed yet [81]. Most of the patients diagnosed with subfertility and/or infertility present problems in sperm function rather than lack of sperm [82] and therefore the study of the molecular mechanisms responsible for sperm functioning is pivotal. Both glycolysis and mitochondrial respiration are active in mammalian sperm and thus the ATP levels in sperm are maintained through glycolytic and non-glycolytic substrates [83] although sperm capacitation is stimulated by glucose [84], which supports that this sugar has an important regulatory role in the overall male reproductive function.

Moreover, the process of sperm capacitation is associated with several metabolic alterations that may induce OS which is often followed by sperm loss of functional competence [85]. As mentioned, spermatozoa progress from a non-motile to a motile status after spermatogenesis and during that process they lose most of their cytoplasm becoming dependent of external energy substrates [86]. Indeed, sperm cells are differentiated cells well known for their motility, a process that requires high amounts of energy. Besides, these cells undergo several energy requiring changes such as tyrosine phosphorylation, hyperactivated motility, calcium movements and acrosome reaction, that allow them to acquire their fertilization potential [87, 88].

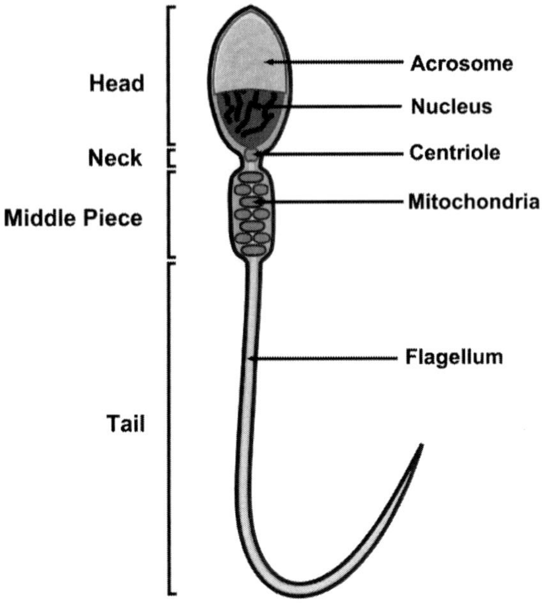

Figure 1. Schematic diagram of the structure of mammalian sperm.

Glucose has been reported to be essential for spontaneous acrosome reaction as a medium lacking glucose inhibited this process that was rapidly restored after glucose addition [89].

Glucose uptake and metabolization by testicular cells and sperm has a preponderant role in the fertility potential of male individuals. Glucides are polar molecules capable to cross the lipidic bilayers in a very slow and inefficient manner. Consequently, the cells need to uptake them through carriers, down a chemical gradient. There are two different families of glucose transport proteins: the Sodium Dependent Glucose Transporters (SGLTs), also known as Solute Carrier Family 5 (SLC5), and the Glucose Transporters (GLUTs), also known as Solute Carrier Family 2 (SLC2) [86, 90, 91]. These two families act in a very distinct manner to transport the hexoses: while the SGLTs are active sugar transporters with energy needs for functioning, the GLUTs passively transport the sugars without energy consumption [91]. The SGLTs family is composed by six different active transporters (SGLT1 to 6) while the GLUTs family is composed by fourteen glucose transporter isoforms (GLUT1 to 14) [91]. Several studies were conducted to identify and characterize GLUTs expression in testicular tissues and sperm of several species (for extensive review see [86]). Briefly, the first GLUT identified in human testis and ejaculated spermatozoa was the GLUT5 [92]. Later GLUT3 was reported in human testis and sperm cells [93]. Still, the first wide study on GLUTs in human, rat and bull sperm cells and testicular tissues was presented a few years later [72]. In human spermatozoa, it was reported that GLUT1 and GLUT2 were present in the acrosomal region and in the principal and endpieces of the tail while GLUT3 was found in the midpiece [72]. Noteworthy, GLUT4 was not found and GLUT5 was detected in subequatorial region and in the mid and principal pieces [72]. GLUT8 was also found in human and mouse mature spermatozoa, especially in the acrosomal membrane [94, 95]. Later, others [96] reported that

GLUT8 was located in the sperm midpiece and also reported the presence of GLUT9 in testis, mature sperm tail and in the apical ridge of sperm acrosome. Finally, GLUT14 was found in human testes and interestingly, it was reported that it presented a 95% homology with GLUT3 [97]. This brief description clearly shows that there is a high degree of functional flexibility in sperm glucose uptake since it presents a distinct and complex distribution of GLUTs isoforms. Energy metabolism is a key feature in supporting spermatozoa function (Figure 2). Spermatozoa require energy, generated as intracellular ATP, mostly for cell motility [98, 99]. Whenever the supply of ATP or ADP is exhausted, motility stops and if spermatozoa cannot swim by using its flagellar motion, it cannot fertilize the egg [100]. Hence, for maintenance of motility, ATP or ADP must be restored. Both oxidative phosphorylation and glycolysis can provide the required energy, together or independently of one another [101, 102]. In the absence of O_2, rebuilding of ATP occurs by anaerobic glycolysis or fructolysis [101, 103]. Substrates for glycolysis and fructolysis provided by the seminal plasma are glucose and fructose, respectively, although mannose and maltose can also be used. Depending on the species, spermatozoa might prefer glucose or fructose as their main energy source [103, 104]. For instance, bull sperm uses fructose and glucose at the same rate, although the seminal plasma has remarkably more fructose than glucose [105].

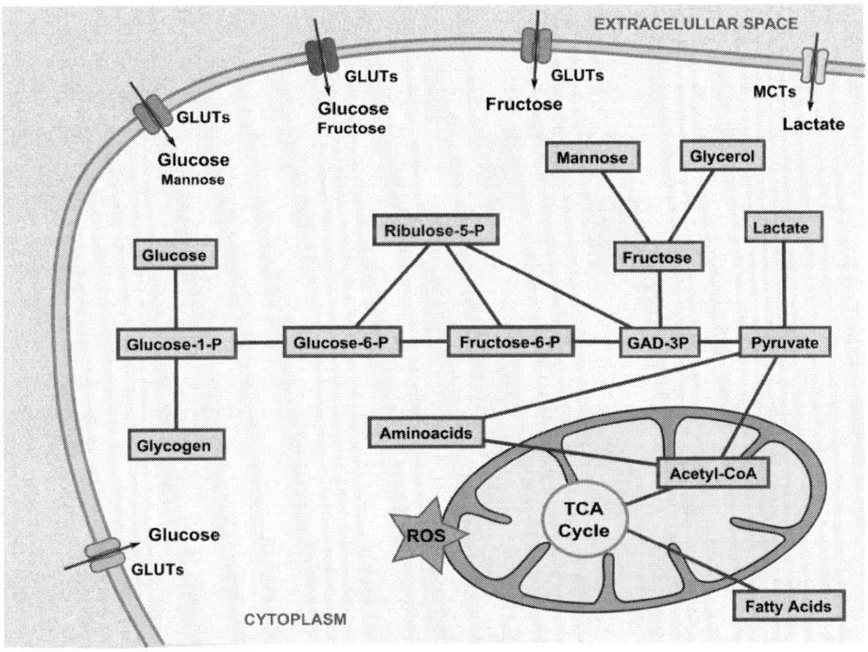

Figure 2. Diagram of the pathways involved in sperm cells energy metabolism. Spermatozoa are capable of consuming a variety of fuels, including fructose, glucose, mannose, lactate, amino and fatty acids. Spermatozoa preferentially metabolize fructose or glucose, which is converted to lactate or oxidized via the TCA cycle in the mitochondrial matrix. The mitochondrial oxidation of these substrates is coupled with ADP phosphorylation, via the electron transport chain to form ATP. Glucose on sperm can also be metabolized via the pentose phosphate pathway, in which ribulose-5-phosphate is a key intermediate. Lactate is transported into cells via the family of proton-linked plasma membrane transporters known as MCTs, while hexoses are imported via the GLUTs family of membrane proteins. Abbreviations: GLUTs, glucose transporters; MCTs, monocarboxylate transporters; ROS, reactive oxygen species; TCA, tricarboxylic acid.

In glycolysis, once glucose gets inside the cell taken up from the seminal plasma, it enters the glycolytic pathway and is decomposed to pyruvate in a sequence of reactions that yields two molecules of ATP, two molecules of pyruvate and two "high energy" electron carrying molecules of NADH. The flux through the glycolytic pathway is adjusted in response to meet energy cellular requirements and the reactions catalyzed by hexokinase, phosphofructokinase, and pyruvate kinase are effectively irreversible in most organisms and potential sites for glycolysis control [106, 107]. Under anaerobic conditions, glycolysis further progresses to produce lactic acid, by lactate dehydrogenase action, which accumulates in spermatozoa [103].

Similarly to what happens to glucose in glycolysis, during fructolysis, fructose suffers a series of transformations in order to be decomposed to pyruvate and later converted into lactate. It shares an initial common step, as fructose enters the cell using specific hexose membrane transporters (GLUTs) [92, 108]. Once inside the cell it is phosphorylated to fructose 1-phosphate by fructokinase. Interestingly, besides sperm and liver no other organs or cells have fructokinase activity which can actively metabolize this hexose. Fructose-1-phosphate undergoes hydrolysis by fructose-1-phosphate aldolase to form dihydroxyacetone phosphate (DHAP) and glyceraldehyde. DHAP can be isomerized to glyceraldehyde-3-phosphate by triosephosphate isomerase and the glyceraldehyde produced may also be converted to glyceraldehyde-3-phosphate by glyceraldehyde kinase [103, 104, 109]. At this point, the metabolism of fructose yields intermediates of the glycolytic pathway, which can be converted to pyruvate and finally reduced to lactate. As happens with glycolysis, fructolysis is a component of both the aerobic and anaerobic metabolism of semen, but whereas in aerobic metabolism it does not constitute the sole source of energy, owing to the occurrence of other aerobic processes, in anaerobic conditions it plays a key role and must be regarded as indispensable for spermatozoa survival [103, 104].

Under aerobic conditions, a variety of products can be utilized by spermatozoa including lactic and pyruvic acid for production of energy in a process far more energetically efficient than anaerobic glycolysis or fructolysis [101, 103]. Pyruvate transported to the mitochondrial matrix is oxidized and decarboxylated by the enzyme pyruvate dehydrogenase (PDH) forming a two carbon intermediate, Acetyl-CoA, which can enter the TCA cycle (Krebs cycle) to combine with oxaloacetate (OAA). OAA is then further oxidized to reduce the mobile electron carrier NAD^+ to NADH and the Complex II prosthetic group FAD_2^+ to $FADH_2$ generating carbon dioxide. The oxidation of these substrates is coupled with adenosine diphosphate (ADP) phosphorylation via the mitochondrial electron transport chain with the consequent production of ATP [103]. Availability of O_2 is not an absolute governing force with respect to sperm metabolism. There are factors that control the degree to which oxidative phosphorylation provides energy for motility [101]. Nevertheless, spermatozoa possess all the required enzymatic machinery for energy metabolism in presence or absence of oxygen, pointing towards a high degree of metabolic plasticity. This makes fertilization possible under divergent conditions. Indeed, intracellular lipid reserves are often used by spermatozoa when other energy sources are absent [110, 111]. Acyl ester bonds of plasmalogen are broken during endogenous respiration resulting in freeing of fatty acids which can then be used as substrate for endogenous respiration [112]. Glycerol can also be utilized by spermatozoa, being converted to fructose and then metabolized in fructolysis [103, 113]. Spermatozoa are capable of oxidizing aminoacids as well, resulting in the formation of H_2O_2. Nevertheless, as they lack the peroxidase/catalase enzymatic machinery, the production of H_2O_2 eventually leads to loss of sperm motility [114]. The effect of glucose on sperm, especially during sperm entry in the oocyte, is also associated to the pentose phosphate pathway (PPP) [115, 116]. In this pathway, the glucose-6-phosphate produced from glucose is converted to NADPH and pentoses. In this process, glucose-6-phosphate dehydrogenase (G6PDH) is the key rate limiting enzyme and is reported to be very active in human spermatozoa [117] regulating not only NADH production but also glucose metabolism [116]. Finally, sperm has glycogen and endogenous sources of glucose that allows sperm to survive in glucose-free conditions [118].

Sperm metabolism and glucose uptake are crucial endpoints for the fertility potential of male individuals. The complexity of these processes is a challenge for all those that are dedicated to study the impact of diseases in the molecular mechanisms of glucose transport and metabolization in sperm. However, it is imperative to understand the dynamics of sperm metabolism and glucose uptake since it may reveal new therapeutic sites for subfertility/infertility treatment.

Influence of Diabetes in Sperm Glucose Uptake and Metabolism

As discussed above, DM has long been associated with reproductive impairments and multiple mechanisms are expected to be responsible for such alterations. The glucose sensing machinery that testicular cells and sperm possess is sensitive to the hormonal fluctuations caused by DM and, thus, the specific mechanisms to counteract hyper- and hypoglycaemia are also key players in the subfertility and/or infertility associated to DM [58, 119]. One of the key features of DM is the endocrine system dysregulation that is reflected in the endocrine control of testicular function [120, 121]. Diabetic individuals have not only altered levels of gonadotropins [122], FSH and luteinizing hormone (LH), [123, 124] but also abnormal sexual steroids feedback [125, 126]. Several reports point towards strong hormonal dysregulation caused by DM, particularly in sex hormones levels [127, 128], which has a direct effect on spermatogenesis, sperm release and sperm metabolism. In fact, besides the ultrastructural alterations in the testis [129] and sexual disorders such as erectile dysfunction [130], retrograde ejaculation, impotence or decreased libido [131], alterations in human sperm motility or morphology are well known in diabetic men [132]. Sperm motility, morphology, count and semen volume are some of the most affected sperm parameters [133-135]. These alterations have been also reported to be associated with hexoses metabolism since semen from diabetic individuals have higher glucose and fructose levels which is a direct evidence of an ineffective metabolic control in semen with drastic consequences in sperm parameters [134]. However, some studies reported an increase in sperm count and sperm concentration in diabetic men, but sperm motility and semen volume were significantly decreased and sperm morphology remained unaffected [136]. Moreover, there is a high level of nuclear and mitochondrial DNA damage in sperm from diabetic men [3]. All these studies, reported in human biopsies or sperm are somewhat limited to reveal the real impact of DM in sperm glucose metabolism and uptake. Therefore, data from DM animal models are very valuable to study those molecular mechanisms. There are numerous studies that not only confirm reduced fertility in males with DM [8, 11, 129, 137-139], but also allow the study of the glycaemic control impact [138, 140]. For instance, in STZ-treated rats the insulin treatment was able to completely restore sperm counts and motility [140] and also some deleterious effects in regions of the epididymis of diabetic males [138].

The sperm from diabetic individuals has not been widely studied in terms of metabolic modulation induced by the disease. This is reasonable since sperm metabolism has not a direct demonstrated effect in natural and assisted reproduction as other endpoints such as DNA integrity and OS. Nevertheless, the disclosure of the glucose metabolism molecular mechanisms in diabetic individual's sperm may open new insights on possible pharmacological approaches. The incubation of spermatozoa in the absence of glucose revealed that there is a progressive loss of spermatozoa velocity that is rapidly restored with glucose addition in a dose-independent manner [141], evidencing that glucose concentration is essential to maintain the male reproductive health. Interestingly, insulin and glucagon failed to affect sperm glucose metabolism and/or sperm motility [141]. Seminiferous tubules and sperm CO_2 production from glucose was found to be decreased in diabetic animals although lactate production was maintained [142] and the presence/absence of insulin had little effect in glucose consumption and lactate production [142]. Nevertheless, diabetic individuals are well known for insulin impairment [143]. Insulin and insulin analogues are usually used by diabetic individuals and although they are crucial for glucose management, it is not clear that they are totally effective and safe [144-146]. Insulin is known to stimulate several SCs functions such as the uptake of free nucleosides, transferrin secretion, DNA and protein synthesis, glycine metabolism and lactate production [18, 65, 147-151]. Leydig cells functions have also been reported to be controlled by the presence or absence of insulin [152]. Noteworthy, insulin can promote the differentiation of spermatogonia into primary spermatocytes through insulin-like growth factor-I (IGF-I) receptor [153] evidencing the need for a tight control of this hormone throughout spermatogenesis. The insulin action is also required after spermatozoa production. Earlier studies reported that insulin stimulates hexoses metabolism [154] and that both, plasma membrane and spermatozoa acrosome, are under hormonal control by insulin [155]. Insulin intratesticular injection also decreases spermatozoa motility and increases the percentage of spermatozoa motility [156]. Washed human spermatozoa from normozoospermic donors treated with insulin increased total and progressive motility, acrosome reaction and also nitric oxide production enhancing the spermatozoa fertilization capacity [157]. The male reproductive health is severely altered by insulin fluctuations. Recently, it was reported that erectile dysfunction is related to IR [158] and that human ejaculated spermatozoa not only secrete insulin but also that this hormone has a crucial role in autocrine glucose metabolism regulation [159]. The imunocytochemical study of insulin expression in human ejaculated spermatozoa reported that the expression pattern of this hormone is very heterogeneous, evidencing that sperm has different energetic status degrees under insulin regulation [159]. Besides, insulin is reported to increase G6PDH activity, the rate limiting enzyme of PPP, *in vivo* and *in vitro* [160] and when this hormone release or intracellular signaling are blocked, there is a decrease in G6PDH activity [159]. Interestingly, the insulin importance goes far beyond the sperm metabolism control and fertilization capacity. The fact that spermatozoa can synthesize insulin, points toward an autocrine regulation of this hormone in sperm, which can be a two-edge sword. If glucose metabolism in sperm can be independent of systemic insulin to sustain the energetic needs, on the other hand sperm may become vulnerable to metabolic diseases such as DM that induce IR and/or insulin insensitivity.

Intimately related with sperm metabolism and DM is OS. DM is related with increased OS as a consequence of ROS overproduction and decreased levels of antioxidant defenses [161]. This is mainly due to glycolysis stimulation by high glucose availability. ROS are very detrimental to the fertility potential of diabetic individuals [58]. Sperm concentration and motility are reported to be sensitive to energy production and free radical concentration [12]. Besides, diabetic men are reported to possess increased sperm DNA damage as a result of the high OS motivated by the hyperglycaemia [162], compromising these individuals fertility potential. ROS production in sperm involves the leakage of electrons from mitochondria. In spermatozoa, mitochondrial complex I and complex III are the major sites of ROS production [163] and spermatozoa are highly vulnerable to ROS because they are constituted by several polyunsaturated fatty acids (PUFA) that are highly vulnerable to ROS attack [164]. OS is also linked to apoptosis [165, 166] which in turn is associated to DM and is proposed to be one of the main mechanisms to explain the subfertility/infertility detected in diabetic individuals [167].

The molecular mechanisms of sperm glucose uptake and metabolism in diabetic conditions are far from being fully disclosed. Moreover, there are important evidences that point toward a crucial role for insulin in these processes. Although from the clinical point of view the study of sperm OS and DNA damage is more appellative due to their direct implication in natural and assisted reproduction, one cannot forget that these phenomena are intimately related to sperm metabolism and glucose uptake and, therefore, the molecular mechanisms of glucose transport, glucose metabolism and hormonal control should be priority to understand the subfertility/infertility known to occur in male diabetics.

Edible and Medicinal Plants as Antidiabetic and Antioxidant Therapy

Traditional herbal medicines used to treat diabetic conditions have aroused considerable interest in recent years [168]. There are a multitude of plants which have insulin mimetic or insulin secretory activity, hypoglycaemic or anti-hyperglycaemic potential, capacity to increase glucose utilization or glucose uptake by cells and combat secondary complications [169, 170]. Conventionally, T1DM is treated with exogenous insulin [171] and T2DM with synthetic oral hypoglycaemic agents such as sulphonylureas and biguanides [170]. Although these drugs are effective in reducing glycaemia, many of them fail as a curative agent for diabetic complications and have a number of serious adverse effects such as weight gain, hypoglycaemia, edema and gastrointestinal disturbances that can discourage patient compliance [24]. Natural compounds are considered to be less toxic and relatively cheaper than synthetic ones and large amounts can be consumed in everyday diet [1, 172]. Therefore, the search for more effective and safer hypoglycaemic agents has become one important area of investigation [173]. There are a large number of plants that have been reported in literature for their antioxidant and antidiabetic effects (Table 1) [1, 170, 174, 175].

As discussed above, GLUTs, which are often stored in vesicles located in the cytoplasm, facilitate glucose transport in and out of the cells [176]. Upon excitation, specific GLUTs are translocated to the plasma membrane and perform the required transport function [176, 177]. In the case of DM the translocation of those GLUTs to the plasma membrane does not occur or occurs deficiently. A group of medicinal plants has been described to help in the effective upregulation or translocation of GLUTs to the plasma membrane, mimicking the action of insulin, and as a result, glucose is more effectively transported into the cells and its blood concentration decreases. For instance, flavonoids present in *Cephalotaxus sinensis* facilitate the translocation of GLUT4, significantly lowering blood glucose levels in STZ-induced diabetic rats [178]. Similarly, seed extracts of *Trigonella foenum-graecum* (L.) facilitate the control of glucose homeostasis in STZ-induced diabetic rats, and this effect is compared to that of insulin administration. *T. foenum-graecum* induces a rapid and dose dependent stimulatory effect on glucose consumption, by cellular activation of mechanisms that lead to GLUT4 translocation to the cell surface, suggesting that the extracts contain factors that can act independently of insulin to increase glucose uptake mediated by GLUT4 [179]. Similarly, administration of *Aloe vera* extract lowers blood glucose and also lowered total cholesterol in STZ-induced diabetic rats. The lyophilized aqueous Aloe extract upregulated the GLUT4 mRNA synthesis in mouse embryonic NIH/3T3 cells [180]. Resveratrol also mediate GLUT4 translocation in the STZ-induced diabetic myocardium by triggering some of the intracellular insulin signaling components [181]. Studies with *Aegles marmelos* (L.) and *Syzygium cumini* extracts, using L6 myotubes as an *in vitro* model, showed an elevation of GLUT4, associated with an upregulation of glucose uptake. Extracts of those plants were found to be significantly more active when compared with insulin [182]. Similar results were reported when using extracts of *Pterocarpus marsupium* (L.), isolated *P. marsupium* isoflavone and extracts of *Momordica charantia* [183]. In another study, alloxan-induced diabetic rats that were administered with *Toona sinensis* leaf extracts showed lower levels of plasma glucose, improvement in plasma insulin levels and a significant increase in both GLUT4 mRNA and protein levels in brown and white adipose tissues. Moreover, the expression of GLUT4 mRNA in red and white muscles were not significantly altered in diabetic rats administered with *Toona sinensis* leaf extracts, suggesting that this plant possesses an hypoglycaemic effect through an increment of insulin that mediates a specific adipose GLUT4 glucose uptake mechanism [184].

Additionally, much attention has been focused on OS, which has been proposed to play a major role in the pathogenesis of both types of DM [185]. Free radicals, commonly known as ROS, are continuously produced in the body as a result of normal metabolic processes and interaction with environmental stimuli. OS arises when oxidant production exceeds antioxidant activity in cells and plasma [186]. Chronic exposure to ROS and the simultaneous decline of antioxidant defense mechanisms can lead to damage of DNA, cellular organelles and enzymes, increased lipid peroxidation, and development of IR [187-189]. These consequences can promote the development of diabetic complications including diabetic retinopathy, nephropathy, peripheral neuropathy, and cardiovascular disease [190, 191].

Antioxidants that scavenge ROS may be of great value in preventing the onset and/or the progression of these oxidative-induced diseases [192]. Polyphenols are well known as antioxidants but were also reported to regulate glucose transport across the intestine by modulating sodium-glucose co-transporter-1 (SGLT-1), altering postprandial hyperglycaemia to normalcy [193].

Polyphenols present in the methanolic extracts of tea (*Camellia sinensis* (L.)), like (+)-catechin and (-)-epigallocatechin 3-gallate, have been reported to have anti-hyperglycemic activity, by enhancing insulin activity and possibly by preventing damage to β-cells [194] and a inhibitory action on SGLT-1 mediated glucose transport [195, 196].

Table 1. List of edible and medicinal plants with potential antidiabetic and antioxidant activities

Family	Plant botanical name	Mechanism of action	References
Alangiaceae	*Alangium salvifolium*	Reduction in serum glucose levels; Anti-hyperlipidemic effect	[198]
Alliaceae	*Allium sativum*	Insulin-like activity	[199]
Annonaceae	*Annona squamosa*	Increase in plasma insulin activity; Stimulation of antioxidant enzymes	[186]
	Annona muricata	Reduction of oxidative stress of pancreatic β-cells	[174]
Araliaceae	*Panax ginseng*	Reduction in serum glucose levels; Stimulation of insulin secretion; Increase in liver glycogen level	[200]
Cephalotaxaceae	*Cephalotaxus sinensis*	Anti-hyperglycemic effect	[178]
Cucurbitaceae	*Momordica charantia*	Reduction in serum glucose levels; Increase in plasma insulin levels	[170]
		Improvement of glucose tolerance	[183, 201]
Cupressaceae	*Juniperus communis*	Reduction in serum glucose levels	[202]
Fabaceae	*Pterocarpus marsupium*	Reduction in serum glucose levels	[173]
		Protective and restorative effects of pancreatic β-cells	[183, 203]
	Cajanus cajan	Reduction in serum glucose levels	[204]
	Trigonella foenum-graecum	Improvement on glucose homeostasis control	[179]
		Stimulation of insulin secretion	[205]
Gentianaceae	*Swertia chirayita*	Reduction in serum glucose levels; Increase in plasma insulin levels	[206]
Leguminosae	*Acacia Arabica*	Stimulation of insulin secretion	[214]
Liliaceae	*Aloe vera*	Reduction in serum glucose levels; Anti-hypocholesterolemic effect	[180]
	Aloe barbadensis	Stimulation of insulin secretion	[207]
Lycopodiaceae	*Selaginella tamariscina*	Reduction in serum glucose levels; Increase in plasma insulin levels	[208]
Malvaceae	*Hibiscus rosa sinensis*	Stimulation of insulin secretion	[206]
Meliaceae	*Toona Sinensis*	Reduction in serum glucose levels; Increase in plasma insulin levels	[184]
Menispermaceae	*Tinospora crispa*	Reduction in serum glucose levels; Stimulation of insulin secretion	[209]
Myrtaceae	*Eucalyptus globulus*	Reduction in serum glucose levels	[210]
	Syzygium cumini	Increased glucose uptake	[182]
Nyctaginaceae	*Boerhaavia diffusa*	Increase in plasma insulin levels; Improvement in glucose tolerance	[199]

Table 1. (Continued)

Family	Plant botanical name	Mechanism of action	References
Oxalidaceae	*Averrhoa bilimbi*	Reduction in serum glucose levels; Anti-hyperlipidemic effect	[211]
Palmae	*Acrocomia mexicana*	Reduction in serum glucose levels	[212]
Rosaceae	*Agrimonia eupatoria*	Stimulation of insulin secretion; Insulin-like activity	[213]
Rutaceae	*Aegle marmelos*	Reduction in serum glucose levels	[214]
		Increased glucose uptake	[182]
Solanaceae	*Capsicum frutescens*	Increased insulin secretion	[215]
Sterculiaceae	*Helicteres isora*	Anti-hyperlipidemic effect; Insulin sensitizing activity	[199]
Theaceae	*Camellia sinensis*	Reduction in serum glucose levels; Increase in plasma insulin levels; Prevention of oxidative damages	[194-196, 199]

Although the exact mechanisms by which tea polyphenols ameliorate diabetes-related dysfunctions are not clear at the moment, all the studies suggest that the high phenolic content of tea leaves have not only a lowering effect on OS but also an anti-hyperglycaemic potential, by decreasing insulin resistance and improving insulin sensitivity [197].

Nevertheless, until now there are no reports about the effect of medicinal plants on DM-induced alterations in glucose uptake and metabolism of spermatozoa. More investigations must be carried out to evaluate the exact mechanism of medicinal plants action with antidiabetic and insulin mimicking activity.

Despite plants are mostly believed to be safe, there are many botanical materials that are not safe for the human being and, therefore, toxicity studies of these plants must be performed before consumption of these plant materials in a human setting [170]. This is an area of research that should deserve special merit from researchers in the next years.

Conclusion

The increasing incidence of DM is closely related with failing birth rates and low fertility. This is alarming since DM is an epidemic disease that is expected to grow in the next decades. The several comorbidities associated with DM highly increase the complexity of this disease. Concerning the subfertility and/or infertility associated with DM, it is clear that not all diabetic men are infertile. Nonetheless, the alterations in glucose uptake and metabolism in testicular cells and sperm induces several changes that are reflected in the male reproductive health and potential. The glucose transport through GLUTs, glucose metabolization, OS, nuclear and mitochondrial DNA fragmentation as well as apoptosis are some of the mechanisms that are severely altered by DM. Although some of the major problems related to DM-induced subfertility and/or infertility are identified, the molecular bases governing such disruptions remain undisclosed. Unveiling of the insulin actions as well as the alterations induced by the hormonal fluctuations that occur in diabetic individuals are priorities to point towards ways to counteract the unwanted effects of DM in male reproductive system. Moreover, in several cases the diabetic patients use several drugs and the real effect of a polypharmacy regimen to the male reproductive health has been completely neglected. Finally, the antioxidant and antidiabetic activity of several edible and medicinal plants is an emerging field of research for those studying DM. Their potential to avoid the deleterious effects of DM in male fertility potential should deserve special attention in the forthcoming years.

References

[1] Coman C, Rugină OD, Socaciu C (2012) Plants and Natural Compounds with Antidiabetic Action. *Notulae Botanicae Horti Agrobotanici Cluj-Napoca* 40(1): 314-325.

[2] Al-Attar AM Zari TA (2010) Influences of crude extract of tea leaves, *Camellia sinensis*, on streptozotocin diabetic male albino mice. Saudi *Journal of Biological Sciences* 17(4): 295-301.

[3] Agbaje IM, Rogers DA, Mcvicar CM, Mcclure N, Atkinson AB, et al. (2007) Insulin dependant diabetes mellitus: implications for male reproductive function. *Human Reproduction* 22(7): 1871-7.

[4] Bener A, Al-Ansari AA, Zirie M, Al-Hamaq AO (2009) Is male fertility associated with type 2 diabetes mellitus? *International Urology and Nephrology* 41(4): 777-84.

[5] Wild S, Roglic G, Green A, Sicree R, King H (2004) Global prevalence of diabetes estimates for the year 2000 and projections for 2030. Diabetes Care 27(5): 1047-1053.

[6] *Diagnosis and Classification of Diabetes Mellitus*, in *Diabetes Care* 2013, American Diabetes Association. p. S67.

[7] Ballester J, Munoz MC, Dominguez J, Rigau T, Guinovart JJ, et al. (2004) Insulin-dependent diabetes affects testicular function by FSH-and LH-linked mechanisms. *Journal of andrology* 25(5): 706.

[8] Frenkel G, Homonnai Z, Drasnin N, Sofer A, Kaplan R, et al. (1978) Fertility of the Streptozotocin-Diabetic Male Rat. *Andrologia* 10(2): 127-136.

[9] Murray FT, Cameron DF, Orth JM (1983) Gonadal dysfunction in the spontaneously diabetic BB rat. *Metabolism* 32(7): 141-147.

[10] Cameron D, Rountree J, Schultz R, Repetta D, Murray F (1990) Sustained hyperglycemia results in testicular dysfunction and reduced fertility potential in BBWOR diabetic rats. *American Journal of Physiology-Endocrinology And Metabolism* 259(6): E881-E889.

[11] Scarano W, Messias A, Oliva S, Klinefelter G, Kempinas W (2006) Sexual behaviour, sperm quantity and quality after short-term streptozotocin-induced hyperglycaemia in rats. *International journal of andrology* 29(4): 482-488.

[12] Amaral S, Moreno AJ, Santos MS, Seiça R, Ramalho-Santos J (2006) Effects of hyperglycemia on sperm and testicular cells of Goto-Kakizaki and streptozotocin-treated rat models for diabetes. *Theriogenology* 66(9): 2056-2067.

[13] Bogan JS (2012) Regulation of glucose transporter translocation in health and diabetes. *Annu. Rev. Biochem.* 81: 507-532.

[14] Mancine R, Penhos J, Izquierdo I, Heinrich J. *Effects of acute hypoglycemia on rat testis*. in *Proceedings of the Society for Experimental Biology and Medicine. Society for Experimental Biology and Medicine (New York, NY)*. 1960. Royal Society of Medicine.

[15] Zysk J, Bushway A, Whistler R, Carlton W (1975) Temporary sterility produced in male mice by 5-thio-D-glucose. *Journal of reproduction and fertility* 45(1): 69-72.

[16] Frenkel G, Peterson R, Freund M. *Changes in the metabolism of guinea pig sperm from different segments of the epididymis*. in *Proceedings of the Society for Experimental Biology and Medicine. Society for Experimental Biology and Medicine (New York, NY)*. 1973. Royal Society of Medicine.

[17] Rato L, Alves MG, Socorro S, Carvalho RA, Cavaco JE, et al. (2012) Metabolic modulation induced by oestradiol and DHT in immature rat Sertoli cells cultured in vitro. *Bioscience Reports* 32(1): 61-9.

[18] Oliveira PF, Alves MG, Rato L, Laurentino S, Silva J, et al. (2012) Effect of insulin deprivation on metabolism and metabolism-associated gene transcript levels of in vitro cultured human Sertoli cells. *Biochim. Biophys. Acta* 1820(2): 84-9.

[19] Oliveira PF, Alves MG, Rato L, Silva J, Sa R, et al. (2011) Influence of 5alpha-dihydrotestosterone and 17beta-estradiol on human Sertoli cells metabolism. *International journal of andrology* 34(6 Pt 2): e612-20.

[20] Boussouar F Benahmed M (2004) Lactate and energy metabolism in male germ cells. *Trends in Endocrinology and Metabolism* 15(7): 345-350.

[21] Lin C-Y, Hung P-H, Vandevoort CA, Miller MG (2009) 1H NMR to investigate metabolism and energy supply in rhesus macaque sperm. *Reprod. Toxicol.* 28(1): 75-80.

[22] Klip A, Tsakiridis T, Marette A, Ortiz PA (1994) Regulation of expression of glucose transporters by glucose: a review of studies in vivo and in cell cultures. *The FASEB journal* 8(1): 43-53.

[23] Handberg A, Vaag A, Damsbo P, Beck-Nielsen H, Vinten J (1990) Expression of insulin regulatable glucose transporters in skeletal muscle from type 2 (non-insulin-dependent) diabetic patients. *Diabetologia* 33(10): 625-627.

[24] Vasconcelos C, Maranhão H, Batista T, Carneiro E, Ferreira F, et al. (2011) Hypoglycaemic activity and molecular mechanisms of *Caesalpinia ferrea* Martius bark extract on streptozotocin-induced diabetes in Wistar rats. *J. Ethnopharmacol.* 137(3): 1533-1541.

[25] Hays NP, Galassetti PR, Coker RH (2008) Prevention and treatment of type 2 diabetes: current role of lifestyle, natural product, and pharmacological interventions. *Pharmacol. Ther.* 118(2): 181-191.

[26] Ali SS, Kasoju N, Luthra A, Singh A, Sharanabasava H, et al. (2008) Indian medicinal herbs as sources of antioxidants. *Food Res. Int.* 41(1): 1-15.

[27] Larson RA (1988) The antioxidants of higher plants. *Phytochemistry* 27(4): 969-978.

[28] Ugochukwu N, Babady N, Cobourne M, Gasset S (2003) The effect ofGongronema latifolium extracts on serum lipid profile and oxidative stress in hepatocytes of diabetic rats. *J. Biosci.* (Bangalore) 28(1): 1-5.

[29] Kesavulu M, Giri R, Rao BK, Apparao C (2008) Lipid peroxidation and antioxidant enzyme levels in type 2 diabetics with microvascular complications.

[30] Baliga V Sapsford R (2009) Diabetes mellitus and heart failure—an overview of epidemiology and management. *Diabetes and Vascular Disease Research* 6(3): 164-171.

[31] Hamden K, Jaouadi B, Carreau S, Aouidet A, Elfeki A (2011) Therapeutic effects of soy isoflavones on α-amylase activity, insulin deficiency, liver–kidney function and metabolic disorders in diabetic rats. *Nat. Prod. Res.* 25(3): 244-255.

[32] Burul-Bozkurt N, Pekiner C, Kelicen P (2010) Diabetes alters aromatase enzyme levels in gonadal tissues of rats. *Naunyn Schmiedebergs Arch. Pharmacol* 382(1): 33-41.

[33] Grieco FA, Sebastiani G, Spagnuolo I, Patti A, Dotta F (2012) Immunology in the clinic review series; focus on type 1 diabetes and viruses: how viral infections modulate beta cell function. *Clin. Exp. Immunol.* 168(1): 24-9.

[34] Heltianu C, Guja C, Manea SA, *Genetic Determinants of Microvascular Complications in Type 1 Diabetes*, ed. D. Wagner 2011: *Tech. Open Access Publisher.*

[35] Defronzo RA, Bonadonna RC, Ferrannini E (1992) Pathogenesis of NIDDM: a balanced overview. *Diabetes Care* 15(3): 318-368.

[36] Berg JM, Tymoczko JL, Stryer L, *Biochemistry*. 5th ed2002: New York: WH Freeman.

[37] Edelstein SL, Knowler WC, Bain RP, Reubin A, Barrett-Connor EL, et al. (1997) Predictors of progression from impaired glucose tolerance to NIDDM: an analysis of six prospective studies. *Diabetes* 46(4): 701.

[38] Tuomilehto J, Lindström J, Eriksson JG, Valle TT, Hämäläinen H, et al. (2001) Prevention of type 2 diabetes mellitus by changes in lifestyle among subjects with impaired glucose tolerance. *New Engl. J. Med.* 344(18): 1343-1350.

[39] Aroda VR Ratner R (2008) Approach to the patient with prediabetes. *J. Clin. Endocrinol. Metab.* 93(9): 3259-65.

[40] Hamman RF (2009) Genetic and environmental determinants of non-insulin-dependent diabetes mellitus (NIDDM). *Diabetes/metabolism reviews* 8(4): 287-338.

[41] Carneiro FS, Giachini FR, Carneiro ZN, Lima VV, Ergul A, et al. (2010) Erectile dysfunction in young non-obese type II diabetic Goto-Kakizaki rats is associated with decreased eNOS phosphorylation at Ser1177. *J. Sex Med.* 7(11): 3620-34.

[42] Golay A Ybarra J (2005) Link between obesity and type 2 diabetes. *Best Pract Res. Clin. Endocrinol. Metab.* 19(4): 649-63.

[43] Patel D, Kumar R, Prasad S, Sairam K, Hemalatha S (2011) Antidiabetic and in vitro antioxidant potential of Hybanthus enneaspermus (Linn) F. Muell in streptozotocin-induced diabetic rats. *Asian Pac. J. Trop. Biomed.* 1(4): 316-322.

[44] Mallidis C, Agbaje I, O'neill J, Mcclure N (2009) The influence of type 1 diabetes mellitus on spermatogenic gene expression. *Fertil Steril* 92(6): 2085-7.

[45] Pauli EM, Legro RS, Demers LM, Kunselman AR, Dodson WC, et al. (2008) Diminished paternity and gonadal function with increasing obesity in men. *Fertil Steril* 90(2): 346-51.

[46] Kasturi SS, Tannir J, Brannigan RE (2008) The metabolic syndrome and male infertility. *J. Androl.* 29(3): 251-9.

[47] Zegers-Hochschild F, Adamson GD, De Mouzon J, Ishihara O, Mansour R, et al. (2009) International Committee for Monitoring Assisted Reproductive Technology (ICMART) and the World Health Organization (WHO) revised glossary of ART terminology, 2009. *Fertil Steril* 92(5): 1520-4.

[48] Dube E, Hermo L, Chan PT, Cyr DG (2008) Alterations in gene expression in the caput epididymides of nonobstructive azoospermic men. *Biol. Reprod.* 78(2): 342-51.

[49] Shukla KK, Mahdi AA, Rajender S (2012) Apoptosis, spermatogenesis and male infertility. *Front Biosci.* (Elite Ed) 4: 746-54.

[50] Du Plessis SS, Kashou AH, Benjamin DJ, Yadav SP, Agarwal A (2011) Proteomics: a subcellular look at spermatozoa. *Reprod Biol. Endocrinol.* 9:36.

[51] Iammarrone E, Balet R, Lower AM, Gillott C, Grudzinskas JG (2003) Male infertility. *Best Pract. Res. Clin. Obstet. Gynaecol* 17(2): 211-29.

[52] Lavizzo-Mourey R (2007) Childhood obesity: what it means for physicians. *JAMA* 298(8): 920-2.

[53] Harjutsalo V, Sjoberg L, Tuomilehto J (2008) Time trends in the incidence of type 1 diabetes in Finnish children: a cohort study. *Lancet* 371(9626): 1777-82.

[54] Nguyen RH, Wilcox AJ, Skjaerven R, Baird DD (2007) Men's body mass index and infertility. *Hum. Reprod.* 22(9): 2488-93.

[55] Delfino M, Imbrogno N, Elia J, Capogreco F, Mazzilli F (2007) Prevalence of diabetes mellitus in male partners of infertile couples. *Minerva Urol Nefrol* 59(2): 131-5.

[56] Sexton WJ Jarow JP (1997) Effect of diabetes mellitus upon male reproductive function. *Urology* 49(4): 508-13.

[57] La Vignera S, Calogero A, Condorelli R, Lanzafame F, Giammusso B, et al. (2009) Andrological characterization of the patient with diabetes mellitus. *Minerva endocrinologica* 34(1): 1.

[58] Amaral S, Oliveira PJ, Ramalho-Santos J (2008) Diabetes and the impairment of reproductive function: possible role of mitochondria and reactive oxygen species. *Curr. Diabetes Rev.* 4(1): 46-54.

[59] Kort HI, Massey JB, Elsner CW, Mitchell-Leef D, Shapiro DB, et al. (2006) Impact of body mass index values on sperm quantity and quality. *J. Androl.* 27(3): 450-2.

[60] La Vignera S, Condorelli R, Vicari E, D'agata R, Calogero AE (2011) Diabetes Mellitus and Sperm Parameters: A Brief Review. *Journal of andrology.*

[61] Kandeel FR, Koussa VKT, Swerdloff RS (2001) Male sexual function and its disorders: physiology, pathophysiology, clinical investigation, and treatment. *Endocr. Rev.* 22(3): 342-388.

[62] Gómez O, Ballester B, Romero A, Arnal E, Almansa I, et al. (2009) Expression and regulation of insulin and the glucose transporter GLUT8 in the testes of diabetic rats. *Hormone and metabolic research* 41(5): 343.

[63] Khan S, Teerds K, Dorrington J (1992) Growth factor requirements for DNA synthesis by Leydig cells from the immature rat. *Biol. Reprod.* 46(3): 335-341.

[64] Hurtado De Catalfo GE De Gómez Dumm INT (1998) Lipid dismetabolism in Leydig and Sertoli cells isolated from streptozotocin-diabetic rats. *The international journal of biochemistry and cell biology* 30(9): 1001-1010.

[65] Borland K, Mita M, Oppenheimer C, Blinderman L, Massague J, et al. (1984) The actions of insulin-like growth factors I and II on cultured Sertoli cells. *Endocrinology* 114(1): 240-246.

[66] Mita M, Borland K, Price JM, Hall PF (1985) The influence of insulin and insulin-like growth factor-I on hexose transport by Sertoli cells. *Endocrinology* 116(3): 987-992.

[67] Schoeller EL, Schon S, Moley KH (2012) The effects of type 1 diabetes on the hypothalamic, pituitary and testes axis. *Cell Tissue Res.* 1-9.

[68] Kapoor D, Aldred H, Clark S, Channer KS, Jones TH (2007) Clinical and biochemical assessment of hypogonadism in men with type 2 diabetes: correlations with bioavailable testosterone and visceral adiposity. *Diabetes Care* 30(4): 911-7.

[69] Alves MG, Rato L, Carvalho RA, Moreira PI, Socorro S, et al. (2013) Hormonal control of Sertoli cells metabolism regulates spermatogenesis. *Cell Mol. Life Sci.*
[70] Boussouar F Benahmed M (2004) Lactate and energy metabolism in male germ cells. *Trends Endocrinol. Metab.* 15(7): 345-50.
[71] Mita M Hall PF (1982) Metabolism of round spermatids from rats: lactate as the preferred substrate. *Biol. Reprod.* 26(3): 445-55.
[72] Angulo C, Rauch MC, Droppelmann A, Reyes AM, Slebe JC, et al. (1998) Hexose transporter expression and function in mammalian spermatozoa: cellular localization and transport of hexoses and vitamin C. *J. Cell Biochem.* 71(2): 189-203.
[73] Miki K (2007) Energy metabolism and sperm function. *Soc. Reprod. Fertil Suppl* 65: 309-25.
[74] Scott TW, White IG, Annison EF (1962) Glucose and acetate metabolism by ram, bull, dog and fowl spermatozoa. *Biochem. J.* 83(398-404.
[75] Setchell BP, Hinks NT, Voglmayr JK, Scott TW (1967) Amino acids in ram testicular fluid and semen and their metabolism by spermatozoa. *Biochem. J.* 105(3): 1061-5.
[76] Terner C (1960) Oxidation of exogenous substrates by isolated human spermatozoa. *American Journal of Physiology* 198: 48-50.
[77] Eddy EM, Toshimori K, O'brien DA (2003) Fibrous sheath of mammalian spermatozoa. *Microsc Res. Tech.* 61(1): 103-15.
[78] Miki K, Qu W, Goulding EH, Willis WD, Bunch DO, et al. (2004) Glyceraldehyde 3-phosphate dehydrogenase-S, a sperm-specific glycolytic enzyme, is required for sperm motility and male fertility. *Proceedings of the National Academy of Sciences of the United States of America* 101(47): 16501-6.
[79] Ford WC (2006) Glycolysis and sperm motility: does a spoonful of sugar help the flagellum go round? *Hum. Reprod. Update* 12(3): 269-74.
[80] Krisfalusi M, Miki K, Magyar PL, O'brien DA (2006) Multiple glycolytic enzymes are tightly bound to the fibrous sheath of mouse spermatozoa. *Biol. Reprod.* 75(2): 270-8.
[81] Piomboni P, Focarelli R, Stendardi A, Ferramosca A, Zara V (2011) The role of mitochondria in energy production for human sperm motility. *International journal of andrology.*
[82] Hull MG, Glazener CM, Kelly NJ, Conway DI, Foster PA, et al. (1985) Population study of causes, treatment, and outcome of infertility. British *Medical Journal* (Clinical Research Ed.) 291(6510): 1693-7.
[83] Goodson SG, Qiu Y, Sutton KA, Xie G, Jia W, et al. (2012) Metabolic substrates exhibit differential effects on functional parameters of mouse sperm capacitation. *Biol. Reprod.* 87(3): 75.
[84] Cappello AR, Guido C, Santoro A, Santoro M, Capobianco L, et al. (2012) The mitochondrial citrate carrier (CIC) is present and regulates insulin secretion by human male gamete. *Endocrinology* 153(4): 1743-54.
[85] Gomez E, Buckingham DW, Brindle J, Lanzafame F, Irvine DS, et al. (1996) Development of an image analysis system to monitor the retention of residual cytoplasm by human spermatozoa: correlation with biochemical markers of the cytoplasmic space, oxidative stress, and sperm function. *Journal of andrology* 17(3): 276-87.
[86] Bucci D, Rodriguez-Gil JE, Vallorani C, Spinaci M, Galeati G, et al. (2011) GLUTs and mammalian sperm metabolism. *Journal of andrology* 32(4): 348-55.

[87] Flesch FM Gadella BM (2000) Dynamics of the mammalian sperm plasma membrane in the process of fertilization. *Biochimica et Biophysica Acta* 1469(3): 197-235.
[88] Harrison RA Gadella BM (2005) Bicarbonate-induced membrane processing in sperm capacitation. *Theriogenology* 63(2): 342-51.
[89] Urner F Sakkas D (1996) Glucose participates in sperm-oocyte fusion in the mouse. *Biol. Reprod.* 55(4): 917-22.
[90] Joost HG Thorens B (2001) The extended GLUT-family of sugar/polyol transport facilitators: nomenclature, sequence characteristics, and potential function of its novel members (review). *Mol. Membr Biol.* 18(4): 247-56.
[91] Scheepers A, Joost HG, Schurmann A (2004) The glucose transporter families SGLT and GLUT: molecular basis of normal and aberrant function. JPEN. *Journal of Parenteral and Enteral Nutrition* 28(5): 364-71.
[92] Burant CF, Takeda J, Brot-Laroche E, Bell GI, Davidson NO (1992) Fructose transporter in human spermatozoa and small intestine is GLUT5. *J. Biol. Chem.* 267(21): 14523-6.
[93] Haber RS, Weinstein SP, O'boyle E, Morgello S (1993) Tissue distribution of the human GLUT3 glucose transporter. *Endocrinology* 132(6): 2538-43.
[94] Schurmann A, Axer H, Scheepers A, Doege H, Joost HG (2002) The glucose transport facilitator GLUT8 is predominantly associated with the acrosomal region of mature spermatozoa. *Cell Tissue Res.* 307(2): 237-42.
[95] Gomez O, Romero A, Terrado J, Mesonero JE (2006) Differential expression of glucose transporter GLUT8 during mouse spermatogenesis. *Reproduction* 131(1): 63-70.
[96] Kim ST Moley KH (2007) The expression of GLUT8, GLUT9a, and GLUT9b in the mouse testis and sperm. *Reprod. Sci.* 14(5): 445-55.
[97] Wu X Freeze HH (2002) GLUT14, a duplicon of GLUT3, is specifically expressed in testis as alternative splice forms. *Genomics* 80(6): 553-7.
[98] Perchec G, Jeulin C, Cosson J, Andre F, Billard R (1995) Relationship between sperm ATP content and motility of carp spermatozoa. *J. Cell Sci.* 108(2): 747-753.
[99] Williams AC Ford W (2001) The role of glucose in supporting motility and capacitation in human spermatozoa. *Journal of andrology* 22(4): 680.
[100] Calamera J, Brugo S, Vilar O (1982) Relation between motility and adenosinetriphosphate (ATP) in human spermatozoa. *Andrologia* 14(3): 239-241.
[101] Storey BT (2008) Mammalian sperm metabolism: oxygen and sugar, friend and foe. *Int. J. Dev. Biol.* 52(5): 427.
[102] Pasupuleti V, *Role of glycolysis and respiration in sperm metabolism and motility*, 2007, Kent State University.
[103] Mann T (2009) Metabolism of semen. *Advances in Enzymology and Related Areas of Molecular Biology* 9: 329-390.
[104] Mann T (1946) Studies on the metabolism of semen: 3. Fructose as a normal constituent of seminal plasma. Site of formation and function of fructose in semen. *Biochem. J.* 40(4): 481.
[105] Voglmayr J, Larsen L, White I (1970) Metabolism of spermatozoa and composition of fluid collected from the rete testis of living bulls. *Journal of reproduction and fertility* 21(3): 449-460.

[106] Scrutton MC Utter MF (1968) The regulation of glycolysis and gluconeogenesis in animal tissues. *Annu. Rev. Biochem.* 37(1): 249-302.
[107] Berg JM, Tymoczko JL, Stryer L (2002) Glycolysis and gluconeogenesis.
[108] Douard V Ferraris RP (2008) Regulation of the fructose transporter GLUT5 in health and disease. *American Journal of Physiology-Endocrinology And Metabolism* 295(2): E227-E237.
[109] Hoskins DD, Stephens D, Casillas E (1971) Enzymic control of fructolysis in primate spermatozoa. *Biochimica et Biophysica Acta (BBA)-General Subjects* 237(2): 227-238.
[110] Hartree E Mann T (1961) Phospholipids in ram semen: metabolism of plasmalogen and fatty acids. *Biochem. J.* 80(3): 464.
[111] Misro M Ramya T (2012) Fuel/Energy Sources of Spermatozoa. *Male Infertility* 209-223.
[112] Hartree E Mann T (1959) Plasmalogen in ram semen, and its role in sperm metabolism. *Biochem. J.* 71(3): 423.
[113] Mohri H Masaki J (1967) Glycerokinase and its possible role in glycerol metabolism of bull spermatozoa. *Journal of reproduction and fertility* 14(2): 179-194.
[114] Cabrita E, Ma S, Diogo P, Martínez-Páramo S, Sarasquete C, et al. (2011) The influence of certain aminoacids and vitamins on post-thaw fish sperm motility, viability and DNA fragmentation. *Anim. Reprod. Sci.* 125(1): 189-195.
[115] Urner F Sakkas D (1999) A possible role for the pentose phosphate pathway of spermatozoa in gamete fusion in the mouse. *Biol. Reprod.* 60(3): 733-9.
[116] Urner F Sakkas D (1999) Characterization of glycolysis and pentose phosphate pathway activity during sperm entry into the mouse oocyte. *Biol. Reprod.* 60(4): 973-8.
[117] Travis AJ, Jorgez CJ, Merdiushev T, Jones BH, Dess DM, et al. (2001) Functional relationships between capacitation-dependent cell signaling and compartmentalized metabolic pathways in murine spermatozoa. *J. Biol. Chem.* 276(10): 7630-6.
[118] Travis AJ, Tutuncu L, Jorgez CJ, Ord TS, Jones BH, et al. (2004) Requirements for glucose beyond sperm capacitation during in vitro fertilization in the mouse. *Biol. Reprod.* 71(1): 139-45.
[119] Alves MG, Oliveira PF, Socorro S, Moreira PI (2012) Impact of diabetes in blood-testis and blood-brain barriers: resemblances and differences. *Curr. Diabetes Rev.* 8(6): 401-12.
[120] Ballester J, Munoz MC, Dominguez J, Rigau T, Guinovart JJ, et al. (2004) Insulin-dependent diabetes affects testicular function by FSH- and LH-linked mechanisms. *J. Androl.* 25(5): 706-19.
[121] Chiodini I, Di Lembo S, Morelli V, Epaminonda P, Coletti F, et al. (2006) Hypothalamic-pituitary-adrenal activity in type 2 diabetes mellitus: role of autonomic imbalance. *Metabolism: Clinical and Experimental* 55(8): 1135-40.
[122] Bestetti G, Locatelli V, Tirone F, Rossi GL, Muller EE (1985) One month of streptozotocin-diabetes induces different neuroendocrine and morphological alterations in the hypothalamo-pituitary axis of male and female rats. *Endocrinology* 117(1): 208-16.
[123] Wright AD, London DR, Holder G, Williams JW, Rudd BT (1976) Luteinizing release hormone tests in impotent diabetic males. *Diabetes* 25(10): 975-7.

[124] Distiller LA, Sagel J, Morley JE, Seftel HC (1975) Pituitary responsiveness to luteinizing hormone-releasing hormone in insulin-dependent diabetes mellitus. *Diabetes* 24(4): 378-80.

[125] Baccetti B, La Marca A, Piomboni P, Capitani S, Bruni E, et al. (2002) Insulin-dependent diabetes in men is associated with hypothalamo-pituitary derangement and with impairment in semen quality. *Human Reproduction* 17(10): 2673-7.

[126] Dong Q, Lazarus RM, Wong LS, Vellios M, Handelsman DJ (1991) Pulsatile LH secretion in streptozotocin-induced diabetes in the rat. *J. Endocrinol.* 131(1): 49-55.

[127] Stanworth RD, Kapoor D, Channer KS, Jones TH (2011) Dyslipidaemia is associated with testosterone, oestradiol and androgen receptor CAG repeat polymorphism in men with type 2 diabetes. *Clinical Endocrinology* 74(5): 624-30.

[128] Maric C, Forsblom C, Thorn L, Waden J, Groop PH (2010) Association between testosterone, estradiol and sex hormone binding globulin levels in men with type 1 diabetes with nephropathy. *Steroids* 75(11): 772-8.

[129] Cameron DF, Murray FT, Drylie DD (1985) Interstitial compartment pathology and spermatogenic disruption in testes from impotent diabetic men. *Anatomical Record* 213(1): 53-62.

[130] De Young L, Yu D, Bateman RM, Brock GB (2004) Oxidative stress and antioxidant therapy: their impact in diabetes-associated erectile dysfunction. *Journal of andrology* 25(5): 830-6.

[131] Schoeffling K, Federlin K, Ditschuneit H, Pfeiffer EF (1963) Disorders of Sexual Function in Male Diabetics. *Diabetes* 12: 519-27.

[132] Bartak V, Josifko M, Horackova M (1975) Juvenile diabetes and human sperm quality. *International Journal of Fertility* 20(1): 30-2.

[133] Bartak V (1979) Sperm quality in adult diabetic men. *International Journal of Fertility* 24(4): 226-32.

[134] Padron RS, Dambay A, Suarez R, Mas J (1984) Semen analyses in adolescent diabetic patients. *Acta Diabetologica Latina* 21(2): 115-21.

[135] Ranganathan P, Mahran AM, Hallak J, Agarwal A (2002) Sperm cryopreservation for men with nonmalignant, systemic diseases: a descriptive study. *Journal of andrology* 23(1): 71-5.

[136] Ali ST, Shaikh RN, Siddiqi NA, Siddiqi PQ (1993) Semen analysis in insulin-dependent/non-insulin-dependent diabetic men with/without neuropathy. *Archives of Andrology* 30(1): 47-54.

[137] Hassan AA, Hassouna MM, Taketo T, Gagnon C, Elhilali MM (1993) The effect of diabetes on sexual behavior and reproductive tract function in male rats. *Journal of Urology* 149(1): 148-54.

[138] Soudamani S, Malini T, Balasubramanian K (2005) Effects of streptozotocin-diabetes and insulin replacement on the epididymis of prepubertal rats: histological and histomorphometric studies. *Endocrine Research* 31(2): 81-98.

[139] Kim ST Moley KH (2008) Paternal effect on embryo quality in diabetic mice is related to poor sperm quality and associated with decreased glucose transporter expression. *Reproduction* 136(3): 313-22.

[140] Seethalakshmi L, Menon M, Diamond D (1987) The effect of streptozotocin-induced diabetes on the neuroendocrine-male reproductive tract axis of the adult rat. *Journal of Urology* 138(1): 190-4.

[141] Gorus FK Pipeleers DG (1986) Glucose metabolism in human spermatozoa: lack of insulin effects and dissociation from alloxan handling. *Journal of Cellular Physiology* 127(2): 261-6.
[142] Ford WC Hamilton DW (1984) The effect of experimentally induced diabetes on the metabolism of glucose by seminiferous tubules and epididymal spermatozoa from the rat. *Endocrinology* 115(2): 716-22.
[143] *Diagnosis and Classification of Diabetes Mellitus*, in *Diabetes Care*, A.D. Association, Editor 2012. p. S64-71.
[144] Cryer PE (2008) The barrier of hypoglycemia in diabetes. *Diabetes* 57(12): 3169-76.
[145] Cardoso S, Carvalho C, Santos R, Correia S, Santos MS, et al. (2011) Impact of STZ-induced hyperglycemia and insulin-induced hypoglycemia in plasma amino acids and cortical synaptosomal neurotransmitters. *Synapse* 65(6): 457-66.
[146] Cardoso S, Santos MS, Seica R, Moreira PI (2010) Cortical and hippocampal mitochondria bioenergetics and oxidative status during hyperglycemia and/or insulin-induced hypoglycemia. *Biochim Biophys Acta* 1802(11): 942-51.
[147] Griswold MD Merryweather J (1982) Insulin stimulates the incorporation of 32Pi into ribonucleic acid in cultured sertoli cells. *Endocrinology* 111(2): 661-7.
[148] Skinner MK Griswold MD (1982) Secretion of testicular transferrin by cultured Sertoli cells is regulated by hormones and retinoids. *Biol. Reprod.* 27(1): 211-21.
[149] Oonk RB, Grootegoed JA, Van Der Molen HJ (1985) Comparison of the effects of insulin and follitropin on glucose metabolism by Sertoli cells from immature rats. *Mol. Cell Endocrinol.* 42(1): 39-48.
[150] Guma FC, Wagner M, Martini LH, Bernard EA (1997) Effect of FSH and insulin on lipogenesis in cultures of Sertoli cells from immature rats. *Braz J. Med. Biol. Res.* 30(5): 591-7.
[151] Alves MG, Socorro S, Silva J, Barros A, Sousa M, et al. (2012) In vitro cultured human Sertoli cells secrete high amounts of acetate that is stimulated by 17beta-estradiol and suppressed by insulin deprivation. Biochimica et Biophysica Acta (BBA) - *Molecular Cell Research* 1823(8): 1389-1394.
[152] Perrard-Sapori MH, Chatelain PC, Rogemond N, Saez JM (1987) Modulation of Leydig cell functions by culture with Sertoli cells or with Sertoli cell-conditioned medium: effect of insulin, somatomedin-C and FSH. *Mol. Cell Endocrinol.* 50(3): 193-201.
[153] Nakayama Y, Yamamoto T, Abe SI (1999) IGF-I, IGF-II and insulin promote differentiation of spermatogonia to primary spermatocytes in organ culture of newt testes. *Int. J. Dev. Biol.* 43(4): 343-7.
[154] Hicks JJ, Rojas L, Rosado A (1973) Insulin regulation of spermatozoa metabolism. *Endocrinology* 92(3): 833-9.
[155] Silvestroni L, Modesti A, Sartori C (1992) Insulin-sperm interaction: effects on plasma membrane and binding to acrosome. *Archives of Andrology* 28(3): 201-11.
[156] Sliwa L (1994) Effects of selected hormones on the motility of spermatozoa in the mouse vas deferens. *Archives of Andrology* 33(3): 145-9.
[157] Lampiao F Du Plessis SS (2008) Insulin and leptin enhance human sperm motility, acrosome reaction and nitric oxide production. *Asian J. Androl.* 10(5): 799-807.

[158] Bansal TC, Guay AT, Jacobson J, Woods BO, Nesto RW (2005) Incidence of metabolic syndrome and insulin resistance in a population with organic erectile dysfunction. *J. Sex Med.* 2(1): 96-103.

[159] Aquila S, Gentile M, Middea E, Catalano S, Ando S (2005) Autocrine regulation of insulin secretion in human ejaculated spermatozoa. *Endocrinology* 146(2): 552-7.

[160] Stumpo DJ Kletzien RF (1984) Regulation of glucose-6-phosphate dehydrogenase mRNA by insulin and the glucocorticoids in primary cultures of rat hepatocytes. *Eur. J. Biochem.* 144(3): 497-502.

[161] Tabak O, Gelisgen R, Erman H, Erdenen F, Muderrisoglu C, et al. (2011) Oxidative lipid, protein, and DNA damage as oxidative stress markers in vascular complications of diabetes mellitus. Clinical and Investigative Medicine. *Medecine Clinique et Experimentale* 34(3): E163-71.

[162] Aitken RJ Baker MA (2006) Oxidative stress, sperm survival and fertility control. *Mol. Cell Endocrinol.* 250(1-2): 66-9.

[163] Koppers AJ, De Iuliis GN, Finnie JM, Mclaughlin EA, Aitken RJ (2008) Significance of mitochondrial reactive oxygen species in the generation of oxidative stress in spermatozoa. *J. Clin. Endocrinol. Metab.* 93(8): 3199-207.

[164] Aitken RJ, Jones KT, Robertson SA (2012) Reactive Oxygen Species and Sperm Function--in Sickness and in Health. Journal of andrology. DOI:10.2164/ jandrol.112.016535.

[165] Grunewald S, Said TM, Paasch U, Glander HJ, Agarwal A (2008) Relationship between sperm apoptosis signalling and oocyte penetration capacity. *International journal of andrology* 31(3): 325-30.

[166] Aitken RJ Koppers AJ (2011) Apoptosis and DNA damage in human spermatozoa. *Asian J. Androl.* 13(1): 36-42.

[167] Roessner C, Paasch U, Kratzsch J, Glander HJ, Grunewald S (2012) Sperm apoptosis signalling in diabetic men. *Reprod Biomed Online* 25(3): 292-9.

[168] Ayyanar M, Sankarasivaraman K, Ignacimuthu S (2008) Traditional herbal medicines used for the treatment of diabetes among two major tribal groups in south Tamil Nadu, India. *Ethnobotanical leaflets* 2008(1): 32.

[169] Bhushan MS, Rao C, Ojha S, Vijayakumar M, Verma A (2010) An analytical review of plants for anti diabetic activity with their phytoconstituent and mechanism of action. *International journal of pharmaceutical sciences and research* 1(1): 29-45.

[170] Patel D, Prasad S, Kumar R, Hemalatha S (2012) An overview on antidiabetic medicinal plants having insulin mimetics property. *Asian Pac J Trop Biomed* 2(4): 320-330.

[171] Felig P, Bergman M, Felig C (1995) The endocrine pancreas: diabetes mellitus. Endocrinology and Metabolism, 3: Momin A. *Role of indigenous medicine in primary health care.* in *Proceedings of first international seminar on Unani medicine. New Delhi, India.* 1987.

[172] Saxena A Vikram NK (2004) Role of selected Indian plants in management of type 2 diabetes: a review. *The Journal of Alternative and Complementary Medicine* 10(2): 369-378.

[173] Malviya N, Jain S, Malviya S (2010) Antidiabetic potential of medicinal plants. *Acta Pol. Pharm* 67(2): 113-118.

[174] Bnouham M, Ziyyat A, Mekhfi H, Tahri A, Legssyer A (2006) Medicinal plants with potential antidiabetic activity-A review of ten years of herbal medicine research (1990-2000). *International Journal of Diabetes and Metabolism* 14(1): 1.

[175] Pessin JE, Thurmond DC, Elmendorf JS, Coker KJ, Okada S (1999) Molecular Basis of Insulin-stimulated GLUT4 Vesicle Trafficking LOCATION! LOCATION! LOCATION! *J. Biol. Chem.* 274(5): 2593-2596.

[176] James DE Piper RC (1994) Insulin resistance, diabetes, and the insulin-regulated trafficking of GLUT-4. *The Journal of cell biology* 126(5): 1123-1126.

[177] Li W, Dai R-J, Yu Y-H, Li L, Wu C-M, et al. (2007) Antihyperglycemic Effect of *Cephalotaxus sinensis* Leaves and GLUT-4 Translocation Facilitating Activity of Its Flavonoid Constituents. *Biological and Pharmaceutical Bulletin* 30(6): 1123-1129.

[178] Vijayakumar MV, Singh S, Chhipa RR, Bhat MK (2005) The hypoglycaemic activity of fenugreek seed extract is mediated through the stimulation of an insulin signalling pathway. *Br. J. Pharmacol.* 146(1): 41-48.

[179] Kumar R, Sharma B, Tomar N, Roy P, Gupta A, et al. (2011) In Vivo Evalution of Hypoglycemic Activity of Aloe spp. and Identification of Its Mode of Action on GLUT-4 Gene Expression In Vitro. *Appl. Biochem. Biotechnol.* 164(8): 1246-1256.

[180] Penumathsa SV, Thirunavukkarasu M, Zhan L, Maulik G, Menon V, et al. (2008) Resveratrol enhances GLUT-4 translocation to the caveolar lipid raft fractions through AMPK/Akt/eNOS signalling pathway in diabetic myocardium. *J. Cell Mol. Med.* 12(6a): 2350-2361.

[181] Anandharajan R, Jaiganesh S, Shankernarayanan N, Viswakarma R, Balakrishnan A (2006) In vitro glucose uptake activity of *Aegles marmelos* and *Syzygium cumini* by activation of Glut-4, PI3 kinase and PPARg in L6 myotubes. *Phytomedicine* 13(6): 434-441.

[182] Anandharajan R, Pathmanathan K, Shankernarayanan NP, Vishwakarma RA, Balakrishnan A (2005) Upregulation of Glut-4 and PPARg by an isoflavone from Pterocarpus marsupium on L6 myotubes: a possible mechanism of action. *J. Ethnopharmacol* 97(2): 253-260.

[183] Wang PH, Tsai MJ, Hsu CY, Wang CY, Hsu HK, et al. (2008) *Toona sinensis* Roem (Meliaceae) leaf extract alleviates hyperglycemia via altering adipose glucose transporter 4. *Food Chem. Toxicol.* 46(7): 2554-2560.

[184] Valko M, Leibfritz D, Moncol J, Cronin MTD, Mazur M, et al. (2007) Free radicals and antioxidants in normal physiological functions and human disease. *International Journal of Biochemistry and Cell Biology* 39(1): 44-84.

[185] Kaleem M, Asif M, Ahmed Q, Bano B (2006) Antidiabetic and antioxidant activity of Annona squamosa extract in streptozotocin-induced diabetic rats. *Singapore medical journal* 47(8): 670.

[186] Halliwell B (1997) Antioxidants and human disease: a general introduction. *Nutr. Rev.* 55(1): S44-S49.

[187] Sohal RS Weindruch R (1996) Oxidative stress, caloric restriction, and aging. *Science* (New York, NY) 273(5271): 59.

[188] Maritim AC, Sanders RA, Watkins JB (2003) Diabetes, oxidative stress, and antioxidants: A review. *J. Biochem. Mol. Toxicol.* 17(1): 24-38.

[189] Cunningham JJ (1998) Micronutrients as nutriceutical interventions in diabetes mellitus. *Journal of the American College of Nutrition* 17(1): 7-10.

[190] Henriksen EJ, Diamond-Stanic MK, Marchionne EM (2011) Oxidative stress and the etiology of insulin resistance and type 2 diabetes. *Free Radical Biol. Med.* 51(5): 993-999.

[191] Willett WC (1994) Diet and health: what should we eat. *Science* 264(5158): 532-537.

[192] Hanhineva K, Torronen R, Bondia-Pons I, Pekkinen J, Kolehmainen M, et al. (2010) Impact of dietary polyphenols on carbohydrate metabolism. *International journal of molecular sciences* 11(4): 1365-1402.

[193] Anderson RA Polansky MM (2002) Tea enhances insulin activity. *J. Agric. Food Chem.* 50(24): 7182-7186.

[194] Chen CH, Hsu HJ, Huang YJ, Lin CJ (2007) Interaction of flavonoids and intestinal facilitated glucose transporters. *Planta medica* 73(4): 348.

[195] Johnston K, Sharp P, Clifford M, Morgan L (2005) Dietary polyphenols decrease glucose uptake by human intestinal Caco-2 cells. *FEBS Lett.* 579(7): 1653-1657.

[196] Islam M (2011) Effects of the aqueous extract of white tea (*Camellia sinensis*) in a streptozotocin-induced diabetes model of rats. *Phytomedicine* 19(1): 25-31.

[197] Takale V, Kshirsagar RP, Darade SS (2010) Effect of Alangium salvifolium (Alangiaceae) on dexamethasone induced insulin resistance in rats. *Journal of Pharmacy Research* 3(11):

[198] Ayodhya S, Kusum S, Anjali S (2010) Hypoglycaemic activity of different extracts of various herbal plants. *International Journal* 1(Kim K Kim HY (2008) Korean red ginseng stimulates insulin release from isolated rat pancreatic islets. *J. Ethnopharmacol.* 120(2): 190-195.

[199] Leatherdale B, Panesar R, Singh G, Atkins T, Bailey C, et al. (1981) Improvement in glucose tolerance due to Momordica charantia (karela). *British Medical Journal* (Clinical Research Ed.) 282(6279): 1823-1824.

[200] De Medina FS, Gamez M, Jimenez I, Jimenez J, Osuna J, et al. (2007) Hypoglycemic activity of juniper "berries". *Planta medica* 60(03): 197-200.

[201] Manickam M, Ramanathan M, Jahromi MaF, Chansouria J, Ray A (1997) Antihyperglycemic activity of phenolics from Pterocarpus marsupium. *J. Nat. Prod.* 60(6): 609-610.

[202] Amalraj T Ignacimuthu S (1998) Hypoglycemic activity of Cajanus cajan (seeds) in mice. *Indian J. Exp. Biol.* 36(10): 1032.

[203] Singh LW (2011) Traditional medicinal plants of Manipur as antidiabetics. *J. Med. Plants Res.* 5(5): 677-87.

[204] Grover J, Yadav S, Vats V (2002) Medicinal plants of India with anti-diabetic potential. *J. Ethnopharmacol.* 81(1): 81.

[205] Ajabnoor MA (1990) Effect of aloes on blood glucose levels in normal and alloxan diabetic mice. *J. Ethnopharmacol.* 28(2): 215-220.

[206] Miao N, Tao H, Tong C, Xuan H, Zhamg G (1996) The Selaginella tamariscina (Beauv.) Spring Complex Treatment for Experimental Diabetes and Its Effect on Blood Rheology. *China journal of chinese materia medica* 21: 493-495.

[207] Noor H Ashcroft SJH (1998) Pharmacological characterisation of the antihyperglycaemic properties of *Tinospora crispa* extract. *J. Ethnopharmacol.* 62(1): 7-13.

[208] Swatson-Flatt S, Day C, Bailey C, Flatt P (1990) Traditional plant treatments for diabetes: Studies in normal streptozotocin diabetic mice. *Diabetologia* 33: 462-464.

[209] Pushparaj P, Tan C, Tan B (2000) Effects of Averrhoa bilinbi leaf extract on blood glucose and lipids in Streptozotocin-diabetic rats. *J. Ethnopharmacol.* 72: 69-76.

[210] Pérez G S, Pérez G RM, Pérez G C, Zavala S MA, Vargas S R (1997) Coyolosa, a new hypoglycemic from *Acrocomia mexicana. Pharmaceutica Acta Helvetiae* 72(2): 105-111.

[211] Gray AM Flatt PR (1998) Actions of the traditional anti-diabetic plant, Agrimony eupatoria (agrimony): effects on hyperglycaemia, cellular glucose metabolism and insulin secretion. *British Journal of Nutrition* 80(1): 109-114.

[212] Seema P, Sudha B, Padayatti PS, Abraham A, Raghu K, et al. (1996) Kinetic studies of purified malate dehydrogenase in liver of streptozotocin-diabetic rats and the effect of leaf extract of Aegle marmelose (L) Correa ex Roxb. *Indian J. Exp. Biol.* 34: 600-602.

[213] Islam MS Choi H (2008) Dietary red chilli (Capsicum frutescens L.) is insulinotropic rather than hypoglycemic in type 2 diabetes model of rats. *Phytother Res.* 22(8): 1025-1029.

[214] Singh LW (2011) Traditional medicinal plants of Manipur as anti-diabetics. *J Med Plant Res.* 5(5): 677-687.

In: Glucose Uptake
Editors: Carter C. Johnson and Davis B. Williams
ISBN: 978-1-62618-670-5

Chapter 6

Beneficial Role of Estrogen Signaling in Glucose Homeostasis and Energy Expenditure

Zsuzsanna Suba[*]
National Institute of Oncology,
Budapest, Hungary

Abstract

Glucose is the most important fuel of mammalian cells and its uptake is crucial for cellular metabolism, differentiation, survival and proliferation. The insulin signaling pathway regulates the uptake of glucose and whole-body metabolic homeostasis by transducing extracellular signals through the insulin receptor (IR) to downstream intracellular targets. Dysregulation of insulin secretion or alterations of IR signal transduction pathways is associated with self-generating, progressive insulin resistance in thorough interplay with an imbalance of sexual steroid production. Insulin resistance and the associated metabolic and hormonal alterations predispose patients to a variety of life threatening diseases; such as metabolic syndrome, type-2 diabetes, cardiovascular lesions and malignancies at different sites. Estrogens have beneficial effects on energy metabolism and glucose homeostasis by several pathways. In the central nervous system, hypothalamic nuclei are the pivotal regulators of food intake and energy expenditure by means of their estrogen receptors (ERs). In the pancreatic islet cells, ERs advantageously regulate the biosynthesis and secretion of insulin and maintain the equilibrium of glycogen synthesis and glycogenolysis in the liver by the balanced activation of glycogen synthase and glycolytic enzymes. In the peripheral tissues, ERs advantageously modulate the insulin stimulated glucose uptake through regulation of the phosphorylation of IR protein and increase the expression and translocation of intracellular glucose transporters (GLUTs). Considering the regulatory effects of ERs on food intake, insulin secretion, glucose uptake and metabolic processes, estrogen administration seems to be a

[*] Correspondent author: Prof. Dr. Zsuzsanna Suba. National Institute of Oncology, Surgical and Molecular Tumor Pathology Centre. Address: H-1122 Ráth György str. 7-9, Budapest, Hungary. Tel: 00 36 1 224 86 00; Fax: 0036 1 224 86 20; E-mail: subazdr@gmail.com.

therapeutic avenue to repair insulin sensitivity in patients with dysmetabolism and diabetes mellitus.

Introduction

A famous hypothesis of Reaven reported in 1988 established a causal association between cellular insulin resistance and compensatory hyperinsulinemia, which were than regarded as basic disorders for several human diseases [1]. Earlier, cardiovascular lesions, hypertension, dyslipidemia, obesity and elevated fasting glucose level were attributed to be the complications of type-2 diabetes. Nevertheless, the sequence of these alterations was frequently inconsequent and contradictory. The new theory revealed that many divergent symptoms and findings all have a common soil; the defective glucose uptake of mammalian cells [2].

Actions of female sexual steroids are much wider than just having crucial roles in female physiology and reproduction. Estrogen receptor signals are fairly involved in the beneficial regulation of metabolic processes and in the gene regulation of cell growth, differentiation, proliferation and apoptosis both in men and women [3]. Physiological estrogen levels in healthy premenopausal women supply protection against insulin resistance, hypertension, cardiovascular diseases and malignancies as compared with men of the same age [4, 5]. However, after menopause a decreased ovarian estrogen synthesis will rapidly deteriorate the glucose uptake and increases the prevalence of ischemic stroke, myocardial infarction and other life threatening diseases among women.

New findings on functions of tissular estrogen receptors indicate that estradiol plays important role in the maintenance of glucose homeostasis and energy expenditure [6]. It beneficially modulates the expression of genes that are involved in insulin secretion, cellular insulin sensitivity and glucose uptake at different sites [6, 7]. Nevertheless, there are conflicting data concerning estradiol action in women and its relation to glucose homeostasis and insulin sensitivity in pre- and postmenopausal cases.

Recent studies suppose that a functional imbalance between the signals of estrogen receptor (ER) isoforms and gene polymorphism of ERs may have important implications for the development of dysmetabolism, such as metabolic syndrome and type-2 diabetes [8]. Moreover, in pathologic pregnancy, hyperestrogenism is erroneously supposed to disrupt the glucose uptake in the adipose tissue mass in obese women and is presumed to contribute to insulin resistance and the associated co-morbidities [9, 10].

Results of animal experiments support that estradiol administration improves the metabolic functions in insulin resistance and obesity [11, 12]. Pregnancy analogue elevated estrogen level advantageously improves the metabolic disorders and decreases the risk for associated diseases. In experimental animals, estradiol administration counteracts to the growth of chemically induced or transplanted malignant tumors [13, 14, 15].

In human metabolic diseases the beneficial effects of estradiol administration justify that ERs conduct advantageous intracellular signals if sufficient estrogen supply is available [16]. By contrast, in estrogen deficient states or in case of defective ER signals an increased prevalence of metabolic syndrome, type-2 diabetes and their comorbidities is characteristic [7].

The purpose of this analytical study is to examine the real significance of estradiol supply in cellular metabolic functions. Experimental studies on disabled ER knockout mice help to evaluate the unique and apparently controversial functions of each ER isoform. Nevertheless, understanding the crosstalk and interplay between them may illuminate the fact that there is no good or bad ER isoform, but they may construe a complex system so as to achieve an ideal internal milieu.

Beneficial Impact of Estrogen Signaling in the Clinical Phases of Insulin Resistance

Reactive hyperinsulinemia in the first compensated phase of insulin resistance transitorily maintains serum glucose level within the normal range by means of an increased secretory capacity of pancreatic β-cells [17]. Hyperinsulinemia favors the luteinizing hormone (LH) secretion of pituitary and stimulates androgen biosynthesis in the adrenal gland and ovarian theca cells at the expense of estrogen loss [18, 19]. This phase may be asymptomatic for a long time though the associated dysmetabolism and hyperandrogenism prepare for many alterations at cellular level.

Insulin resistance, hyperinsulinemia and excessive IGF-I activity may decrease estrogen synthesis by counteraction to aromatase enzyme gene expression. The aromatase enzyme complex catalyzes the conversion of androgens to estrogens in a wide variety of tissues [20].

In young women there are different alterations associated with insulin resistance, hyperinsulinism and hyperandrogenism. In premenopausal, highly insulin resistant women, ovaries exhibit increased androgen synthesis at the expense of estrogen loss attributed to a reduction of ovarian aromatase activity [21]. The most frequent endocrine disorder among adolescents and premenopausal cases is polycystic ovarian syndrome (PCOS) with insulin resistance and hyperandrogenism [17]. An artificial hormonal cycle created by oral contraceptive use improves dysmetabolism and decreases the risk of its comorbidities [22, 23]. In postmenopausal women, each year yields decreasing insulin sensitivity with ovarian senescence [24] and increasing male to female sexual hormone ratio [16]. Estradiol substitution increases insulin sensitivity and yields favorable changes in plasma lipid levels [6, 8].

Insulin has diverse metabolic effects and functions as a growth factor too with strong mitogenic capacity. High insulin level in itself is an excessive stimulation of multiple cellular signaling cascades [25]. Hyperinsulinemia increases the hepatic synthesis and mitogenic activity of other, insulin-like growth factors, such as IGF-I. In clinical studies estradiol lowered, whereas testosterone increased total IGF-I level and estradiol specifically suppressed free IGF-I level [26].

Metabolic syndrome (prediabetes) develops in the second, phase of insulin resistance, when the increased insulin synthesis of pancreatic islet cells is not enough to maintain euglycemia. This phase is characterized by a quartet of elevated fasting glucose, high serum triglyceride, low HDL cholesterol level and hypertension, which is frequently associated with visceral obesity [27]. The metabolic syndrome is a high risk for both cardiovascular diseases and malignancies [2, 28].

Estrogen administration counteracts to *hyperglycemia* as it has beneficial effects on the glucose homeostasis. Estrogens advantageously regulate the insulin production capacity of pancreatic islet cells [29, 30]. In the liver, estrogen regulates insulin sensitivity by the balanced activation of hepatic enzymes to maintain the equilibrium of glycogen synthesis and glycogenolysis [31]. In the peripheral tissues estrogen improves the insulin stimulated glucose uptake through regulation of the phosphorylation of insulin receptor protein [32]. Moreover, estrogen has a crucial role in cellular glucose uptake by regulation of glucose transporters (GLUTS) enhancing both GLUT4 expression and translocation [33].

Dyslipidemia has important role in the development of cardiovascular diseases in insulin resistant states [34]. Estrogen, in healthy premenopausal women has crucial role in the maintenance of normal serum lipid levels. By contrast, postmenopausal women have elevated total cholesterol, LDL cholesterol and triglyceride levels and lower HDL cholesterol levels. These changes may be regarded as a shift toward a more atherogenic lipid profile in an estrogen deficient milieu [35].

Estradiol has a cardiovascular protective impact by *its anti-hypertensive activity* as well. Estrogens can downregulate components of the renin-angiotensin system (RAS) and reduce the expression and activity of angiotensin 1-converting enzyme [36, 37]. Estradiol inhibits the excessive synthesis of vasoconstrictor endothelin and improves endothelial dysfunction in ovariectomized, spontaneously hypertensive female rats [38]. In the pathogenesis of postmenopausal hypertension increased androgen to estrogen ratio may be associated with the activation of renin-angiotensin and endothelin systems [39].

Obesity and overweight are important concomitants of insulin resistance and are strongly associated with disturbed equilibrium of male to female sexual hormone concentrations in women [40]. Healthy estrogen predominance in women induces subcutaneous gluteofemoral adipose tissue deposition [41]. Conversely, androgen excess and deficient estrogen synthesis exhibit close correlation with visceral obesity both in pre- and postmenopausal cases [42]. In obese young women, oral contraceptive treatment has advantageous effect against adiposity, whereas in obese postmenopausal cases HRT may reduce abdominal adiposity and insulin resistance [22, 43].

Type-2 diabetes is the uncompensated phase of insulin resistance when the secretory capacity of insular β-cell becomes exhausted and the insufficient serum insulin level results in disturbance of cellular glucose uptake and excessive hyperglycemia. The disrupted glucose homeostasis, the abundant formation of free radicals and the protein glycation depress the activities of antioxidant scavengers and enzymes [44]. These noxious processes cause serious damages in all biological structures even at a molecular level [45, 46].

Type-2 diabetes may develop insidiously, without conspicuous symptoms but the metabolic breakdown may be a sword of Damocles hanging over the patient. Stroke, myocardial infarct and malignant tumors are the most frequent health catastrophes associated with the final phase of insulin resistance.

Estrogens have beneficial effects on energy metabolism and glucose homeostasis [6], as well as on immune surveillance [42]. Clinical and experimental studies support that estradiol also has antioxidant activities and protects against oxidative stress [47, 48]. Estrogen administration may be a new therapeutic avenue for diabetic patients if the misbelief of estrogen excess related diseases could be shattered.

Role of Estradiol and Its Receptors in the Regulation of Cellular Glucose Uptake and Energy Homeostasis

The metabolic state of the body is controlled by a central regulation of the brain through signals arriving from the pancreas, liver, adipose tissue, skeletal muscle, and gut [8]. A wide variety of these signals includes hormones (insulin, leptin, adiponectin, etc.), cytokines (TNF-α, IL-6] and nutrients [glucose, free fatty acids, lipids]. The central nervous system may induce metabolic and behavioral changes by anorexigenic and orexigenic stimuli so as to maintain the serum glucose level and energy homeostasis in the different organs [8].

Role of Estrogen Signaling in the Central Regulation of Metabolism and Energy Homeostasis

The *hypothalamus* is a pivotal regulator of food intake and maintenance of energy homeostasis [49]. Several hypothalamic nuclei express estrogen receptors. Central actions of estrogen on metabolism induce changes in feeding behavior rather than modulation of the metabolism of carbohydrates and lipids. In animal experiments, direct actions of estrogen in the hypothalamus lead to anorexia [50]. In ovariectomized mice, increase in food intake, body weight and abdominal fat mass are characteristic, which are reversible by estradiol administration [51].

Energy homeostasis and feeding behavior follows the menstrual cycle in women conducted by the hypothalamic nuclei. In women, highest estradiol levels during the periovulatory period result in decreased food intake and lower fat accumulation [52].

Leptin is a crucial hormone in the central regulation of metabolism and transfers catabolic signals to the hypothalamic nuclei to inhibit food intake and increase energy expenditure [53, 54]. Leptin is an adipocytokine secreted by adipose tissue in direct proportion to body fat mass. Estrogen modulates leptin synthesis and secretion via ER-dependent transcriptional mechanisms [55]. Increased estrogen level has been associated with raised leptin sensitivity in the brain [56]. Ovariectomy reduces the sensitivity to leptin and this effect can be reversed by estradiol substitution [57].

In rats and mice the ERα-selective agonist, PPT, rapidly results in a decrease in food intake [58], whereas silencing of ERα leads to hyperphagia, obesity, decreased glucose tolerance and reduced energy expenditure [59]. In ovariectomized rodents treatment with an ERα-specific ligand (PPT) resulted in an inhibition of eating and reduced body weight gain.

The exact role of ERβ in the central regulation of feeding is less known [8]. ERβ exhibits the highest expression in the paraventricular nucleus (PVN). In ovariectomized rats, estradiol and ERβ anti-sense oligodeoxynucleotides (ODN) administered into the third ventricle in the brain reduced the inhibitory effect of estrogen on food intake. By contrast, ERα antisense ODN had no effect in this experiment [60].

The hypothalamus has a wide control over insulin, glucocorticoid and gonadal hormone functions, which have important role in defining insulin sensitivity and consequently in peripheral carbohydrate and lipid metabolism [61]. The ratio of male to female sexual hormone levels particularly affects peripheral fat distribution attributed to a modulation of insulin sensitivity. Estrogen deficiency or defective estrogen signaling results in insulin resistance and increase in the adipose tissue mass, preferentially in visceral location [62].

Estrogen Receptor Actions on Pancreatic Islets and Insulin Production

Pancreatic islets are responsible for insulin synthesis and secretion. Through the insulin signaling pathway, cellular transmembrane insulin receptors (IRs) regulate whole-body glucose homeostasis by transducing extracellular signals to downstream intracellular targets [8]. Estradiol and its receptors are key players in the physiology and insulin production capacity of the β cells of pancreatic islets [29].

Estradiol administration is associated with pancreatic islet hypertrophy and increased insulin release from the β cells in rats. Islet cells isolated from ovariectomized mice respond to glucose with a smaller insulin release than islet cells from intact mice. Estradiol replacement in ovariectomized mice normalizes the insulin response to glucose ingestion [6]. Estrogen receptor alpha (ER-α) activation promotes β-cell mass proliferation and insulin biosynthesis in diabetic and obese cases, whereas estrogen receptor beta (ER-β) activation improves glucose stimulated insulin secretion [30].

In postmenopausal women, estrogen loss decreases the insulin sensitivity causing a higher insulin demand. After menopause, reduced insulin secretion is transitorily compensated by the reduced elimination of insulin [63]. During postmenopausal hormone replacement therapy or contraceptive use estradiol improves insulin secretion, sensitivity and elimination.

Estrogen signaling pathways seem to be essential for structural and functional β cell adaptation, especially during high metabolic demand and insulin resistant periods. Recognition of these correlations will lead to novel therapies for β-cell related diseases [8]. Estrogen administration supplies a therapeutic possibility to preserve functional β-cell mass in patients with diabetes mellitus.

Estrogen Receptors and Energy Homeostasis in the Liver

Liver function disturbances are in close correlation with insulin resistance, hyperglycemia and dyslipidemia. Estrogen plays a pivotal role in the regulation of glucose homeostasis and insulin sensitivity of mouse liver [31]. ERα is the predominant receptor isoform in human hepatocytes and ERα signal is essential for the glucose tolerance of the liver [64].

Glucose homeostasis in the liver is not dependent on intracellular glucose transporters (GLUTs). Insulin modulates hepatic carbohydrate metabolism through direct effects on enzymatic activities. Glucose uptake is controlled by the activation of glycogen synthase and glycogen phosphorylase leading to glucose storage as glycogen in the liver; whereas insulin may stimulate glycolysis through the activation of several hepatic enzymes [31].

PPT, a selective ERα agonist, improved glucose tolerance and insulin sensitivity in genetically obese mice suggesting that estradiol has antidiabetogenic impact via ERα [65]. In the liver of ERα knock out (ERαKO) mice, hepatic insulin resistance, increased glucose production and lipid synthesis as well as decreased lipid transport were observed [31]. By contrast ERβ might be diabetogenic as ERβ knock out (ERβKO) mice with increased body weight exhibited improved hepatic and muscular insulin sensitivity due to reduced accumulation of triglycerides [66].

ERα confers estradiol mediated protection of the liver from inflammatory injuries [67]. Ovariectomized mice were treated by IL-18 to induce hepatic inflammation and estradiol hampered the expression of cytokines in ERβKO but not in ERαKO mice. These data suggest that ERα is the chief regulator of defense against experimental hepatitis.

Estradiol replacement in postmenopausal women increased HDL and decreased LDL, total cholesterol, lipoprotein, fasting insulin and glucose levels, exhibiting antidiabetogenic and antiatherogenic impacts [68]. Nevertheless, estrogen deficiency, such as antiestrogen (tamoxifen) therapy or ovariectomy resulted in atherogenic lipid profile and hepatic steatosis increasing the risk of metabolic syndrome and cardiovascular diseases [69].

Estradiol Actions on the Energy Homeostasis of Skeletal Muscles

Skeletal muscle mass is responsible for 75% of the insulin-mediated glucose uptake in the body and consequently, physical activity is in close correlation with insulin sensitivity [8].

Peripheral glucose uptake requires active transport through the double lipid layer of the cell membrane. Insulin receptors have outer α-subunits with binding sites for insulin, whereas the two transmembrane β-subunits are responsible for intracellular signal transduction. Insulin interaction with the external α-subunits induces auto-phosphorylation of the β subunits at multiple tyrosines resulting in an activation of signal transduction [70]. The phosphorylation cascade provokes translocation of glucose transporter (GLUT4) containing cytoplasmic vesicles to the cell membrane. GLUT4 anchored to the cell membrane enables the facilitated diffusion of glucose from the extracellular space into the cell [71]. Alterations in this mechanism, such as insulin deficiency, disturbance in insulin signal transduction, GLUT4 expression and/or translocation may result in insulin resistance.

Estradiol stimulates the phosphorylation of Akt, AMPK and the Akt substrate in soleus muscle [72]. Estradiol administration to insulin resistant rats or mice increases the insulin receptor substrate content and the concentration of the phosphorylated form of Akt in muscles, restoring the action of insulin [12, 73].

Estrogen receptors are important players in the balance of glucose uptake and energy expenditure [8]. They advantageously modulate insulin stimulated glucose uptake through regulation of the tyrosine phosphorylation of insulin receptor protein [32]. ERs participate in GLUT4 expression and translocation as well. Estradiol treatment improves glucose homeostasis through its ability to increase muscle GLUT4 content on the cell membrane [74]. In ovariectomized rats, the decreased amount of GLUT1 protein in the blood-brain barrier was increased after estradiol substitution [75].

In aromatase knockout (ARKO) mice with inactivation of the enzyme for estrogen synthesis, the importance of estradiol for glucose homeostasis has been justified [76]. Both male and female estrogen deficient ARKO mice exhibit reduced glucose oxidation, increased adiposity and hyperinsulinemia [77]. Glucose intolerance and insulin resistance can be reversed by estradiol administration even in male ARKO mice [78].

Skeletal muscle expresses both ERs, and in mice ERβ is the predominant isoform. Knowledge on the role of each receptor has been obtained from receptor knockout mice and from studies with selective estrogen receptor modulators [8]. Treatment with the ERα selective agonist, PPT increased GLUT4 translocation to the cell membrane of L6 myoblasts, and when ERα was silenced a decreased translocation was noted [79]. ERα knockout (ERαKO) mice are glucose intolerant and insulin resistant [80] as the absence of ERα involves a reduced glucose uptake in muscles [31]. ERβ is a repressor of GLUT4 expression and translocation. In ERβKO mice both glucose tolerance and insulin release remains normal or better than in wild type mice [31, 66]. These data exhibit that a steadily balanced activation of both receptors may ensure the ideal glucose tolerance and energy expenditure.

During estrogen loss in the perimenopausal and postmenopausal periods, muscle strength exhibits a striking decline that can be reversed by hormone replacement therapy (HRT), suggesting that estrogens are important modulators of muscle physiology [81, 82].

Pathologic hyperestrogenism is extremely rare in human practice and the results concerning its correlation with insulin sensitivity are thoroughly controversial. Some authors have established that hyperestrogenism is also related to insulin resistance similarly, to estrogen deficiency [83]. In women with irregular menstrual cycles and gestational diabetes, hyperestrogenism was presumed to be a contributor to insulin resistance [9, 10]. Nevertheless, today these pathologic insulin resistant states are attributed to hyperandrogenism and deficient estrogen synthesis rather than to estrogen excess [7, 16].

Treatment with pregnancy equivalent estradiol seems to have antidiabetogenic, antiobesity and anticancer effects in animal experiments [14, 15]. Moreover, high dose estrogen administration as ovulation provocation and good hormonal equilibrium in multiparous women prove to be metabolically advantageous and exhibit anticancer capacities [16].

Taken together, estrogens are beneficial for insulin sensitivity and an estrogen deficient milieu endangers the balanced glucose uptake and energy expenditure of skeletal muscles leading to insulin resistance [7].

Role of Estrogen Receptors in the Deposition and Energy Homeostasis of Adipose Tissue

Increased *adipose tissue in visceral location* has been linked to a self-generating process of insulin resistance. Adipose tissue participates in a variety of metabolic, endocrine and immunologic processes and interacts with CNS and peripheral organs by means of adipokine secretion. The size of the fatty tissue compartment clearly reflects the balance between whole-body energy intake and expenditure [84].

In rats, ovariectomy increased body weight, intra-abdominal fat, fasting glucose, insulin levels and insulin resistance. Estradiol substitution restored normoglycemia, increased the expression of adiponectin and decreased resistin expression resulting in improvement of insulin resistance [85].

In healthy premenopausal women, estradiol counteracts to the development of obesity and insulin resistance [57]. Estradiol prevents the accumulation of visceral fat, and decreases the lipogenic activity of lipoprotein lipase in adipose tissue. By contrast, irregular or long menstrual cycles in young women are associated with insulin resistance and predict occurrence of type-2 diabetes attributed to the defective estrogen synthesis [86, 87]. After menopause, with decreasing estradiol production and deepening insulin resistance there is an increased lipid deposition and decreased lipid utilization resulting in visceral fat mass accumulation [88].

Human subcutaneous and visceral adipose tissues express both ERα and ERβ. The presence of ERα and ERβ isoforms was confirmed in adipocytes from both subcutaneous and intra-abdominal fat, with clear predominance of ERα [89]. Although the separated functions of the two ERs in adipose tissue may be studied on disabled ERαKO and ERβKO mice, these results may hardly be extrapolated to human practice. The only clinical case was a 28-year-old man with defective ERα who presented with glucose intolerance, hyperinsulinemia and obesity [90]. Hyperestrogenism in this patient seemed to be a contra-regulatory effect. Failure of his ERα signal leaded to premature coronary artery disease and decreased levels of total, LDL and HDL cholesterol [91].

In male mice deletion of ERα induces insulin resistance and progressive increase in adipose tissue with advancing age. In female ERαKO mice insulin resistance, increased adiposity, higher leptin and cholesterol levels and smaller LDL particles are characteristic [80].

Estradiol substitution in ovariectomized mice kept on high-fat diet preserved glucose tolerance and insulin sensitivity [12]. Studies on ERα and ERβ knockout mice suggest that ERα is the main regulator of GLUT4 expression in adipose tissue [8] and the two ER isoforms seem to have opposite functions on fat metabolism [92]. In human adipocytes GLUT4 abundance is highly correlated with insulin responsiveness. In PCOS cases with ovarian overproduction of testosterone and defective estrogen synthesis, insulin stimulated glucose uptake was reduced due to decreased amounts of GLUT4 on adipocyte membrane [93].

Presumed Role of Gene Polymorphism in the Alterations of Estrogen Signaling and in the Development of Metabolic Syndrome

Epidemiological and prospective studies suggest controversial associations between estrogen signaling and several aspects of the metabolic syndrome and its comorbidities [8]. Rapid development of new methods for genetic investigations enabled scientists to analyze the polymorphism of the ESR1 gene coding for ERα. Gene polymorphism may induce either enhancing or silencing of the signal transduction of ERs, and these observations apparently yielded possibilities for the rational explanation of puzzling links among estrogen, metabolic disorders and the associated morbidities.

Of the many hypotheses trying to explain the health advantage of women over men, none has received more attention than the estrogen hypothesis [94, 95]. Nevertheless, some of the large studies have indicated a lack of metabolic benefit and an early increase in CVD and breast cancer risk from the use of hormone replacement therapy in postmenopausal cases [96]. As a plausible possibility, genetic variations in estrogen mediated pathways emerged as contributors to interindividual and sex-specific variation in health risk [97]. Gene polymorphisms of ESR1 and their correlations with each component of metabolic syndrome and insulin resistance associated comorbidities were thoroughly investigated but the results proved to be contradictory.

Apparently controversial results were published in relation to ESR1 polymorphism and the prevalence of myocardial infarct. In the Framingham Heart study, male carriers of the common ESR1 IVS1-397T>C C/C genotype, were found to have a substantial increase in the risk of myocardial infarction and serious atherosclerosis [98]. Conversely, in the Rotterdam study, postmenopausal female carriers of the ESR1 haplotype 1 (IVS1-397T>C*T allele and IVS1-351A> G*A allele), were shown to have an increased risk of myocardial infarct and ischemic heart disease, whereas in men an absence of this association was reported [99].

Several other studies also demonstrated conflicting results. The presence of the IVS1-397T>C*C allele was protective against the risk of in-stent restenosis [100]. In postmenopausal women with HRT use, homozygous for the C allele at the IVS1-397T>C site, the severity of atherosclerosis was milder and less progressive [101]. By contrast, negative effects of the same allele have also been described; it was associated with risk of aortic valve sclerosis and coronary atherosclerosis at angiography in women [102]. Nevertheless, age distribution, body weight, HRT use and other factors of the examined population of women may thoroughly influence the CVD risk besides the ER genotype.

Associations of polymorphism in the estrogen receptor β gene (ESR2) with recurrent spontaneous abortion were examined in the Chinese population. There were no significant differences in gene polymorphisms between women with full term pregnancy and with recurrent spontaneous abortion [103].

Correlations between genetic polymorphisms in the ERα gene and breast cancer risk are highly inconclusive. Pvull polymorphism was associated with a moderately increased risk, whereas, Xbal polymorphism was related to non-significantly elevated risk confined to older postmenopausal women in Shanghai [104]. Conversely, in a Korean study Pvull genotype distribution did not show any differences between breast cancer cases and controls, while the adjusted odds ratio for the Xbal allele containing genotypes was advantageously decreased (OR=0.4), particularly in postmenopausal cases [105]. These results suggest that even in genetic investigations, mainly postmenopausal state and ageing associated estrogen deficiency define the altered estrogen signaling leading to breast cancer [106]. Moreover, these studies cannot clarify the ligand binding strength and the negative or positive changes in the signaling pathways of estrogen attributed to gene polymorphism.

Although endometrial physiology is closely dependent on estrogen exposure, there were no significant associations between estrogen receptor-alpha gene polymorphisms and endometriosis in Japanese women [107]. Results on ESR1 polymorphism in relation to endometrial cancer risk are controversial as well. Some studies indicate that common genetic variants within ER cofactor genes are unlikely to play significant role in endometrial cancer risk in the European population [108]. In a further study the c.454-397C → T CC and c.454-351A → G GG genotypes were associated with a decreased risk for cancers of the endometrium [109].

In women with PCOS, Pvull and Xbal polymorphisms of ERα, as well as Alul and Rsal polymorphisms of ERβ were genotyped, and no difference was found in the distribution of these gene variants between patients and healthy controls [110]. Nevertheless, in PCOS women, carriers of TC and TT genotypes of Pvull polymorphism had lower fasting glucose to insulin ratio compared with carriers of CC genotype. These results suggest that there are some associations of ERα polymorphisms with insulin resistance in PCOS.

The correlation between hormone replacement therapy (HRT) and mortality of women was examined to determine whether the risk of HRT use associated mortality varies depending on the genetic variability of estrogen receptors [111]. These studies suggested that some women are genetically more vulnerable to the effects of HRT in terms of their estrogen receptor genotype.

Lifelong Changes in the Sex Hormone Levels of Women and Their Correlation with Insulin Resistance, Type-2 Diabetes and Obesity

In premenopausal women, the good equilibrium of sexual steroid synthesis defines somatic health and reproductive capacity, whereas good, symptom-free adaptation to the estrogen deficient environment is a prerequisite of postmenopausal health (16, 106). Changes in the sexual hormone equilibrium during women's lives strongly influence insulin sensitivity and the associated risk for type-2 diabetes and obesity (Table 1).

Changes in Sexual Hormone Levels and Insulin Sensitivity in Puberty

In puberty, the extreme somatic growth and explosion-like sexual development mean a great challenge for the entire metabolic and hormonal systems. In this period there is a higher risk for development of insulin resistance, particularly in overweight cases, induced either by inherited inclination or by exogenous noxae. Increased insulin resistance does not return to prepubertal values at the end of puberty and represents high risk for adult metabolic syndrome, type-2 diabetes and their complications [112, 113].

Table 1. Lifelong changes in the sex hormone levels of women and their correlations with insulin resistance, type-2 diabetes and obesity

Hormonal changes in the life periods of women	Estrogen level	Androgen level	Insulin resistance	Risk for type-2 DM and obesity
Adolescence				
menstrual disorder	↓	↑	↑	↑
anovulatory cycles	↓	↑	↑	↑
Premenopausal women				
anovulatory infertility	↓	↑	↑	↑
nulliparity	↓	↑	↑	↑
PCOS	↓	↑	↑	↑
hyperprolactinemia	↓	↑	↑	↑
antiestrogen treatment	↓	↑	↑	↑
contraceptive use	↑	↓	↓	↓
Postmenopausal women				
HRT use	↑	↓	↓	↓
non HRT use	↓	↑	↑	↑
antiestrogen treatment	↓↓	↑↑	↑↑	↑↑
hysterectomy	↓↓	↑↑	↑↑	↑↑

In adolescent girls, the developing insulin resistance leads to abnormal ovarian sexual steroidgenesis as well, resulting in excessive androgen and defective estrogen production and a higher frequency of menstrual irregularity and anovulatory cycles [114, 115, 116]. Elevated serum androgen concentrations associated with insulin resistance in adolescents are preserved into adulthood and are reflected by defective fertility patterns at least until 30 years of age [115, 116].

Development of pathologic dysmetabolism and hyperandrogenism in the critical period of puberty might be a defining, dangerous factor for later metabolic syndrome, type-2 diabetes and their comorbidities [117].

Correlations between Changes in Estrogen Levels and Insulin Sensitivity in Premenopausal Women

In premenopausal young women, defective estrogen synthesis and decreased estrogen level are frequently associated with anovulation and infertility. Clinical signs of ovarian insufficiency are the long and/or irregular menstrual cycles [118, 119]. These reproductive dysfunctions are related to hyperinsulinism and excessive androgen production.

Among premenopausal hormonal disorders with insulin resistance and fertility failure, the polycystic ovarian syndrome (PCOS) is the most prevalent, presumably caused by a large number of different genetic abnormalities [17]. It seems to be a pathologic model of hormonal and metabolic alterations of postmenopausal status in young women. PCOS may usually be manifested by menstrual disorders, anovulatory infertility, hirsutism and obesity or overweight. Nevertheless, polycystic ovaries are common findings in symptom-free cases with normal menstrual cycles as well; only the laboratory findings of hyperinsulinemia and hyperandrogenism reveal the early phase of metabolic and hormonal disturbances [120, 121].

PCOS is not only an infertility disease but represents also a high systemic health risk for the affected women. Increased prevalence of type-2 diabetes, hypertension and cardiovascular complications were observed in a follow up study of a Dutch population of women with PCOS [86]. Close associations between PCOS and premature coronary and aortic atherosclerosis were revealed in middle-aged women [122, 123]. A retrospective Swedish study found a 7.4-fold risk of myocardial infarction among women suffering of PCOS [124].

In young infertile, nulliparous women with or without PCOS, an elevated risk for endometrial cancer was observed [125]. The high prevalence of endometrial cancer is frequently associated with synchronous primary cancers of the ovary or breast [125, 126]. This female organ triad has the highest estrogen demand, showing peculiar cancer risk even in a slightly estrogen deficient environment [7, 16, 127].

Earlier, some authors presumed that elevated estrogen levels unopposed by progestin continuously stimulate estrogen receptors in women with PCOS. This concept seemed to be a plausible explanation for the high risk of endometrial and breast cancers observed in these cases based on the concept of the carcinogenic capacity of estrogen [128]. Recently, insulin resistance and hyperinsulinemia in PCOS patients are regarded as concomitants of high ovarian and adrenal androgen synthesis at the expense of defective estrogen production [16, 129].

Hyperprolactinemia is also associated with insulin resistance, obesity, cycle disorders, reproductive dysfunction and hyperandrogenism in women, however, it should be delineated from PCOS and other disorders related to androgen excess. Glucose intolerance and obesity are characteristic in hyperprolactinemia, suggesting that prolactin or the associated hormonal disturbances might also be modulators of insulin sensitivity and body weight [130, 131]. In a population based cohort study the overall cancer risk was elevated in patients with hyperprolactinemia [132].

Oral contraceptive (OC) use replaces the natural menstrual cycle with relatively steady levels and fluctuations of artificial sex hormones. In PCOS cases, hormone treatment by oral contraceptives reduces the volume of cystic ovaries, decreases testosterone secretion and improves the carbohydrate and lipid metabolism as well [23]. Conversely, administration of the insulin sensitizing metformin in PCOS cases primarily lowers the high insulin level and at the same time improves menstrual abnormalities, ovulatory dysfunction and hirsutism [129].

Epidemiologic studies have confirmed that combined oral contraceptives provide substantial protection against endometrial and ovarian cancers in endangered anovulatory women [133]. A recent patent disclosed a method for treating hyperandrogenism and associated conditions, including PCOS by an estrogen derivative compound [134]. This therapy seems to be more advantageous against the dangerous dysmetabolism of PCOS cases than oral contraceptive administration.

Tamoxifen is a nonsteroidal anti-estrogenic drug used for adjuvant therapy of breast cancer and recently as a chemopreventive agent for breast cancer and also, on an investigational basis, for other cancers [135]. Nevertheless, worldwide administration of antiestrogen compounds yielded thorough disappointment [136].

Antiestrogens are cytostatic agents blocking the most important regulatory mechanisms associated with estrogen signaling. They have ambiguous impacts on mammary tumor development, however, the estrogen deprivation induces several life-threatening side-effects and exhibits strong carcinogenic capacity, particularly in the highly estrogen dependent endometrium [137]. Results of case-control studies demonstrated an increased prevalence of fatty liver, intraabdominal fat accumulation and type-2 diabetes in breast cancer cases receiving tamoxifen [135, 138, 139]. Artificial blocking of estrogen signaling pathways seems to confer serious insulin resistance. Metabolic syndrome and type-2 diabetes are well-known promoters of cancer initiation and progression affecting preferentially the endometrium [140, 141], which is the most vulnerable site in an estrogen deficient milieu.

Correlations between Estrogen Loss and the Risk of Increasing Insulin Resistance in Postmenopausal Women

Menopause at 50-52 years of age means a sudden loss of ovarian estrogen synthesis and confers further decline in the circulating hormone level.

Postmenopausal women never using HRT are obviously insulin resistant and exhibit increasing inclination to the associated comorbidities. With ageing, every year after menopause is associated with continuous estrogen loss and parallel advancing insulin resistance [24]. For women aged 55-65 years, weight gain and obesity are their major health risk [142]. In postmenopausal women, deepening dysmetabolism, obesity and disturbance of male to female sexual steroid levels are associated with increased prevalence of metabolic syndrome, type-2 diabetes, cardiovascular disease and malignancies.

In HRT user postmenopausal cases, the protective effect of estrogen substitution may counteract to the developing insulin resistance and their metabolic and hormonal equilibrium becomes reminiscent to that of young women with preserved circulatory estrogen levels [16]. Estradiol administration increases insulin sensitivity [6], yields favorable changes in plasma lipid levels [143] and its anti-obesity effect decreases fat accumulation, particularly in visceral location [144, 145]. All these impacts justify that HRT use is beneficially protective against insulin resistance and its comorbidities in postmenopausal women.

Hysterectomy and bilateral oophorectomy mean an abrupt, shocking hormone deprivation as compared with natural menopause. These patients are highly endangered as they have no possibility for gradual adaptation to estrogen loss by a compensatory hormone synthesis of the peripheral tissues [106, 127]. Bilateral oophorectomy is used as a risk reduction strategy in BRCA1/2 mutation carriers, although data on long-term side effects are not yet available. In the US population, oophorectomy, particularly at a young age, has been associated with highly increased overall and cardiovascular disease (CVD) mortality [146].

In 2011 the WHI Randomized Controlled Trial substantiated that estrogen treatment in women with prior hysterectomy resulted in a significantly lower breast cancer risk than in untreated controls [147]. Hysterectomy seems to be a near uniformly high breast cancer risk for women, thus HRT studies on these homogenously selected cases proved to be methodologically strong and yielded unexpectedly correct results [7].

In postmenopausal women, antiestrogen administration is a worldwide practice as either therapeutic or cancer preventive agent. Both ER-blocker and aromatase inhibitor types of antiestrogens further aggravate the estrogen deficiency and insulin resistance of aged female patients [136].

Conclusion

Correlations between estrogen signaling and human morbidity seems to be very unclear and difficult topic. Nowadays, the opinions of scientists have been partially changed regarding "estrogen induced diseases" however; the overwhelming literary data are still contradictory.

Individual and population specific ER associated gene polymorphisms were presumed to illuminate the highly controversial clinical and experimental results on the role of estrogens in health and disease. Nevertheless, the puzzling biology behind estrogen action was not reassuringly clarified by studies on either disabled ERKO animals or patients with genotyping of their ERs.

Diverse disorders associated with insulin resistance are usually well treatable by estradiol substitution both in pre- and postmenopausal women as well as in animal experiments. ERs seem to have balanced interplay in the maintenance of ideal glucose uptake and energy expenditure with continuous adaptation to the momentarily changing intra and extracellular stimuli. This equilibrium may be shattered in case of a defective estrogen supply or by the derangement of ER signaling pathways.

Recently, new therapeutic options are planned based on the development of tissue specific estrogen receptor modulators (SERMs) by the pharmaceutical industry.

Nevertheless, considering the beneficial results of estradiol substitution in diverse diseases, we should understand the marvelous network of ER signal transduction pathways instead of trying to manipulate it.

References

[1] Reaven, G. M. Banting lecture 1988: Role of insulin resistance in human disease. *Diabetes* 1988; 37:1595-1607.

[2] DeFronzo, R. A., Ferrannini, E. Insulin resistance: a multifaceted syndrome responsible for NIDDM, obesity, hypertension, dyslipidemia and atherosclerotic cardiovascular disease. *Diabetes Care* 1991; 14:173-194.

[3] Björnström, L., Sjöberg, M. Mechanisms of estrogen receptor signaling: Convergence of genomic and nongenomic actions on target genes. *Mol. Endocrinol.* 2005; 19:833-842.

[4] Reckelhoff, J. F. Sex steroids, cardiovascular disease, and hypertension. *Hypertension* 2005; 45:170-174.

[5] Suba, Zs. Estrogen deficiency as cancer risk factor. In: Suba, Zs Ed. *Estrogen versus Cancer.* Chapter 5. Nova Science Publishers Inc. New York, 2013

[6] Barros, R. P. A., Machadon, U. F., Gustafsson, J. A. Estrogen receptors: new players in diabetes mellitus. *Trends Molecular Med.* 2006; 9:425-431.

[7] Suba, Z. Interplay between insulin resistance and estrogen deficiency as co-activators in carcinogenesis. *Pathol. Oncol. Res.* 2012; 18:123-33.

[8] Barros, R. P., Gustafsson, J. Å. Estrogen receptors *Cell Metab.* 2011; 14:289-99.

[9] Garvey, T. L., Maianu, J. H., Zhu, W., Hancock, J. A., Golichowski, A. M. Multiple defects in the adipocyte glucose transport system cause cellular insulin resistance in gestational diabetes. Heterogeneity in the number and a novel abnormality in subcellular localization of GLUT4 glucose transporters. *Diabetes* 1993; 42:1773–1785.

[10] Okuno, S., Akazawa, S.,Yasuhi, I., Kawasaki, E., Matsumoto, K., Yamasaki, H., Matsuo, H., Yamaguchi, Y., Nagataki, S. Decreased expression of the GLUT4 glucose transporter protein in adipose tissue during pregnancym. *Horm. Metab. Res.* 1995; 27: 231–234.

[11] Stubbins, R. E., Najjar, K., Holcomb, V. B., Hong, J., Núnez, N. P. Estrogen alters adypocyte biology and protects female mice from adipocyte inflamation and insulin resistance. *Diabetes Obes. Metab.* 2012; 14:58-66.

[12] Riant, E., Waget, A., Cogo, H., Arnal, J. F., Burcelin, R., Gourdy, P. Estrogens protect against high-fat diet-induced insulin resistance and glucose intolerance in mice. *Endocrinology* 2009; 150:2109-17.

[13] Deguchi, J., Miyamoto, M., Okada, S. Sex hormone-dependent renal cell carcinogenesis induced by ferric nitrilotriacetate in Wistar rats. *Jpn. J. Cancer Res.* 1995; 86:1068-1071.

[14] Nkhata, K. J., Ray, A., Dogan, S., Grande, J. P., Cleary, M. P. Mammary tumor development from T47-D human breast cancer cells in obese ovariectomized mice with and without estradiol supplements. *Breast Cancer Res. Treat.* 2009; 114:71-83.

[15] Hong, J., Holcomb, V. B., Kushiro, K., Núñez, N. P. Estrogen inhibits the effects of obesity and alcohol on mammary tumors and fatty liver. *Int. J. Oncol.* 2011; 39:1443-53.

[16] Suba, Z. Circulatory Estrogen Level Protects Against Breast Cancer in Obese Women. *Recent Pat. Anticancer Drug Discov.* 2012 Oct. 11. [Epub. ahead of print]

[17] Bloomgarden, Z. T. Second World Congress on the Insulin Resistance Syndrome: Mediators, pediatric insulin resistance, the polycystic ovary syndrome, and malignancy. *Diab. Care* 2005; 28:1821-1830.

[18] Moghetti, P., Castello, R., Negri, C., Tosi, F., Spiazzi, G. G., Brun, E., *et al.* Insulin infusion amplifies 17 alpha-hydroxycorticosteroid intermediates response to adrenocorticotropin in hyperandrogenic women: apparent relative impairment of 17,20-lyase activity. *J. Clin. Endocrinol. Metab.* 1996; 81:881-6.

[19] Cara, J. F., Fan, J., Azzarello, J., Rosenfield, R. L. Insulin-like growth factor-I enhances luteinizing hormone binding to rat ovarian theca-interstitial cells. *J. Clin. Invest.* 1990; 86:560-5.

[20] Simpson, E. R. Sources of estrogen and their importance. *J. Steroid Biochem. Mol. Biol.* 2003; 86:225-30.

[21] Stamataki, K. E., Spina, J., Rangou, D. B., Chlouverakis, C. S., Piaditis, G. P. Ovarian function in women with non-insulin dependent diabetes mellitus. *Clin. Endocrinol.* (Oxf.) 1996; 45:615-21.

[22] Diamanti-Kandarakis, E., Baillargeon, J. P., Iuorno, M. J., Jakubowicz, D. J., Nestler, J. E. A modern medical quandary: polycystic ovary syndrome, insulin resistance, and oral contraceptive pills. *J. Clin. Endocrinol. Metab.* 2003; 88:1927-32.

[23] ESHRE Capri Workshop Group. Ovarian and endometrial function during hormonal contraception. *Hum. Reprod.* 2001; 16:1527-35.

[24] Schneider, J. G., Tompkins, C., Blumenthal, R. S., Mora, S. The metabolic syndrome in women. *Cardiol. Rev.* 2006; 14:286-291.

[25] Gupta, K., Krishnaswamy, G., Karnad, A., Peiris, A. Insulin: A novel factor in carcinogenesis. *Am. J. Med. Sci.* 2002; 323:140-145.

[26] Veldhuis, J. D., Frystyk, J., Iranmanesh, A., Ørskov, H. Testosterone and estradiol regulate free insulin-like growth factor I (IGF-I), IGF binding protein 1 (IGFBP-1), and dimeric IGF-I/IGFBP-1 concentrations. *J. Clin. Endocrinol. Metab.* 2005; 90:2941-7.

[27] Reaven, G. M. Insulin resistance, cardiovascular disease, and the metabolic syndrome: How well do the emperor's clothes fit? *Diabetes Care* 2004; 27:1011-1012.

[28] Doyle, S. L., Donohoe, C. L., Lysaght, J., Reynolds, J. V. Visceral obesity, metabolic syndrome, insulin resistance and cancer. *Proc. Nutr. Soc.* 2012; 71:181-9.

[29] Choi, S. B., Jang, J. S., Park, S. Estrogen and exercise may enhance beta-cell function and mass via insulin receptor substrate-2 induction in ovariectomized diabetic rats. *Endocrinology* 2005; 146:4786-94.

[30] Tiano, J. P., Mauvais-Jarvis, F. Importance of oestrogen receptors to preserve functional β-cell mass in diabetes. *Nat. Rev. Endocrinol.* 2012; 8:342-51.

[31] Bryzgalova, G., Gao, H., Ahren, B., Zierath, J. R., Galuska, D., Steiler, T. L., *et al.* Evidence that oestrogen receptor-alpha plays an important role in the regulation of glucose homeostasis in mice: insulin sensitivity in the liver. *Diabetologia* 2006; 49:588-97.

[32] Muraki, K., Okuya, S., Tanizawa, Y. Estrogen receptor alpha regulates insulin sensitivity trough IRS-1 thyrosin phosphorylation in mature 3T3-L1 adipocytes. *Endocr. J.* 2006; 53:841-851.

[33] Welch, R. D., Gorski, J., Welch, R. D., Gorski, J. Regulation of glucose transporters by estradiol in the immature rat uterus. *Endocrinology* 1999; 140:3602-8.

[34] Taskinen, M. R. Diabetic dyslipidaemia: from basic research to clinical practice. *Diabetologia* 2003; 46:733-49.

[35] Jensen, J., Nilas, L., Christiansen, C. Influence of menopause on serum lipids and lipoproteins. *Maturitas* 1990; 12:321-31.

[36] Harrison-Bernard, L. M., Schulman, I. H., Raij, L. Postovariectomy hypertension is linked to increased renal AT1 receptor and salt sensitivity. *Hypertension* 2003; 42: 1157-63.

[37] Gallagher, P. E., Li, P., Lenhart, J. R., Chappell, M. C., Brosnihan, K. B. Estrogen regulation of angiotensin-converting enzyme mRNA. *Hypertension* 1999; 33:323-8.

[38] Widder, J., Pelzer, T., von Poser-Klein, C., Hu, K., Jazbutyte, V., Fritzemeier, K. H., et al. Improvement of endothelial dysfunction by selective estrogen receptor-alpha stimulation in ovariectomized SHR. *Hypertension* 2003; 42:991-6.

[39] Yanes, L. L., Reckelhoff, J. F. Postmenopausal hypertension. *Am. J. Hypertens.* 2011; 24: 740-9.

[40] Baghaei, F., Rosmond, R., Westberg, L., Hellstrand, M., Eriksson, E., Holm, G., et al. The CYP19 gene and associations with androgens and abdominal obesity in premenopausal women. *Obes. Res.* 2003; 11:578-85.

[41] Poehlman, E. T., Toth, M. J., Gardner, A. W. Changes in energy balance and body composition at menopause: a controlled longitudinal study. *Ann. Intern. Med.* 1995; 123: 673-5.

[42] Björntorp, P. The regulation of adipose tissue distribution in humans. *Int. J. Obes. Relat. Metab. Disord.* 1996; 20:291-302.

[43] Salpeter, S. R., Walsh, J. M., Ormiston, T. M., Greyber, E., Buckley, N. S., Salpeter, E. E. Meta-analysis: effect of hormone-replacement therapy on components of the metabolic syndrome in postmenopausal women. *Diabetes Obes. Metab.* 2006; 8:538-54.

[44] Pennathur, S., Heinecke, J. W. Mechanisms for oxidative stress in diabetic cardiovascular disease. *Antioxid. Redox. Signal* 2007; 9:955-69.

[45] Baynes, J. W., Thorpe, S. R. Role of oxidative stress in diabetic complications: a new perspective on an old paradigm. *Diabetes* 1999; 48:1-9.

[46] Altomare, E., Vendemiale, G., Chicco, D., Procacci, V., Cirelli, F. Increased lipid peroxidation *Diabetes Metab.* 1992; 18: 264-71.

[47] Dantas, A. P., Tostes, R. C., Fortes, Z. B., Costa, S. G., Nigro, D., Carvalho, M. H. In vivo evidence for antioxidant potential of estrogen in microvessels of female spontaneously hypertensive rats. *Hypertension* 2002; 39:405-11.

[48] Viña, J., Sastre, J., Pallardó, F. V., Gambini, J., Borrás, C. Role of mitochondrial oxidative stress to explain the different longevity between genders: protective effect of estrogens. *Free Radic. Res.* 2006; 40:1359-65.

[49] Morton, G. J., Cummings, D. E., Baskin, D. G., Barsh, G. S., Schwartz, M. W. Central nervous system control of food intake and body weight. *Nature.* 2006;443(7109):289-95.

[50] Roepke, T. A. Oestrogen modulates hypothalamic control of energy homeostasis through multiple mechanisms. *J. Neuroendocrinol.* 2009; 21:141–150.
[51] Laudenslager, M. L., Wilkinson, C. W., Carlisle, H. J., Hammel, H. T. Energy balance in ovariectomized rats with and without estrogen replacement. *Am. J. Physiol.* 1980; 238: R400-5.
[52] Asarian, L., Geary, N. Modulation of appetite by gonadal steroid hormones. *Philos. Trans. R Soc. Lond. B Biol. Sci.* 2006; 361:1251-63.
[53] Ahima, R. S., Bjorbaek, C., Osei, S., Flier, J. S. Regulation of neuronal and glial proteins by leptin: implications for brain development. *Endocrinology.* 1999; 140:2755-62.
[54] Elias, C. F., Aschkenasi, C., Lee, C., Kelly, J., Ahima, R. S., Bjorbaek, C., Flier, J. S., Saper, C. B., Elmquist, J. K. Leptin differentially regulates NPY and POMC neurons projecting to the lateral hypothalamic area. *Neuron* 1999; 23:775-86.
[55] Machinal, F., Dieudonne, M. N., Leneveu, M. C., Pecquery, R., Giudicelli, Y. In vivo and in vitro ob gene expression and leptin secretion in rat adipocytes: evidence for a regional specific regulation by sex steroid hormones. *Endocrinology* 1999; 140:1567-74.
[56] Ainslie, D. A., Morris, M. J., Wittert, G., Turnbull, H., Proietto, J., Thorburn, A. W. Estrogen deficiency causes central leptin insensitivity and increased hypothalamic neuropeptide Y. *Int. J. Obes. Relat. Metab. Disord.* 2001; 25:1680-8.
[57] Clegg, D. J., Brown, L. M., Woods, S. C., Benoit, S. C. Gonadal hormones determine sensitivity to central leptin and insulin. *Diabetes* 2006; 55:978-87.
[58] Thammacharoen, S., Geary, N., Lutz, T. A., Ogawa, S., Asarian, L. Divergent effects of estradiol and the estrogen receptor-alpha agonist PPT on eating and activation of PVN CRH neurons in ovariectomized rats and mice. *Brain Res.* 2009; 1268:88-96.
[59] Musatov, S., Chen, W., Pfaff, D. W., Mobbs, C. V., Yang, X. J., Clegg, D. J., Kaplitt, M. G., Ogawa, S. Silencing of estrogen receptor alpha in the ventromedial nucleus of hypothalamus leads to metabolic syndrome. *Proc. Natl. Acad. Sci. US* 2007; 104:2501–2506.
[60] Liang, Y. Q., Akishita, M., Kim, S., Ako, J., Hashimoto, M., Iijima, K., Ohike, Y., Watanabe, T., Sudoh, N., Toba, K., Yoshizumi, M., Ouchi, Y. Estrogen receptor beta is involved in the anorectic action of estrogen. *Int. J. Obes. Relat. Metab. Disord.* 2002; 26:1103-9.
[61] Genabai, N. K., Briski, K. P. Adaptation of arcuate insulin receptor, estrogen receptor-alpha, estrogen receptor-beta, and type-II glucocorticoid receptor gene profiles to chronic intermediate insulin-induced hypoglycemia in estrogen-treated ovariectomized female rats. *J. Mol. Neurosci.* 2010; 41:304–309.
[62] Pallottini, V., Bulzomi, P., Galluzzo, P., Martini, C., Marino, M. Estrogen regulation of adipose tissue functions: involvement of estrogen receptor isoforms. *Infect. Disord. Drug Targets* 2008; 8:52-60.
[63] Godsland, I. F. Oestrogens and insulin secretion. *Diabetologia* 2005; 48:2213–2220.
[64] Gao, H., Fält, S., Sandelin, A., Gustafsson, J. A., Dahlman-Wright, K. Genome-wide identification of estrogen receptor alpha-binding sites in mouse liver. *Mol. Endocrinol.* 2008; 22:10–22.
[65] Lundholm, L., Bryzgalova, G., Gao, H., Portwood, N., Fält, S., Berndt, K. D., Dicker, A., Galuska, D., Zierath, J. R., Gustafsson, J. A., et al. The estrogen receptor alpha-

selective agonist propyl pyrazole triol improves glucose tolerance in ob/ob mice; potential molecular mechanisms. *J. Endocrinol.* 2008; 199:275–286.

[66] Foryst-Ludwig, A., Clemenz, M., Hohmann, S., Hartge, M., Sprang, C., Frost, N., Krikov, M., Bhanot, S., Barros, R., Morani, A., et al. Metabolic actions of estrogen receptor beta (ERbeta) are mediated by a negative cross-talk with PPARgamma. *PLoS Genet.* 2008; 4: e1000108.

[67] Evans, M. J., Lai, K., Shaw, L. J., Harnish, D. C., Chadwick, C. C. Estrogen receptor alpha inhibits IL-1beta induction of gene expression in the mouse liver. *Endocrinology* 2002; 143:2559–2570.

[68] Espeland, M. A., Marcovina, S. M., Miller, V., Wood, P. D., Wasilauskas, C., Sherwin, R., Schrott, H., Bush, T. L. Effect of postmenopausal hormone therapy on lipoprotein(a) concentration. PEPI Investigators. *Postmenopausal Estrogen/Progestin Interventions Circulation* 1998; 97:979–986.

[69] Redig, A. J., Munshi, H. G. Care of the cancer survivor: metabolic syndrome after hormone-modifying therapy. *Am. J. Med.* 2010; 123:87 e1–e6.

[70] Kasuga, M., Hedo, J. A., Yamada, K. M., Kahn, C. R. The structure of insulin receptor and its subunits. Evidence for multiple nonreduced forms and a 210,000 possible proreceptor. *J. Biol. Chem.* 1982; 257: 10392-9.

[71] Zhou, L., Chen, H., Xu, P., Cong, L. N., Sciacchitano, S., Li, Y., Graham, D., Jacobs, A. R., Taylor, S. I., Quon, M. J. Action of insulin receptor substrate-3 (IRS-3) and IRS-4 to stimulate translocation of GLUT4 in rat adipose cells. *Mol. Endocrinol.* 1999; 13: 505-14.

[72] Rogers, N. H., Witczak, C. A., Hirshman, M. F., Goodyear, L. J., Greenberg, A. S. Estradiol stimulates Akt, AMP-activated protein kinase (AMPK) and TBC1D1/4, but not glucose uptake in rat soleus. *Biochem. Biophys. Res. Commun.* 2009; 382:646–650.

[73] Ordóñez, P., Moreno, M., Alonso, A., Llaneza, P., Díaz, F., González, C. 17beta-Estradiol and/or progesterone protect from insulin resistance in STZ-induced diabetic rats. *J. Steroid Biochem. Mol. Biol.* 2008; 111: 287–294.

[74] Moreno, M., Ordoñez, P., Alonso, A., Díaz, F., Tolivia, J., González, C. Chronic 17beta-estradiol treatment improves skeletal muscle insulin signaling pathway components in insulin resistance associated with aging. *Age* (Dordr.) 2010; 32:1–13.

[75] Shi, J., Simpkins, J. W. 17 beta-Estradiol modulation of glucose transporter 1 expression in blood-brain barrier. *Am. J. Physiol.* 1997; 272:E1016-22.

[76] Fisher, C. R., Graves, K. H., Parlow, A. F., Simpson, E. R. Characterization of mice deficient in aromatase (ArKO) because of targeted disruption of the cyp19 gene. *Proc. Natl. Acad. Sci. US* 1998; 95:6965-70.

[77] Jones, M. E., Thorburn, A. W., Britt, K. L., Hewitt, K. N., Wreford, N. G., Proietto, J., Oz, O. K., Leury, B. J., Robertson, K. M., Yao, S., Simpson, E. R. Aromatase-deficient (ArKO) mice have a phenotype of increased adiposity. *Proc. Natl. Acad. Sci. US* 2000; 97:12735-40.

[78] Takeda, K., Toda, K., Saibara, T., Nakagawa, M., Saika, K., Onishi, T., Sugiura, T., Shizuta, Y. Progressive development of insulin resistance phenotype in male mice with complete aromatase (CYP19) deficiency. *J. Endocrinol.* 2003; 176:237-46.

[79] Galluzzo, P., Rastelli, C., Bulzomi, P., Acconcia, F., Pallottini, V., Marino, M. 17beta-Estradiol regulates the first steps of skeletal muscle cell differentiation via ER-alpha-mediated signals. *Am. J. Physiol. Cell Physiol.* 2009; 297:C1249–C1262.

[80] Heine, P. A., Taylor, J. A., Iwamoto, G. A., Lubahn, D. B., Cooke, P. S. Increased adipose tissue in male and female estrogen receptor-alpha knockout mice. *Proc. Natl. Acad. Sci. US* 2000; 97:12729–12734.

[81] Phillips, S. K., Rook, K. M., Siddle, N. C., Bruce, S. A., Woledge, R. C. Muscle weakness in women occurs at an earlier age than in men, but strength is preserved by hormone replacement therapy. *Clin. Sci.* 1993; 84:95–98.

[82] Greising, S. M., Baltgalvis, K. A., Lowe, D. A., Warren, G. L. Hormone therapy and skeletal muscle strength: a meta-analysis. *J. Gerontol. A Biol. Sci. Med. Sci.* 2009; 64: 1071–1081.

[83] Livingstone, C., Collison, M. Sex steroids and insulin resistance. *Clin. Sci.* 2002; 102: 151–166.

[84] Frayn, K. N., Karpe, F., Fielding, B. A., Macdonald, I. A., Coppack, S. W. Integrative physiology of human adipose tissue. *Int. J. Obes. Relat. Metab. Disord.* 2003; 27:875–888.

[85] Kim, J. Y., Jo, K. J., Kim, O. S., Kim, B. J., Kang, D. W., Lee, K. H., H. W. Baik, M. S. Han, S. K. Lee. Parenteral 17beta-estradiol decreases fasting blood glucose levels in non-obese mice with short-term ovariectomy. *Life Sci.* 2010; 87:358–366.

[86] Elting, M. W., Korsen, T. J., Bezemer, P. D., Schoemaker, J. Prevalence of diabetes mellitus, hypertension and cardiac complaints in a follow-up study of a Dutch PCOS population. *Hum. Reprod.* 2001;16:556-60.

[87] Suba, Zs. Insulin resistance, estrogen deficiency and cancer risk. In: Ed. Suba, Zs. *Estrogen versus Cancer*. Chapter 7. Nova Science Publishers Inc. New York. 2009.

[88] Wohlers, L. M., Spangenburg, E. E. 17beta-estradiol supplementation attenuates ovariectomy-induced increases in ATGL signaling and reduced perilipin expression in visceral adipose tissue. *J. Cell Biochem.* 2010; 110:420–427.

[89] Dieudonné, M. N., Leneveu, M. C., Giudicelli, Y., Pecquery, R. Evidence for functional estrogen receptors alpha and beta in human adipose cells: regional specificities and regulation by estrogens. *Am. J. Physiol. Cell Physiol.* 2004; 286:C655-61.

[90] Smith, E. P., Boyd, J., Frank, G. R., Takahashi, H., Cohen, R. M., Specker, B., Williams, T. C., Lubahn, D. B., Korach, K. S. Estrogen resistance caused by a mutation in the estrogen-receptor gene in a man. *N Engl. J. Med.* 1994; 331:1056–1061.

[91] Sudhir, K., Chou, T. M., Chatterjee, K., Smith, E. P., Williams, T. C., Kane, J. P., Malloy, M. J., Korach, K. S., Rubanyi, G. M. Premature coronary artery disease associated with a disruptive mutation in the estrogen receptor gene in a man. *Circulation* 1997; 96:3774–3777.

[92] Naaz, A., Zakroczymski, M., Heine, P., Taylor, J., Saunders, P., Lubahn, D., Cooke, P. S. Effect of ovariectomy on adipose tissue of mice in the absence of estrogen receptor alpha (ERalpha): a potential role for estrogen receptor beta (ERbeta). *Horm. Metab. Res.* 2002; 34:758–763.

[93] Rosenbaum, D., Haber, R. S., Dunaif, A. Insulin resistance in polycystic ovary syndrome: decreased expression of GLUT-4 glucose transporters in adipocytes. *Am. J. Physiol.* 1993; 264:E197-202.

[94] Austad, S. N.: Why women live longer than men: Sex differences in longevity. *Gender Medicine* 2006; 3:79-92.

[95] Suba, Z. The role of estrogen in health and disease. In: Ed. Suba, Z. *Estrogen versus Cancer*. Chapter 4. Nova Science Publishers Inc. New York. 2009.

[96] Rossouw, J. E., Anderson, G. L., Prentice, R. L., LaCroix, A. Z., Kooperberg, C., Stefanick, M. L., et al. Wrightin Group for the Women's Health Initiative Investigators. Risks and benefits of estrogen plus progestin in healthy postmenopausal women: principal results from the Women's Health Initiative randomized controlled trial. *JAMA* 2002; 288:321-33.
[97] Newton-Cheh, C., O'Donnell, C. J. Sex differences and genetic associations with myocardial infarction. *JAMA* 2004; 291:3008-10.
[98] Shearman, A. M., Cupples, L. A., Demissie, S., et al. Association between estrogen receptor alpha gene variation and cardiovascular disease. *JAMA* 2003; 290:2263-2270.
[99] Schuit, S. C., Oei, H. H., Witteman, J. C., et al. Estrogen receptor alpha gene polymorphisms and risk of myocardial infarction. *JAMA* 2004; 291:2969-2977.
[100] Ferrero, V., Ribichini, F., Matullo, G., et al. Estrogen receptor-alpha polymorphisms and angiographic outcome after coronary artery stenting. *Arterioscler. Thrombosis Vascular Biol.* 2003; 23:2223-2228.
[101] Koivu, T. A., Fan, Y-M., Mattila, K. M., et al.The effect of hormone replacement therapy on atherosclerotic severity in relation to ESR1 genotype in postmenopausal women. *Maturitas* 2003; 44:29-38.
[102] Nordström, P., Glader, C. A., Dahlen, G., et al. Oestrogen receptor alpha gene polymorphism is related to aortic valve sclerosis in postmenopausal women. *J. Int. Med.* 2003; 254:140-146.
[103] Hu, J., Wang, J., Xiang, H., Li, Z., Wang, B., Cao, Y., Ma, X. Association of polymorphisms in the estrogen receptor β (ESR2) with unexplained recurrent spontaneous abortion (URSA) in Chinese population. *J. Matern. Fetal Neonatal. Med.* 2012; 25:1727-9.
[104] Cai, Q., Shu, X. O., Jin, F., Dai, Q., Wen, W., Cheng, J. R., Gao, Y. T., Zheng, W. Genetic polymorphisms in the estrogen receptor alpha gene and risk of breast cancer: results from the Shanghai Breast Cancer Study. *Cancer Epidemiol. Biomarkers Prev.* 2003; 12:853-9.
[105] Shin, A., Kang, D., Nishio, H., Lee, M. J., Park, S. K., Kim, S. U., Noh, D. Y., Choe, K. J., Ahn, S. H., Hirvonen, A., Kim, J. H., Yoo, K. Y. Estrogen receptor alpha gene polymorphisms and breast cancer risk. *Breast Cancer Res. Treat.* 2003; 80:127-31.
[106] Suba, Z. Common soil of smoking-associated and hormone-related cancers: estrogen deficiency. *Oncol. Rev.* 2010; 4:73-87.
[107] Matsuzaka, Y., Kikuti, Y. Y., Izumi, S., Goya, K., Suzuki, T., Cai, L. Y., Oka, A., Inoko, H., Kulski, J. K., Kimura, M. Failure to detect significant association between estrogen receptor-alpha gene polymorphisms and endometriosis in Japanese women. *Environ. Health Prev. Med.* 2012; 17:423-8.
[108] Li, Y., Low, H. Q., Foo, J. N., Darabi, H., Einarsdóttir, K., Humphreys, K., Spurdle, A.; ANECS Group, Easton, D. F., Thompson, D. J., Dunning, A. M., Pharoah, P. D., Czene, K., Chia, K. S., Hall, P., Liu, J. Genetic variants in ER cofactor genes and endometrial cancer risk. *PLoS One*. 2012; 7:e42445.
[109] Weiderpass, E., Persson, I., Melhus, H., Wedren, S., Kindmark, A., Baron, J. A. Estrogen receptor alpha gene polymorphisms and endometrial cancer risk. *Carcinogenesis* 2000; 21:623–627.
[110] Nectaria, X., Leandros, L., Ioannis, G., Agathocles, T. The importance of ERα and ERβ gene polymorphisms in PCOS. *Gynecol. Endocrinol.* 2012; 28:505-8.

[111] Ryan, J., Canonico, M., Carcaillon, L., Carrière, I., Scali, J., Dartigues, J. F., Dufouil, C., Ritchie, K., Scarabin, P. Y., Ancelin, M. L. Hormone treatment, estrogen receptor polymorphisms and mortality: a prospective cohort study. *PLoS One*. 2012;7:e34112.

[112] Theintz, G. From obesity to type-2 diabetes in children and adolescents. *Rev. Med. Suisse* 2005; 1:477-80.

[113] Vanhala, M. J., Vanhala, P. T., Keinänen-Kiukaanniemi, S. M., Kumpusalo, E. A., Takala, J. K. Relative weight gain and obesity as a child predict metabolic syndrome as an adult. *Int. J. Obes. Relat. Metab. Disord.* 1999; 23:656-9.

[114] Stoll, B. A. Teenage obesity in relation to breast cancer risk. *Int. J. Obes. Relat. Metab. Disord.* 1998; 22:1035-40.

[115] Apter, D., Vihko, R. Endocrine determinants of fertility: serum androgen concentrations during follow-up of adolescents into the third decade of life. *J. Clin. Endocrinol. Metab.* 1990; 71:970-4.

[116] Baer, H. J., Colditz, G. A., Willett, W. C., Dorgan, J. F. Adiposity and sex hormones in girls. *Cancer Epidemiol. Biomarkers Prev.* 2007; 16: 1880-8.

[117] Lewy, V. D., Danadian, K., Witchel, S. F., Arslanian, S., : Early metabolic abnormalities in adolescent girls with polycystic ovarian syndrome. *J. Pediatrics* 2001; 138: 38-44.

[118] Zain, M. M., Norman, R. J. Impact of obesity on female fertility and fertility treatment. *Women's Health* (Lond. Engl.) 2008; 4:183-94.

[119] Rose, D. P., Vona-Davis, L. Interaction between menopausal status and obesity in affecting breast cancer risk. *Maturitas* 2010; 66:33-8.

[120] Polson, D. W., Wadsworth, J., Adams, J., et al. Polycystic ovaries: a common finding in normal women. *Lancet* 1988; 1:870-872.

[121] Carmina, E., Lobo, R. A. Polycystic ovaries in hirsute women with normal menses. *Am. J. Med.* 2001; 111:602-606.

[122] Talbott, E., Guzick, D., Clerici, A., et al. Coronary heart disease risk factors in women with polycystic ovary syndrome. *Arteroscler. Thromb. Vasc. Biol.* 1995; 15:821-826.

[123] Talbott, E. O., Zborowski, J. V., Rager, J. R., Boudreaux, M. Y., Edmundowicz, D. A., Guzick, D. S. Evidence for an association between metabolic cardiovascular syndrome and coronary and aortic calcification among women with polycystic ovary syndrome. *J. Clin. Endocrinol. Metab.* 2004; 89:5454-5461.

[124] Dahlgren, E., Johansson, S., Lindstedt, G., et al. Women with polycystic ovary syndrome wedge resected in 1956 to 1975: a long-term follow-up focusing on natural history and circulating hormones. *Fertil. Steril.* 1992; 57:505-513.

[125] Soliman, P. T., Oh, J. C., Schmeler, K. M., Sun, C. C., Slomovitz, B. M., Gershenson, D. M., *et al.* Risk factors for young premenopausal women with endometrial cancer. *Obstet. Gynecol.* 2005; 10:575-80.

[126] Uccella, S., Cha, S. S., Melton, L. J. 3rd, Bergstralh, E. J., Boardman, L. A., Keeney, G. L., *et al.* Risk factors for developing multiple malignancies in patients with endometrial cancer. *Int. J. Gynecol. Cancer* 2011; 21:896-901.

[127] Suba, Zs. Discovery of estrogen deficiency as common cancer risk factor for highly and moderately estrogen dependent organs. In: Ed. Suba, Zs. *Estrogen prevention for breast cancer*. Chapter 1. Nova Science Publishers Inc. New York. 2013.

[128] Gadducci, A., Gargini, A., Palla, E., Fanucchi, A., Genazzani, A. R. Polycystic ovary syndrome and gynecological cancers: is there a link? *Gynecol. Endocrinol.* 2005; 20: 200-208.

[129] Nestler, J. E. Obesity, insulin, sex steroids and ovulation. *Int. J. Obes. Relat. Metab. Disord.* 2000; 24 Suppl. 2: S71-3.

[130] Tuzcu, A., Yalaki, S., Arikan, S., Gokalp, D., Bahcec, M., Tuzcu, S. Evaluation of insulin sensitivity in hyperprolactinemic subjects by euglycemic hyperinsulinemic clamp technique. *Pituitary* 2009; 12:330-334.

[131] Shibli-Rahhal, A., Schlechte, J. The effects of hyperprolactinemia on bone and fat. *Pituitary*. 2009; 12:96-104.

[132] Berinder, K., Akre, O., Granath, F., Hulting, A. L. Cancer risk in hyperprolactinemia patients: a population-based cohort study. *Eur. J. Endocrinol.* 2011; 165:209-15.

[133] Deligeoroglou, E., Michailidis, E., Creatsas, G. Oral contraceptives and reproductive system cancer. *Ann. NY Acad. Sci.* 2003; 997:199-208.

[134] Tejura, B. Methods of treating hyperandrogenism and conditions associated therewith by administering a fatty acid ester of an estrogen or an estrogen derivative. *US20100286105* (2010).

[135] Nguyen, M. C., Stewart, R. B., Banerji, M. A., Gordon, D. H., Kral, J. G. Relationships between tamoxifen use, liver fat and body fat distribution in women with breast cancer. *Int. J. Obes. Relat. Metab. Disord.* 2001; 25:296-8.

[136] Suba, Z. Antiestrogen or estrogen as anticancer drug. In: Hoffmann, A. B. Ed. *Sex hormones: development, regulation and disorders.* Chapter 4. Pp. 75-94. New York, Nova Science Publishers Inc., 2011.

[137] Suba, Z. Failures and controversies of the antiestrogen treatment of breast cancer. In: Suba, Z. Ed. *Estrogen prevention for breast cancer.* Chapter 6. Nova Science Publishers Inc. New York, 2013.

[138] Saphner, T., Triest-Robertson, S., Li, H., Holzman, P. The association of nonalcoholic steatohepatitis and tamoxifen in patients with breast cancer. *Cancer.* 2009; 115:3189-95.

[139] Lipscombe, L. L., Fischer, H. D., Yun, L., Gruneir, A., Austin, P., Paszat, L., Anderson, G. M., Rochon, P. A. Association between tamoxifen treatment and diabetes: a population-based study. *Cancer.* 2012; 118:2615-22.

[140] Burzawa, J. K., Schmeler, K. M., Soliman, P. T., Meyer, L. A., Bevers, M. W., Pustilnik, T. L., Anderson, M. L., Ramondetta, L. M., Tortolero-Luna, G., Urbauer, D. L., Chang, S., Gershenson, D. M., Brown, J., Lu, K. H. Prospective evaluation of insulin resistance among endometrial cancer patients. *Am. J. Obstet. Gynecol.* 2011; 204: 355.e1-7.

[141] Friedenreich, C. M., Biel, R. K., Lau, D. C., Csizmadi, I., Courneya, K. S., Magliocco, A. M., Yasui, Y., Cook, L. S. Case-control study of the metabolic syndrome and metabolic risk factors for endometrial cancer. *Cancer Epidemiol. Biomarkers Prev.* 2011; 20:2384-95.

[142] Davis, S. R., Castelo-Branco, C., Chedraui, P., Lumsden, M. A., Nappi, R. E., Shah, D., Villaseca, P.; Writing Group of the International Menopause Society for World Menopause Day 2012. Understanding weight gain at menopause. *Climacteric* 2012; 15: 419-29.

[143] Walsh, B. W., Schiff, I., Rosner, B., Greenberg, L., Ravnikar, V., Sacks, F. M. Effects of postmenopausal estrogen replacement on the concentrations and metabolism of plasma lipoproteins. *N Engl. J. Med.* 1991; 325:1196-204.

[144] Gunter, M. J., Hoover, D. R., Yu, H., Wassertheil-Smoller, S., Rohan, T. E., Manson, J. E., *et al.* Insulin, insulin-like growth factor and risk of breast cancer in postmenopausal women. *J. Natl. Cancer Inst.* 2009; 101:48-60.

[145] Tchernof, A., Calles-Escandon, J., Sites, C. K., Poehlman, E. T. Menopause, central body fatness and insulin resistance: effects of hormone-replacement therapy. *Coron. Artery Dis.* 1998; 9:503-11.

[146] McCarthy, A. M., Menke, A., Ouyang, P., Visvanathan, K. Bilateral oophorectomy, body mass index, and mortality in US women aged 40 years and older. *Cancer Prev. Res.* (Phila.). 2012; 5:847-54.

[147] LaCroix, A. Z., Chlebowski, R. T., Manson, J. E., Aragaki, A. K., Johnson, K. C., Martin, L., et al. Health outcomes after stopping conjugated equine estrogens among postmenopausal women with prior hysterectomy. A randomized controlled trial. *JAMA* 2011; 305:1305-14.

In: Glucose Uptake
Editors: Carter C. Johnson and Davis B. Williams
ISBN: 978-1-62618-670-5

Chapter 7

Glucose Uptake and Androgen Responsiveness of Prostate Cancer Cells

C. V. Vaz, M. G. Alves, R. Marques, P. F. Oliveira, C. J. Maia and S. Socorro*

CICS-UBI - Health Sciences Research Center, University of Beira Interior, Portugal

Abstract

Prostate cancer (PCa) onset and development requires a sequence of genetic changes and additional metabolic adaptations to ensure survival of the modified cells and to finally achieve a malignant phenotype. This concept may be associated with the "Warburg phenomenon", which consists of an increased glycolytic capacity and reduced mitochondrial oxidative phosphorylation, maintained even in the presence of high oxygen concentrations. It has been proposed that this alteration may be a metabolic mechanism of tumor cells to survive and grow under hypoxic conditions through the modulation of key enzymes and transporters involved in glycolytic metabolic pathways. PCa progression is characterized by a transition from an androgen-dependent to an androgen-independent phenotype, which corresponds to early and advanced stages of PCa respectively, and an increase of glucose uptake and consumption is required for rapid proliferation of androgen-dependent cells. In addition, androgen-responsive and androgen-nonresponsive PCa cells are known to have distinct glycolytic characteristics and present different levels of oxidative stress, which have been associated with a deregulated expression of transporters and enzymes related with glucose uptake and utilization. Identifying the metabolic alterations that occur in the transition of androgen-responsive to androgen-nonresponsive stage is a fundamental step for the development of new diagnosis and therapeutic approaches of PCa.

* Corresponding author: Sílvia Socorro, Faculdade de Ciências da Saúde, Av. Infante D. Henrique, 6200-506 Covilhã, Portugal, Phone: +351 275 329 053, FAX: +351-275 329 099, E-mail: ssocorro@fcsaude.ubi.pt

Introduction

Despite intensive scientific research in the field, and remarkable advances in diagnosis and treatment, prostate cancer (PCa) continues to be a source of considerable morbidity and mortality for men around the world. For example, the American Cancer Society estimates that over 215000 new diagnoses and 32000 deaths occur per year in the United States of America [1]. Environmental issues, namely diet and other life style factors, host immune and inflammatory responses, and endogenous hormones along with genetic factors, all concur in the pathogenesis of PCa [2, 3]. Anyway, the disease results from the progression of acquired phenotypes that are continuously generated by aberrant changes in gene expression and function [4]. A large number of genes, including oncogenes, tumor suppressor genes and hormonal receptors, among others, are altered in prostate tumor tissues, and have been implicated in the etiology of PCa [5, 6]. Androgens are important regulators of cell proliferation exerting their actions mainly through the activation of the classical intracellular androgen receptor (AR) [7-9]. The size of adult prostate gland is maintained by a homeostatic balance between the processes of cell renewal and death. This balance is essentially regulated by androgens, which are known to play a key role in the control of prostate cell growth and evolution of PCa [10]. Histologically PCa progresses from precursor lesions, termed prostatic intraepithelial neoplasia (PIN) arising primarily in the peripheral zone of prostate. In contrast, the non-malignant overgrowth of prostate epithelial cells - Benign Prostatic Hyperplasia (BPH) - which is common among aging man occurs almost exclusively within the transition zone [11]. PCa is initially confined to the prostate and at initial diagnosis the majority of tumors show an androgen-dependent proliferation [12]. Therefore, the AR is commonly used as therapeutic target in PCa treatment and androgen deprivation approaches result in tumor regression [13]. However, as disease progresses, PCa is characterized by a transition from the androgen-dependent to the androgen-independent phenotype [14], and these tumor cells are able to survive and progress even in the absence of circulating androgens. Unfortunately, as the new androgen-independent status emerges, androgen ablation therapies fail and tumor recurs to a lethally resistant phenotype [15, 16]. Understanding the pathways that lead to androgen-nonresponsiveness and characterizing the physiology of androgen-nonresponsive PCa cells is a crucial step towards developing new and effective therapies for this lethal form of PCa.

The uncontrolled cell proliferation typical of cancerous tissues is associated with increased energy requirements. Increased glucose consumption is a basic feature of prostate malignant cells, and their glycolytic pathways must provide the energy needed for abnormal cell proliferation and growth [17, 18]. Increased rates of glycolysis may also confer adaptive advantages if the resulting excess of pyruvate is available for lipid synthesis, or provides essential anabolic substrates, such as ribose, for nucleic acid synthesis [19]. In addition, PCa has been considered to have an unique metabolism and the enhanced glycolytic metabolic profile attributed to the advanced stages of PCa results from adaptation of cells to hypoxia, being strongly linked with oncogene activation and silencing of tumor suppressor genes [20, 21]. Also, a relationship between androgens and the metabolic changes associated with progression of PCa has been established [22, 23]. Moreover, it was recently shown that androgen-responsive and androgen-nonresponsive PCa cells display distinct metabolic profiles [24, 25]. Detailing the metabolic reprogramming and the characteristic metabolic profile of PCa cells, and also unraveling the molecular hallmarks associated with transition to androgen-nonresponsive stages will undoubtedly contribute for development of new metabolism-based approaches in diagnosis and treatment of PCa.

Prostate Cancer Development and Androgen-Responsiveness

PCa is a multifactorial disease arising from mutations in cancer susceptibility genes, action of environmental agents that influence the acquisition of somatic genetic changes, and several other systemic or local factors, including behavior/diet, hormones and growth factors, resulting in cell growth or survival advantages [26]. In addition, epigenetic changes contribute also to the development and progression of this disease [27].

PCa progresses from the PIN precursor lesions (Figure 1) and the high-grade PIN is considered to represent the most likely precursor of PCa [11, 28]. It has been identified as traditional histopathology and consists of prostate acini, lined by basal cells and by luminal cells that show dysplastic features including nuclear hyperchromasia, nuclear crowding and prominent nucleoli. An alternative precursor is proliferative inflammatory atrophy (PIA). PIA is defined as discrete foci of proliferative glandular epithelium occurring in association with inflammation [11, 29]. Prostate dysplastic lesions, in turn, may subsequently give rise to a heterogeneous population of cancer epithelial cells showing aberrant differentiation, unlimited division and a decreased rate of apoptotic cell death [30]. PCa can be confined within the prostate, spread into adjacent tissue or lead to a metastatic disease, first to the pelvic lymph nodes and eventually to distant lymph nodes and bones [12]. During the initial stages, PCa cells exhibit unique characteristics with androgens regulating their growth, and thus tumor proliferation. Androgen action is mediated through AR, which appears to have structural and functional alterations during PCa progression [31]. In primary prostate cancer only few *AR* mutations have been found, and at initial stages androgen ablation therapies lead to a tumor regression in the majority of patients [28, 31]. However, PCa cells may acquire additional genetic alterations in comparison with early stages of the disease; among them are mutations in the *AR* gene, which are frequently found in metastatic PCa [31].

In this advanced stage, PCa cells are able to survive and proliferate without the signals that in normal conditions are delivered by circulating androgens [28]. It corresponds to the so-called recurrent androgen-independent PCa phenotype or hormone-refractory PCa, and in these cases androgen ablation therapies eventually fail [15, 16, 28, 31]. Therefore, the loss of androgen responsiveness restricts therapeutic options and is usually associated with lethal forms of disease [32]. This androgen insensitivity refers to the ability of tumor cells to grow even without circulating androgens, through the activation of AR intracellular signaling pathways by mechanisms independent of hormone-binding to receptor [28]. There are several pathways involved in androgen-independent PCa progression, and many antiapoptotic gene products, growth factors and their receptors are differently expressed in the different stages of disease progression, contributing to tumor growth and cell survival, migration and invasiveness.

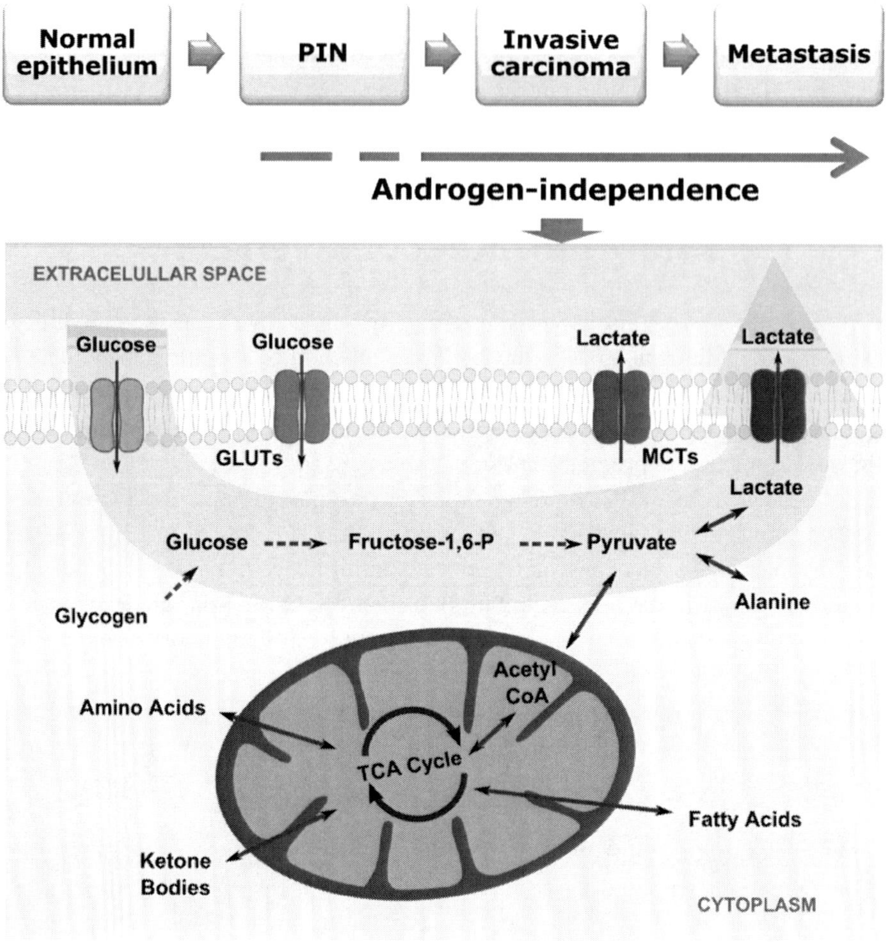

Figure 1. Schematic diagram of progression stages of prostate cancer (PCa) and the overall cellular metabolism. The stages of progression are indicated, together with alterations in the metabolic pathways associated with the androgen-independent advanced stage of PCa. Tumor cells display enhanced glycolytic capacity due to increased glucose uptake from the extracellular media. After transport into the cells via glucose transporters (GLUTs), glucose is metabolized to fructose-1,6-P, which in turn originates pyruvate. Pyruvate can be converted to lactate via lactate dehydrogenase (LDH), or alternatively originate alanine. The produced lactate is transported to the extracellular milieu by monocarboxylate transporters (MCTs). Gray-shaded arrow highlights the metabolic pathway identified in androgen-nonresponsive PCa cells, which exhibit increased lactate production with concomitant increased LDH activity and MCTs levels. Abbreviations: GLUTs, glucose transporters; MCTs, monocarboxylate transporters; PIN, prostatic intraepithelial neoplasia; TCA, tricarboxylic acid.

The activation of the signaling cascades mediated by AR, estrogen receptor (ER), epidermal growth factor receptor (EGFR), Human Epidermal Growth Factor Receptor 2 (HER's-2), hedgehog and Wnt/β-catenin, and the differential expression of cell survival factors, such as Bcl-2, p53, Bax, syndecan-1 and clusterin are among the most frequent genetic alterations associated with aggressiveness of PCa [30, 33, 34]. Moreover, the down-regulation of elements in several apoptotic signaling cascades, such as ceramides and caspases, was also observed in metastatic PCa cells [30]. Indeed, PCa growth occurs by the unbalanced ratio between cell proliferation and apoptosis. Androgens are viewed as the main regulators of this ratio, and in patients who androgen ablation had failed, a high percentage of tumors present AR overexpression, which is rarely observed in the early stages of PCa. The androgen/AR pathway may be hyperactivated through different mechanisms, which were reviewed by [32]: (a) enhanced AR sensitivity at low androgens levels by increasing AR expression or elevation of the local androgen production by increasing 5α-redutase activity; (b) alteration of the normal androgen response by way of mutations in AR gene, that lead to decreased specificity of the ligand binding, with AR becoming responsive to other steroid hormones, growth factors or antiandrogens, or by changes in the expression of co-regulators of the AR; and (c) several growth factors, including EGF, keratinocyte growth factor (KGF), insulin-like growth factor-1 (IGF-1) can activate AR and AR target genes in absence of androgens [30]; HER-2/neu (and eventually other receptor tyrosine kinases) usually overexpressed in PCa, could be an important mechanism for the progression to hormone refractory prostate disease, because it can activate the AR by indirect activation of mitogen-activated protein kinase (MAPK), the phosphatidylinositol 3-kinase (PI3K)/Akt pathway, or inactivating the tumor suppressor gene phosphatase and tensin homolog (PTEN) [35]. Akt is also involved in the inactivation of pro-apoptotic Bad and procaspase-9 proteins [36].

Other alternative or complementary pathways to AR signaling may be activated in the androgen-independent stages, which include Bcl-2 upregulation and/or combined loss of p53/Rb or PTEN/NKx3.1 tumor suppressor genes [37-39]. Others refer that the androgen independence characteristic of the advanced stage of PCa results from a clonal survival and expansion of transformed prostate epithelial stem cells [40]. The stimulation of the AR transcription rate may promote cell proliferation at low androgens levels. On the other hand, the diminution or loss of AR activity promotes growth and selection of cell clones that proliferate in an AR independent manner [35]. A large part of the current knowledge on the physiology of androgen-dependent and androgen-independent PCa stages comes from *in vitro* studies using different PCa cell line models. The majority of studies, classically and routinely have been using PC3 and DU145 cell lines as androgen nonresponsive, and LNCaP cell line as androgen-responsive models, mimicking, respectively, androgen-independent and androgen-dependent PCa stages. PC3 and DU145 cells are derived from bone and brain metastasis, respectively, and are associated with a relatively high proliferation rate [41, 42]. The LNCaP cell line, originated from a lymph node metastasis, expresses a functionally active mutated form of the AR [43] and is androgen-responsive in terms of cell proliferation [44]. In sum, several of the genetic alterations that cause PCa development directly affect the cellular response to hypoxia, the ability of tumor cells to recruit new blood vessels, and glycolysis.

Metabolic Reprogramming and the "Warburg Effect" in Prostate Cancer Cells

The energetic substrate ATP, required for maintenance of cell functions, is produced through aerobic glycolysis by the conversion of glucose via pyruvate into acetyl-CoA, which in non neoplastic cells is completely oxidized in the mitochondria to carbon dioxide and water, through the tricarboxylic acid (TCA) cycle and oxidative phosphorylation [18]. As an adaptive condition cancer cells exhibit alternative metabolic strategies to ensure energy production. The first and the well-known energetic metabolism alteration in tumor cells, was proposed by Warburg in 1920 [17] and it has been known as the ‘‘Warburg effect’’. It assumes that glycolysis is predominantly used for ATP production in detriment of oxidative phosphorylation, producing high levels of lactate, an alteration maintained even in the presence of high oxygen levels (‘‘aerobic glycolysis’’) [17, 45]. Glycolysis is the conversion of glucose to pyruvate, which can be subsequently transformed in lactate. For this, glucose enters into the cell via glucose transporters (GLUTs), then is metabolized to fructose-1,6-bisphosphate, which in turn originates pyruvate (Figure 1). Hexokinase (HK) and phosphofructokinase (PFK) are both important glycolytic enzymes involved in this process. Pyruvate originates lactate via lactate dehydrogenase (LDH) activity, which is extruded to the extracellular medium through monocarboxylate transporters (MCT) [45]. Tumor cells may have to survive in conditions of fluctuating oxygen tension and the increase in glycolytic flux (Figure 1) to generate ATP, may be a metabolic strategy to ensure their survival and growth in environments with low oxygen concentrations [46].

Tumor cells, as highly proliferating cells, consume elevated amounts of glucose and glutamine. Glucose is mostly converted to pyruvate through glycolysis, which is frequently excreted as lactate, an important step for regeneration of nicotinamide adenine dinucleotide (NAD^+) for glycolysis. Lactate production contributes to the acidification of the microenvironment, which favors tumor invasion [47] and suppresses anticancer immune effectors [48]. The lactate produced by tumor cells, eventually enters in stromal cells following oxidative phosphorylation pathway or regenerating pyruvate that in turn, may be used by cancer cells [45, 49]. This process generates complementary metabolic pathways allowing cancer cells survival and growth. On the other hand, glucose can also be used as a source of carbon to produce ribose, used in DNA and RNA synthesis, through the pentose phosphate pathway (PPP). In addition, this process generates nicotinamide adenine dinucleotide phosphate (NADPH) that ensures antioxidant defenses of the cell and contributes to fatty acid (FA) synthesis [20]. Pyruvate can also be converted to acetyl-CoA (Figure 1) that tends to be introduced in a “truncated” TCA cycle in the form of citrate, which is exported back to the cytosol, and acetyl-CoA becomes also available for the synthesis of FA, cholesterol and isoprenoids [19]. *De-novo* FA and sterol synthesis are both characteristics of early and advanced PCa. A number of studies have demonstrated diminished citrate in PCa [19], consistent with the regulation by androgens of a number of key enzymes involved in FA and cholesterol synthesis, which exhibit increased activity in PCa cells [50]. PCa express high levels of the enzyme Fatty Acid Synthase (FASN), a key metabolic enzyme catalyzing the terminal steps in the synthesis of long chain saturated FAs [51]. The PCa cell lines, PC3 and DU145, demonstrated significantly increased mRNA levels of sterol regulatory element-binding proteins SREBP-1a and SREBP-2 in comparison with LNCaP, but diminished mRNA levels of SREBP-1c, the isoform linked to lipogenesis. Although no significant differences in FASN mRNA levels were verified, Acetyl-CoA carboxylase 1 (ACC1), an enzyme that also mediates lipid synthesis, was significantly more expressed in both PC3 and DU145 cells relatively to LNCaP [25]. The increase in lipid synthesis is advantageous for tumor cells in terms of growth advantage, self-survival and drug resistance.

Interestingly, in cancer cells, glycolytic intermediaries are provided for anabolic reactions, including glucose 6-phosphate for the synthesis of glycogen, ribose 5-phosphate essential for sustention of protein and DNA synthesis, and pyruvate to produce non-essential amino acids such as alanine [20]. In cancers with a predisposition to use aerobic glycolysis for producing energy, the emergence of this new metabolic phenotype may arise from the selection of cells that have adapted to the tumor microenvironment and/or from anomalous signaling induced by genetic changes. It is not totally clear how this is an advantage for tumor growth in a different microenvironment, but increased levels of both lactate and alanine were detected in PCa compared with benign prostate tissues [52], which indicates enhanced glycolytic activity or the Warburg effect (Figure 1). Although, the underlying molecular mechanisms need to be detailed, it is evident that the resulting alterations confer adaptive, proliferative, and survival advantages to the tumor cells.

Mechanisms of Metabolic Reprogramming

The mechanisms that underlie tumor metabolic reprogramming have been investigated and one of the main events involved in these metabolic changes resides in the activation of hypoxia-inducible transcription factor (HIF), an important transcription factor that is activated by hypoxic stress, due to different oxygen concentrations in the tumor environment [53]. Also, the activation of oncogenes or inactivation of tumor suppressor genes, inflammatory response, metabolic factors, and oxidative stress contribute to HIF expression [54-56]. HIF controls the transcription of several genes involved in cell survival, increases metabolic resistance to hypoxia and apoptosis, favors tumor invasion, stimulates angiogenesis by upregulating vascular endothelial growth factor (VEGF) and regulates glucose metabolism [57]. HIF-1 enhancement promotes the conversion of glucose to pyruvate by upregulating GLUT1, HK1 and HK2 (catalyzers of the initial step of glycolysis), PFK-1 and PFK-2 (involved in an irreversible reaction in glycolysis), and LDHA and MCT4 (involved in lactate production and transport) [46, 55]. In addition, HIF-1 decreases the conversion of pyruvate to acetyl-CoA by regulating pyruvate dehydrogenase kinase 1, which inhibits pyruvate dehydrogenase (PDH) [58, 59]. All these mechanisms lead to a stimulation of the glycolytic pathway. The majority of cancers, including PCa, present a distinct glycolytic metabolic profile under regulation of HIF [60]. In PCa, HIF-1 expression is upregulated at early phases of the neoplasic development [61] but it is also highly correlated with the risk of metastases [57]. In addition, DU145 and PC3 PCa cells, which represent more aggressive states of this disease, show higher levels of HIF-1, as well as of HIF target genes such as VEGF isoforms, than LNCaP cells [25, 62]. In addition to HIF-1 upregulation, also oncogenes and tumor suppressor genes have been recognized to be important in the metabolic reprogramming at different levels. The tumor suppressor gene *p53*, regulates metabolism at multiple steps, including the enhancement of glutamate synthesis, through upregulation of glutaminase 2 [63], and also the inhibition of FA synthesis and glycolysis. P53 regulates positively the synthesis of SCO2 (synthesis of cytochrome c oxidase protein), which interferes with the mitochondrial electron transport chain [64] and activates TIGAR (TP53-induced glycolysis and apoptosis regulator) inhibiting glycolysis by decreasing the levels of fructose-2,6-bisphosphate (Fru-2,6-P2) [65].

The oncogenes Ras and myc, that are often mutated and hyperactivated in a set of cancers, including PCa, increase glycolysis and lactate production by enhancing the expression of various metabolism related genes [66-68]. Myc is also the main oncogene implicated in glutamine addiction of cancer cells, through the upregulation of glutamate synthesis [69]. The mammalian target of rapamycin (mTOR) pathway is hyperactivated in

PCa regulating glycolysis, through HIF-1 dependent and independent pathways, and lipogenesis through the transcription factors SREBP and Myc. In addition, PI3K/Akt and MAPK pathways are also involved in glycolysis and lipid biosynthesis, and are deregulated and constitutively activated in human PCa [51]. 5-AMP-activated protein kinase (AMPK), other control point in different metabolic pathways is also deregulated in PCa. AMPK-mediated phosphorylation events participate in the regulation of glucose and lipid metabolism, through inhibiting metabolic functions that require ATP consumption, which includes FA, sterols, glycogen and protein synthesis, and increases glucose uptake and mitochondrial activity, controlling lipogenic and PI3K/mTOR pathways [51].

Glycolytic Profile of Androgen-Responsive and Androgen-Nonresponsive Prostate Cancer Cells

Tumor cells need a constant supply of glucose as an important source of energy, as well as a precursor of glycoproteins, triglycerides and glycogen. PCa has been recognized to have unique metabolic features associated with neoplasic development and the rapid proliferation of androgen-nonresponsive PCa cells, which include an increase in lactate and alanine levels and an increase of glucose consumption [52, 70-72]. There are strong evidences that glycolysis is enhanced in tumor cells due to modulation of the expression of key enzymes and transporters involved in this metabolic pathway [73]. In this topic we focus on the description of the glycolytic phenotype of PCa cells and their metabolic strategies to ensure tumor growth and survival. Contrasting data on these topics, regarding androgen-responsive and androgen-nonresponsive PCa cells also will be explored. The low rate of ATP production coupled with high-energy requirements leads to an increased need of glycolysis, a process that starts with glucose uptake. This increased glucose consumption is required for the rapid proliferation of both androgen-responsive and nonresponsive PCa cells [74], and has been proved to be useful and frequently used for cancer detection by positron emission tomography (PET), through measurement of the uptake of a glucose analogue, 2-deoxy-2-fluoro-D-glucose labeled with fluorine-18 (^{18}F-FDG) [75]. Studies have demonstrated a higher FDG consumption in PCa cases with a higher Gleason score [76]. Furthermore, it was reported a decreased uptake of FDG in the prostate tissues of PCa patients undergoing androgen ablation therapy [22], which suggests that glucose metabolism differs between early and advanced stages of PCa. Also *in vitro* studies showed a higher FDG uptake in androgen-responsive PC346C cells derived from a non-progressive prostate primary tumor [23, 77]. Moreover, under androgen deprivation, a diminished FDG uptake was observed in androgen-responsive LNCaP cells [23], which was lower than that observed in androgen-nonresponsive PC3 cells [78]. Recently, our research group demonstrated that both LNCaP and PC3 PCa cells consumed higher amounts of glucose than nonneoplastic PNT1A cells [24]. Others have also described that glucose uptake is increased in PCa allowing rapid cell proliferation, and that androgen-nonresponsive PC3 and DU145 cells present a high rate of glucose consumption [74].

The transport of glucose across the plasma membrane may be the first rate-limiting step for glucose metabolism. Glucose is a hydrophilic compound and crosses the plasma membrane by mediation of GLUTs, which belong to the gene family of solute carriers 2A

(SLC2) [79]. This family consists of 14 glucose transporters, in which GLUT14 appears to be a duplication of GLUT3, and GLUT13 is a proton-coupled myoinisitol transporter (HMIT). GLUTs can be grouped in three different classes based on their sequence similarities. The class I includes the well characterized GLUT1-4, identical 48-63 % to each other in human. GLUT5, 7, 9 and 11 belong to Class II, are 36-40% identical and all demonstrate transport activity for fructose. Class III is composed for the remaining GLUT6, 8, 10, 12 and the HMIT [79]. GLUTs exhibit variable hexose affinity and are metabolic- and hormone-regulated, displaying tissue-specific expression. They mediate the bidirectional transport of glucose (or fructose) between the intracellular and extracellular space, according with the gradient concentration [80, 81]. Given that glucose and (eventually) fructose are required as substrates to an increased energy production necessary for tumor proliferation, several authors speculated that GLUT family might play a fundamental role in cancer progression. In fact, an aberrant expression of GLUT family members has been reported in cancer cases [80]. Although GLUT1, 3, 5, 7 and 10 have been detected in nonneoplastic prostate [82-84], a total loss of GLUT5 expression was observed in PCa [82]. In contrast, GLUT1 is usually upregulated in different cancer tissues, and GLUT1 mRNA expression is correlated with FDG uptake in a rat model of PCa [85]. In PCa, expression of GLUT1 gene was significantly higher in comparison with BPH tissue and correlated with differentiation grade and Gleason score [86, 87]. In addition, over-expression of GLUT1 was observed in some specimens of highly proliferative intraductal PCa [88]. Effert *et al.* investigated the expression of GLUT1 in LNCaP, PC3 and DU145 PCa cell lines and GLUT1 was detected in all cells, but the expression in the poorly-differentiated cell lines, DU145 and PC3, was higher relatively to the androgen-responsive LNCaP cell line [86]. Another report studied GLUTs expression in human PCa cell lines, LNCaP and its derivatives (C4, C4-2 and C4-2B), which are associated with increased tumor progression and androgen independence. All tested cell lines present GLUT1 and GLUT12 mRNA and protein expression, however no relationship was established between GLUT1 and GLUT12 expression of and tumor progression or androgen independence grade [89]. The presence of GLUT12 was demonstrated in malignant prostate tissues and in BPH cells surrounding cancerous ducts, but not confirmed in BPH cells from normal prostate tissues [89]. Recently, we demonstrated an increase in GLUT3 protein expression, but not GLUT1, in LNCaP cells when compared with PNT1A [24]. In addition, others had shown that androgens induce the expression of GLUT1 in LNCaP but not in PC3 cells through activation of the HIF-1 [90, 91]. These findings suggest a role for GLUT1, GLUT3 and GLUT12 for PCa progression.

Once inside the cells glucose is phosphorylated by the enzyme HK. Changes in the pattern of HK isoform expression occur in several tumor cells in comparison to nonneoplastic cells, which seem to be part of the mechanisms involved in the highly glycolytic flux presented by tumor cells [67]. In mammalian cells four different isoforms of HK (HK-1-4, or glucokinase) have been identified, which differ by their kinetic properties and tissue-specific expression [92, 93]. HK-2 is frequently overexpressed in proliferating tumor cells and seems to be important in the initiation and maintenance of high rates of glucose catabolism [67]. Another key control step in the glycolytic flux is the conversion of fructose 6- phosphate into fructose 1,6-bisphosphate by PFK-1, and is often considered a rate-limiting step in glycolysis. The activity of this enzyme is controlled by Fru-2,6-P2, whose biosynthesis and degradation are modulated by the bifunctional PFK-2 [67, 94]. PC3 cells displayed higher mRNA levels for HK-2 [25]. In DU145 cells an increase in expression of the glycolytic enzymes, HK, PFK-

1 and PFK-2 was observed [67]. Our research group demonstrated that PFK-1 is differently modulated in androgen-responsive and nonresponsive PCa cell lines; lower PFK protein levels, accompanied by a significant decrease in PFK enzymatic activity, were detected in PC3 cells relatively to LNCaP cells [24]. Importantly, it was reported that androgens upregulate the expression of HK-2 and PFK-2 in LNCaP cells, which mediated the increased uptake of glucose [95]. In addition, an increased expression of aldolase and piruvate kinase (PK), both important enzymes in glycolysis, was observed in PCa cells [25]. A significant increase in lactate and alanine concentration has been described as one of the metabolic changes associated with PCa development [52, 71, 72, 96]. *In vitro* studies suggested an increase in L-lactate levels in both androgen-responsive (LNCaP) and androgen-nonresponsive (PC3) cells. However, in our studies, a significantly higher lactate concentration was detected in the extracellular medium of both LNCaP and PC3 cells when compared with PNT1A. In addition, PC3 cells significantly produced more lactate than LNCaP [24]. Other studies also described a higher lactate production by PC3 and DU145 cells when compared to LNCaP [25], and higher lactate production/concentration ratios have been associated with more aggressive cancer stages and metastasis [72, 96, 97]. Accordingly, LDHA has been used as a marker of neoplastic transformation [67]. The lactate produced from glucose results from pyruvate conversion, a process achieved by the activity of LDH enzyme, a tetramer of A and B subunits [98]. *In vitro* studies showed an increase in LDH mRNA levels in androgen-nonresponsive PC3 cells comparatively to androgen-responsive LNCaP cells [24, 25], and augmented enzymatic activity of LDH in both PC3 and LNCaP cells in comparison with PNT1A. Moreover, when comparing both cancer cell lines, it was observed a significant higher LDH enzymatic activity in androgen-nonresponsive PC3 cells [24]. MCTs represent a large gene family (*SLC16*) composed by 14 members [99]. MCT1 to MCT4 proteins are proton symporters and mediate the transmembrane transport of lactate, pyruvate and ketone bodies [99]. So, MCTs are responsible for lactate export, associated with hyperglycolytic phenotype found in cancer cells, and for pH regulation, important to acid-resistant phenotype. MCTs have been associated with PCa progression and it has also been reported a correlation between MCT4 expression and a poor prognosis in PCa [100, 101]. MCT1 and MCT4 are expressed in around 90% of PCa cases, while no expression was found in nonneoplastic prostate tissues, PIN lesions or in non-tumor regions adjacent to primary PCa tissues [100]. The expression of MCT1 and MCT4 was associated with high pretreatment PSA levels, high Gleason score, high pathological stage, and nodal involvement [100].

In another study, MCT1 expression was demonstrated in normal samples and significantly less frequently expressed in tumor samples, and its expression was associated with higher PSA levels, absence of perineural invasion, and presence of biochemical recurrence [101]. On the other hand, MCT2 and MCT4 were significantly more expressed in tumor cells than in nonneoplastic tissues [101]. MCT4 expression was associated with higher PSA levels, advanced tumor stage, higher Gleason score, presence of perineural invasion, and presence of biochemical recurrence [101]. *In vitro* studies reported that MCT4 protein expression is increased in both, PC3 and LNCaP cell lines relatively to nonneoplastic PNT1A cells, and a significantly higher expression was observed in androgen-nonresponsive cells, PC3, in comparison with androgen-responsive LNCaP cells, which is indicative of enhanced MCT4 protein expression in more aggressive stages of PCa [24]. Besides the lactate, also alanine results from pyruvate conversion. PCa has been shown to produce higher levels of alanine compared with nonneoplastic prostate tissue [52, 70, 71]. A study performed by proton and 13C HR-MAS spectroscopy using pyruvate as a metabolic subtract demonstrated a high fractional enrichment of alanine in PCa cells [102].

Other *in vitro* studies, demonstrated that both, androgen-responsive, LNCaP, and androgen-nonresponsive, PC3 cells, produce significantly more alanine than the noncancerous PNT1A cells [24]. The lactate/alanine ratio is associated with the $NADH/NAD^+$ equilibrium in the cell because the conversion of pyruvate to lactate and alanine is coupled with NAD^+ and NADH reactions. Thus, lactate/alanine ratio reflects the cell redox state [103]. We reported that lactate/alanine ratio is increased in both PC3 and LNCaP cancerous cell lines, but the ratio is higher in androgen-nonresponsive PC3 cells, which suggests a higher oxidative stress in androgen-nonresponsive cells. This is particularly attractive since it has been suggested that antioxidant therapy prevents the progression from androgen-dependent to androgen-independent PCa [104].

Glucose Uptake and Future Target Therapies/Hallmarks in Prostate Cancer

The development of novel chemopreventive and chemotherapeutic strategies preventing prostate carcinogenesis and PCa progression into more aggressive stages is of great clinical interest, either by reducing costs or improving health outcomes for patients. PCa cancer, similarly to tumors of other tissues, is highly dependent on glucose metabolism presenting a distinct glycolytic profile associated with hallmarks of disease progression (Figure 2), which strongly supports the demand for new anticancer treatments targeting glycolysis [23-25, 74, 76]. Indeed, intense research efforts have targeted the metabolic pathways altered in PCa development and progression. Also, therapeutic strategies aiming at reach nonmetabolic pathways of PCa, including the inhibition of cell-autonomous growth signals (suppressing the PI3K/Akt/mTOR pathway), restoration of apoptosis (one example is the reestablishment of P53 function) and inhibition of angiogenesis (by inhibition of HIF), may also suppress cancer-specific metabolic programs and normalize the metabolic state (Figure 2) [45]. In addition, the inhibition of the processes and enzymes involved in cancer metabolic reprogramming may have a fundamental role in limiting cancer cell-specific bioenergetic requirement and anabolic reactions, and hence reversing the neoplastic phenotype [45, 51]. The prevalence of a high glycolytic state in cancer cells results primarily from overexpression of the high affinity glucose transporters GLUTs and the phosphorylation of glucose by the HK-2 enzyme. It has been demonstrated by clinical application of PET imaging measuring the incorporation of ^{18}F-FDG in cancer tissues, in which 2-deoxyglucose (2DG) competes with glucose for transmembrane transport and is then phosphorylated by HK-2, but due to the lack in 2′hydroxyl group cannot be metabolized, which leads to accumulation in the cells [105]. In addition, treatment with 2DG leads to intracellular ATP depletion and induces autophagy in PCa cells [106]. 2DG has been tested as an anticancer drug, and a recently terminated Phase I clinical study in PCa patients defined a dose of 45 mg/kg to be used in Phase II trials [107]. Targeting glucose metabolism with agents such as 2DG may have a potential to competitively reduce glucose uptake and change the tumor glycolytic profile [106, 108].

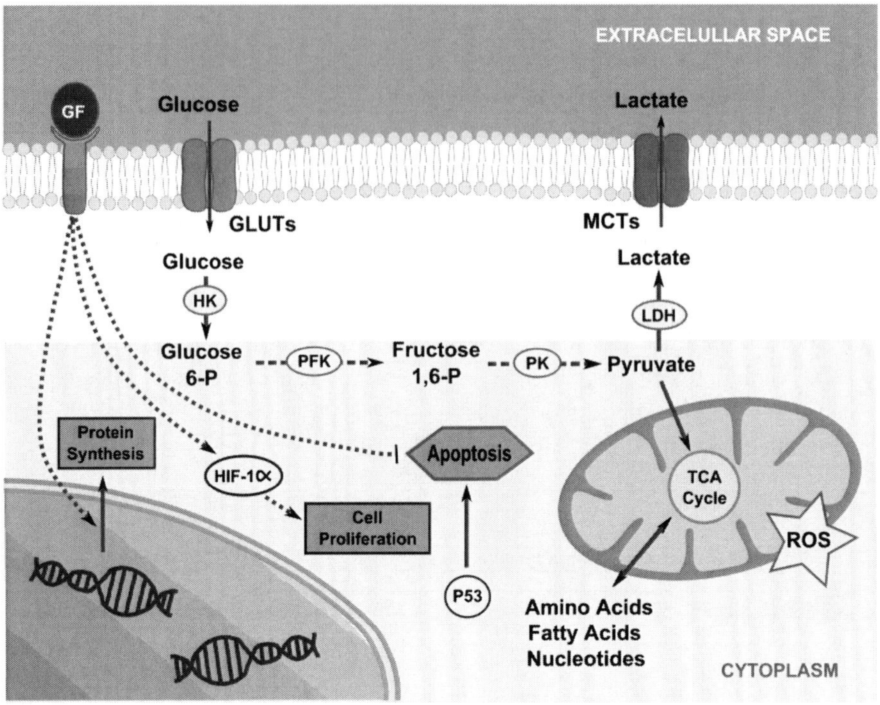

Figure 2. Schematic illustration of the main cellular metabolic pathways in prostate cancer (PCa) with emphasis in the hallmarks of disease development and progression, which could be important therapeutic targets. Prostate tumorigenesis is a multistep process that reflects the progressive transformation of normal cells into malignant derivatives. PCa cells present a distinct glycolytic profile, associated with the stage of progression, which strongly supports the increased cellular energetic demands. The prevalence of high glycolysis in PCa cells results primarily from overexpression of the high affinity glucose transporters (GLUTs), glycolytic enzymes and lactate transporters (MCTs). Cell-autonomous growth signals, increase of angiogenesis (via HIF-1α) and suppression of apoptosis (with loss of P53 function) are also important features of PCa. Elevated oxidative stress in the prostatic microenvironment occupies also a central role in PCa progression by leading to inflammation and induction of lipid, protein and DNA damage that causes genomic instability. Dotted-arrows represent direct or indirect activation and blunt dotted-arrow indicates an inhibitory effect. Abbreviations: GLUTs, glucose transporters; HK, hexokinase; LDH, lactate dehydrogenase; MCTs, monocarboxylate transporters; PFK, phosphofructokinase; PK, pyruvate kinase; ROS, reactive oxygen species; HIF-1α, Hypoxia-Inducible Factor; TCA, tricarboxylic acid.

In addition, studies based on glucose deprivation or using glucose uptake inhibitors (product-based energy restriction-mimetic agents-ERMA) have been reported in a therapeutic/preventive perspective and a link between ERMA antitumor effects [74, 109-111]. Another strategy to inhibit glucose metabolism may be through the inhibition of expression or functionality of GLUT family members (Figure 2). Specially GLUT1 and GLUT3 have a high affinity for glucose and therefore, once expressed, play an essential role ensuring efficient glucose uptake when glucose becomes a limiting resource [112]. As previously described, GLUT1, GLUT3, and more recently GLUT12, were demonstrated to be overexpressed in PCa, with GLUT1 being differentially expressed in tumors with different grades of disease evolution [86, 89]. Thus, the use of GLUT inhibitors may constitute a good anticancer strategy in PCa in different stages of the disease. The silybin/silibinin is a GLUT inhibitor recently identified, which inhibits tumor formation and is actually being tested on ongoing clinical Phase I trials for PCa treatment [113]. Nevertheless, the development of specific inhibitors for GLUTs isoforms, namely for GLUT1 and GLUT3, can be a useful strategy for PCa treatment, targeting stages of disease specifically overexpressing these transporters. Lactate production and extrusion to the extracellular space contributes to the acidification of the microenvironment and is associated with the glycolytic metabolic profile presented by PCa cells (Figs. 1 and 2) [24, 25]. This perturbation selects cells that present an acid-resistant phenotype and it has been considered an essential condition for cancer cell survival [114]. The inhibition of these adaptive mechanisms could be potentially used as a target to kill or to diminish the proliferation rate of cancer cells, increase the immune response against tumor cells and increase sensitivity to chemotherapeutic agents [115]. Other interesting target playing an essential role producing lactate and restoring the NAD for maintenance of the high-rate glycolytic profile is LDHA (Figure 2). Since LDHA has been described to be overexpressed in PCa cells [24], it can be a potential therapeutic option in PCa [116]. These results are particularly interesting knowing that in other human cancer cells, LDHA silencing by RNA interference impairs tumor initiation, maintenance and progression [117, 118].

The transport of lactate across the membrane is mediated through MCTs. MCTs play an essential role in pH regulation acting as transporters of lactate in the metabolic symbiosis created between glycolytic tumor cells and oxidative tumor cells, which produce and consume lactate respectively [119]. Importantly, MCT expression has been correlated with the progression of PCa [100]. MCT1, MCT2 and MCT4 expression was demonstrated to be increased in tumor cells, and specifically, MCT4 levels were correlated with a poor prognosis in PCa [100, 101]. In addition, MCT4 silencing results in decreased cancer cell migration through a mechanism involving interaction of MCT4 with β1-integrin [120]. This suggests that MCT1 and MCT2 should be further explored as tumor marker, and MCT4 as a marker of poor prognosis in PCa [101]. The contribution of lactate to the malignant phenotype, together with the up-regulation of MCTs in PCa cells makes MCT inhibition a tempting strategy for PCa treatment. Moreover, inhibitors of enzymes that occupy a central role in cancer-specific metabolism, such as HK, PFK and PK, all important enzymes in the glycolytic process (Figure 2) and overexpressed in PCa, might be targets for cancer therapy [24, 25, 67]. Other interesting concept is the function of redox metabolism and reactive oxygen species (ROS) in tumor progression. Production and removal of ROS is intimately related with glucose and glutamine metabolism (Figure 2). In cancer, ROS may act as signaling molecules that promote proliferation or as mediators of cell death [121]. Our studies suggested that androgen-nonresponsive PCa cells, representing more aggressive stages of disease, display higher oxidative stress levels compared to androgen-responsive cells [24]. Some of the effects of metabolic reprogramming may due to regulation of ROS, and the use of antioxidant therapies have been suggested to prevent the progression from androgen-dependent to androgen-independent PCa [104].

Other attractive and possible anticancer target is HIF-1, which plays a critical role in upregulating glycolysis as an adaptation of tumor cells to hypoxia and is also upregulated in different stages of PCa [25, 57]. In addition to glucose metabolism, unique genetic and biochemical features in amino acid metabolism, lipid metabolism and nucleotide metabolism of cancer have been reported as potential targets for PCa cancer imaging and therapies [51, 122, 123]. However, the heterogeneity of tumor cells and environment conditions leads to the existence of a range of metabolic pathways used by invasive tumor cells, implying the use of multiple agents that inhibit energy production through aerobic and anaerobic pathways. The most therapeutic strategies in research associate the use of glycolytic inhibitors with chemo-radiotherapeutic agents, increasing the cytotoxicity of such agents [124, 125].

Conclusion

Cancer is a metabolic disease, and not surprisingly an altered glucose metabolism has been described in primary and metastatic PCa. Moreover, an enhanced glycolytic profile is characteristic of androgen-nonreponsive PCa cells, resulting from the necessity of these cells to produce enough energy to survive when resources are limited. Androgen-nonresponsive PCa cells also use metabolic intermediates to the biosynthetic pathways, including lipid and nucleic acid synthesis, which support cell proliferation. Currently, irrefutable and promising clinical and experimental evidences highlight the potential of therapeutic strategies inhibiting the targets that participate in the metabolic reprogramming, in order to suppress growth of established invasive PCa. Targeting the metabolic processes alone or in combination with other anti-cancer drugs may represent an improvement in the treatment of advanced PCa.

Funding

This work was partially supported by the Pluriannual Programme of Portuguese Foundation for Science and Technology (FCT). PF Oliveira was financed by FCT under the program Ciência 2008. Cátia Vaz, Ricardo Marques and Marco G. Alves were funded by FCT fellowships (SFRH/BD/70316/2010, SFRH/BD/66875/2009 and SRFH/BPD /80451 /2011, respectively).

References

[1] American Cancer Society. *Cancer Facts and figures 2010* American Cancer Society: Atlanta, 2010.
[2] Ekman, P. Genetic and environmental factors in prostate cancer genesis: identifying high-risk cohorts. *European Urology*, 1999 35(5-6), 362-369.
[3] Clinton, SK; Giovannucci, E. Diet, nutrition, and prostate cancer. *Annual Review of Nutrition*, 1998 18(1), 413-440.
[4] Grover, PL; Martin, FL. The initiation of breast and prostate cancer. *Carcinogenesis*, 2002 23(7), 1095-1102.
[5] Kallioniemi, OP; Visakorpi, T. Genetic basis and clonal evolution of human prostate cancer. *Advances in Cancer Research*, 1996 68, 225-255.
[6] Rubin, MA; De Marzo, AM. Molecular genetics of human prostate cancer. *Modern Pathology*, 2004 17(3), 380-388.
[7] Yeap, BB; Krueger, RG; Leedman, PJ. Differential posttranscriptional regulation of androgen receptor gene expression by androgen in prostate and breast cancer cells. *Endocrinology*, 1999 140(7), 3282-3291.

[8] DePrimo, SE; Diehn, M; Nelson, JB; Reiter, RE; Matese, J; Fero, M; Tibshirani, R; Brown, PO; Brooks, JD. Transcriptional programs activated by exposure of human prostate cancer cells to androgen. *Genome Biology*, 2002 3(7).

[9] Nelson, PS; Clegg, N; Arnold, H; Ferguson, C; Bonham, M; White, J; Hood, L; Lin, B. The program of androgen-responsive genes in neoplastic prostate epithelium. *Proceedings of the National Academy of Sciences*, 2002 99(18), 11890-11895.

[10] Prins, GS; Jung, MH; Vellanoweth, RL; Chatterjee, B; Roy, AK. Age-dependent expression of the androgen receptor gene in the prostate and its implication in glandular differentiation and hyperplasia. *Developmental Genetics*, 1998 18(2), 99-106.

[11] Joshua, A; Evans, A; Van der Kwast, T; Zielenska, M; Meeker, A; Chinnaiyan, A; Squire, J. Prostatic preneoplasia and beyond. *Biochimica et Biophysica Acta-Reviews on Cancer*, 2008 1785(2), 156-181.

[12] Abate-Shen, C; Shen, M. Mouse models of prostate carcinogenesis. *Trends in Genetics*, 2002 18(5), S1-S5.

[13] Brawer, MK. Hormonal therapy for prostate cancer. *Reviews in Urology*, 2006 8(Suppl 2), S35.

[14] Jenster, G. The role of the androgen receptor in the development and progression of prostate cancer. *Seminars in Oncology*, 1999 26, 407-421.

[15] Thompson, J; Hyytinen, ER; Haapala, K; Rantala, I; Helin, HJ; Janne, OA; Palvimo, JJ; Koivisto, PA. Androgen receptor mutations in high-grade prostate cancer before hormonal therapy. *Laboratory Investigation*, 2003 83(12), 1709-13.

[16] Koivisto, P; Kononen, J; Palmberg, C; Tammela, T; Hyytinen, E; Isola, J; Trapman, J; Cleutjens, K; Noordzij, A; Visakorpi, T; Kallioniemi, OP. Androgen receptor gene amplification: a possible molecular mechanism for androgen deprivation therapy failure in prostate cancer. *Cancer Research*, 1997 57(2), 314-9.

[17] Warburg, O; Wind, F; Negelein, E. Ueber den Stoffwechsel von Tumoren im korper. *Jounal of Molecular Medicine*, 1926 5(19), 829-832.

[18] Costello, LC; Franklin, RB. '~Why do tumour cells glycolyse?': From glycolysis through citrate to lipogenesis. *Molecular and Cellular Biochemistry*, 2005 280(1), 1-8.

[19] Trock, BJ. Application of metabolomics to prostate cancer. *Urologyc Oncology*, 2011 29(5), 572-581.

[20] Gatenby, RA; Gillies, RJ. Why do cancers have high aerobic glycolysis? *Nature Reviews Cancer*, 2004 4(11), 891-899.

[21] Jones, RG; Thompson, CB. Tumor suppressors and cell metabolism: a recipe for cancer growth. *Genes and Development*, 2009 23(5), 537-548.

[22] Oyama, N; Akino, H; Suzuki, Y; Kanamaru, H; Ishida, H; Tanase, K; Sadato, N; Yonekura, Y; Okada, K. FDG PET for evaluating the change of glucose metabolism in prostate cancer after androgen ablation. *Nucleal Medicine Communications*, 2001 22(9), 963-969.

[23] Emonds, KM; Swinnen, JV; van Weerden, WM; Vanderhoydonc, F; Nuyts, J; Mortelmans, L; Mottaghy, FM. Do androgens control the uptake of 18 F-FDG, 11 C-choline and 11 C-acetate in human prostate cancer cell lines? *European Journal of Nuclear Medicine and Molecular Imaging*, 2011 38, 1842-1853.

[24] Vaz, CV; Alves, MG; Marques, R; Moreira, PI; Oliveira, PF; Maia, CJ; Socorro, S. Androgen-responsive and nonresponsive prostate cancer cells present a distinct

glycolytic metabolism profile. *The International Journal of Biochemistry and Cell Biology*, 2012 44(11), 2077-84.

[25] Higgins, L; Withers, H; Garbens, A; Love, H; Magnoni, L; Hayward, S; Moyes, C. Hypoxia and the metabolic phenotype of prostate cancer cells. *Biochimica et Biophysica Acta*, 2009 1787(12), 1433-1443.

[26] Ponder, B. Cancer genetics. *Nature*, 2001 411(6835), 336-341.

[27] Chin, SP; Dickinson, JL; Holloway, AF. Epigenetic regulation of prostate cancer. *Clinical Epigenetics*, 2011 2(2), 151-169.

[28] Nieto, M; Finn, S; Loda, M; Hahn, WC. Prostate cancer: Re-focusing on androgen receptor signaling. *The International Journal of Biochemistry and Cell Biology*, 2007 39(9), 1562-1568.

[29] De Marzo, AM; Nakai, Y; Nelson, WG. Inflammation, atrophy, and prostate carcinogenesis. *Urologic Oncology*, 2007 25(5), 398-400.

[30] Mimeault, M; Batra, SK. Recent advances on multiple tumorigenic cascades involved in prostatic cancer progression and targeting therapies. *Carcinogenesis*, 2006 27(1), 1-22.

[31] Marcelli, M; Ittmann, M; Mariani, S; Sutherland, R; Nigam, R; Murthy, L; Zhao, Y; DiConcini, D; Puxeddu, E; Esen, A; Eastham, J; Weigel, NL; Lamb, DJ. Androgen receptor mutations in prostate cancer. *Cancer Research*, 2000 60(4), 944-9.

[32] Feldman, BJ; Feldman, D. The development of androgen-independent prostate cancer. *Nature Reviews Cancer*, 2001 1(1), 34-45.

[33] Prevarskaya, N; Skryma, R; Shuba, Y. Ca2+ homeostasis in apoptotic resistance of prostate cancer cells. *Biochemical and Biophysical Research Communications*, 2004 322(4), 1326-1335.

[34] Jin-Rong, Z; Lunyin, Y; Luiz, F; Towia, A; George, L. Progression to androgen-independent LNCaP human prostate tumors: cellular and molecular alterations. *International Journal of Cancer. Journal international du cancer*, 2004 110(6), 800.

[35] Heinlein, CA; Chang, C. Androgen receptor in prostate cancer. *Endocrine Reviews*, 2004 25(2), 276-308.

[36] Zhou, H; Li, XM; Meinkoth, J; Pittman, RN. Akt regulates cell survival and apoptosis at a postmitochondrial level. *Science Signalling*, 2000 151(3), 483.

[37] Gao, H; Ouyang, X; Banach-Petrosky, WA; Shen, MM; Abate-Shen, C. Emergence of androgen independence at early stages of prostate cancer progression in Nkx3. 1; Pten mice. *Cancer Research*, 2006 66(16), 7929-7933.

[38] Gleave, M; Tolcher, A; Miyake, H; Nelson, C; Brown, B; Beraldi, E; Goldie, J. Progression to androgen independence is delayed by adjuvant treatment with antisense Bcl-2 oligodeoxynucleotides after castration in the LNCaP prostate tumor model. *Clinical Cancer Research*, 1999 5(10), 2891-2898.

[39] Zhou, Z; Flesken-Nikitin, A; Corney, DC; Wang, W; Goodrich, DW; Roy-Burman, P; Nikitin, AY. Synergy of p53 and Rb deficiency in a conditional mouse model for metastatic prostate cancer. *Cancer Research*, 2006 66(16), 7889-7898.

[40] Craft, N; Chhor, C; Tran, C; Belldegrun, A; DeKernion, J; Witte, ON; Said, J; Reiter, RE; Sawyers, CL. Evidence for clonal outgrowth of androgen-independent prostate cancer cells from androgen-dependent tumors through a two-step process. *Cancer Research*, 1999 59(19), 5030-5036.

[41] Kaighn, ME; Narayan, KS; Ohnuki, JF; Jones, LW. Establishment and characterization of a human prostatic carcinoma cell line (PC-3). *Investigative Urology*, 1979 17(1), 16-23.

[42] Stone, KR; Mickey, DD; Wunderli, H; Mickey, GH; Paulson, DF. Isolation of a human prostate carcinoma cell line (DU 145). *International Journal of Cancer*, 1978 21(3), 274-281.

[43] Horoszewicz, JS; Leong, SS; Kawinski, E; Karr, JP; Rosenthal, H; Chu, TM; Mirand, EA; Murphy, GP. LNCaP model of human prostatic carcinoma. *Cancer research*, 1983 43(4), 1809-18.

[44] Wilding, G; Zugmeier, G; Knabbe, C; Flanders, K; Gelmann, E. Differential effects of transforming growth factor- β on human prostate cancer cells in vitro. *Molecular and Cellular Endocrinology*, 1989 62(1), 79-87.

[45] Kroemer, G; Pouyssegur, J. Tumor cell metabolism: cancer's Achilles' heel. *Cancer cell*, 2008 13(6), 472-482.

[46] Pouyssegur, J; Dayan, F; Mazure, NM. Hypoxia signalling in cancer and approaches to enforce tumour regression. *Nature*, 2006 441(7092), 437-443.

[47] Swietach, P; Vaughan-Jones, RD; Harris, AL. Regulation of tumor pH and the role of carbonic anhydrase 9. *Cancer and Metastasis Reviews*, 2007 26(2), 299-310.

[48] Fischer, K; Hoffmann, P; Voelkl, S; Meidenbauer, N; Ammer, J; Edinger, M; Gottfried, E; Schwarz, S; Rothe, G; Hoves, S. Inhibitory effect of tumor cell-derived lactic acid on human T cells. *Blood*, 2007 109(9), 3812-3819.

[49] Koukourakis, MI; Giatromanolaki, A; Harris, AL; Sivridis, E. Comparison of metabolic pathways between cancer cells and stromal cells in colorectal carcinomas: a metabolic survival role for tumor-associated stroma. *Cancer Research*, 2006 66(2), 632-637.

[50] Swinnen, JV; Heemers, H; de Sande, TV; Schrijver, ED; Brusselmans, K; Heyns, W; Verhoeven, G. Androgens, lipogenesis and prostate cancer. *The Journal of Steroid Biochemistry and Molecular Biology*, 2004 92(4), 273-279.

[51] Flavin, R; Zadra, G; Loda, M. Metabolic alterations and targeted therapies in prostate cancer. *The Journal of Pathology*, 2011 223(2), 284-295.

[52] Swanson, MG; Zektzer, AS; Tabatabai, ZL; Simko, J; Jarso, S; Keshari, KR; Schmitt, L; Carroll, PR; Shinohara, K; Vigneron, DB. Quantitative analysis of prostate metabolites using 1H HR-MAS spectroscopy. *Magnetic Resonance in Medicine*, 2006 55(6), 1257-1264.

[53] Kaelin, WG; Ratcliffe, PJ. Oxygen sensing by metazoans: the central role of the HIF hydroxylase pathway. *Molecular Cell*, 2008 30(4), 393-402.

[54] Harris, AL. Hypoxia-a key regulatory factor in tumour growth. *Nature Reviews Cancer*, 2002 2(1), 38-47.

[55] Semenza, GL. Hypoxia-inducible factor 1 (HIF-1) pathway. *Science Signalling*, 2007 2007(407), cm8.

[56] Taylor, CT; Pouyssegur, J. Oxygen, hypoxia, and stress. *Annals of the New York Academy of Sciences*, 2007 1113(1), 87-94.

[57] Kimbro, K; Simons, J. Hypoxia-inducible factor-1 in human breast and prostate cancer. *Endocrine-Related Cancer*, 2006 13(3), 739-749.

[58] Kim, J; Tchernyshyov, I; Semenza, GL; Dang, CV. HIF-1-mediated expression of pyruvate dehydrogenase kinase: a metabolic switch required for cellular adaptation to hypoxia. *Cell Metabolism*, 2006 3(3), 177-185.

[59] Papandreou, I; Cairns, RA; Fontana, L; Lim, AL; Denko, NC. HIF-1 mediates adaptation to hypoxia by actively downregulating mitochondrial oxygen consumption. *Cell Metabolism*, 2006 3(3), 187-197.

[60] Pan, Z; Chen, Z. Hypoxia-inducible factor 1 and prostate cancer. *National Journal of Andrology*, 2007 13(4), 356.

[61] Pipinikas, CP; Carter, ND; Corbishley, CM; Fenske, CD. HIF-1α mRNA gene expression levels in improved diagnosis of early stages of prostate cancer. *Biomarkers*, 2008 13(7-8), 680-691.

[62] Robey, IF; Lien, AD; Welsh, SJ; Baggett, BK; Gillies, RJ. Hypoxia-inducible factor-1α and the glycolytic phenotype in tumors. *Neoplasia* 2005 7(4), 324.

[63] Hu, W; Zhang, C; Wu, R; Sun, Y; Levine, A; Feng, Z. Glutaminase 2, a novel p53 target gene regulating energy metabolism and antioxidant function. *Science Signalling*, 2010 107(16), 7455.

[64] Matoba, S; Kang, JG; Patino, WD; Wragg, A; Boehm, M; Gavrilova, O; Hurley, PJ; Bunz, F; Hwang, PM. p53 regulates mitochondrial respiration. *Science Signalling*, 2006 312(5780), 1650.

[65] Bensaad, K; Tsuruta, A; Selak, MA; Vidal, M; Nakano, K; Bartrons, R; Gottlieb, E; Vousden, KH. TIGAR, a p53-inducible regulator of glycolysis and apoptosis. *Cell*, 2006 126(1), 107-120.

[66] Ramanathan, A; Wang, C; Schreiber, SL. Perturbational profiling of a cell-line model of tumorigenesis by using metabolic measurements. *Proceedings of the National Academy of Sciences*, 2005 102(17), 5992-5997.

[67] Moreno-Sánchez, R; Rodríguez-Enríquez, S; Marín-Hernández, A; Saavedra, E. Energy metabolism in tumor cells. *FEBS Journal*, 2007 274(6), 1393-1418.

[68] Gordan, JD; Thompson, CB; Simon, MC. HIF and c-Myc: sibling rivals for control of cancer cell metabolism and proliferation. *Cancer Cell*, 2007 12(2), 108-113.

[69] Wise, DR; DeBerardinis, RJ; Mancuso, A; Sayed, N; Zhang, XY; Pfeiffer, HK; Nissim, I; Daikhin, E; Yudkoff, M; McMahon, SB. Myc regulates a transcriptional program that stimulates mitochondrial glutaminolysis and leads to glutamine addiction. *Proceedings of the National Academy of Sciences*, 2008 105(48), 18782-18787.

[70] Cornel, E; Smits, G; Oosterhof, G; Karthaus, H; Deburyne, F; Schalken, J; Heerschap, A. Characterization of human prostate cancer, benign prostatic hyperplasia and normal prostate by in vitro 1H and 31P magnetic resonance spectroscopy. *The Journal of Urology*, 1993 150(6), 2019.

[71] Tessem, MB; Swanson, MG; Keshari, KR; Albers, MJ; Joun, D; Tabatabai, ZL; Simko, JP; Shinohara, K; Nelson, SJ; Vigneron, DB. Evaluation of lactate and alanine as metabolic biomarkers of prostate cancer using 1H HR-MAS spectroscopy of biopsy tissues. *Magnetic Resonance in Medicine*, 2008 60(3), 510-516.

[72] Albers, MJ; Bok, R; Chen, AP; Cunningham, CH; Zierhut, ML; Zhang, VY; Kohler, SJ; Tropp, J; Hurd, RE; Yen, YF. Hyperpolarized 13C lactate, pyruvate, and alanine: noninvasive biomarkers for prostate cancer detection and grading. *Cancer Research*, 2008 68(20), 8607-15.

[73] Atsumi, T; Chesney, J; Metz, C; Leng, L; Donnelly, S; Makita, Z; Mitchell, R; Bucala, R. High expression of inducible 6-phosphofructo-2-kinase/fructose-2,6-bisphosphatase (iPFK-2; PFKFB3) in human cancers. *Cancer Research*, 2002 62(20), 5881-7.

[74] Singh, G; Lakkis, CL; Laucirica, R; Epner, DE. Regulation of prostate cancer cell division by glucose. *Journal of Cell Physiology*, 1999 180(3), 431-438.

[75] Hicks, RJ; Binns, DS; Fawcett, ME; Ware, RE; Kalff, V; McKenzie, AF; Zalcberg, JP; Peters, LJ. Positron emission tomography (PET): experience with a large-field-of-view three-dimensional PET scanner. *The Medical journal of Australia*, 1999 171(10), 529.

[76] Agus, DB; Golde, DW; Sgouros, G; Ballangrud, A; Cordon-Cardo, C; Scher, HI. Positron emission tomography of a human prostate cancer xenograft: association of changes in deoxyglucose accumulation with other measures of outcome following androgen withdrawal. *Cancer Research*, 1998 58(14), 3009-3014.

[77] Marques, RB; Erkens-Schulze, S; de Ridder, CM; Hermans, KG; Waltering, K; Visakorpi, T; Trapman, J; Romijn, JC; van Weerden, WM; Jenster, G. Androgen receptor modifications in prostate cancer cells upon long-term androgen ablation and antiandrogen treatment. *International Journal of Cancer*, 2005 117(2), 221-229.

[78] Hara, T; Bansal, A; DeGrado, TR. Effect of hypoxia on the uptake of [*methyl*-^{3}H]choline, [1-^{14}C] acetate and [^{18}F]FDG in cultured prostate cancer cells. *Nuclear Medicine and Biology*, 2006 33(8), 977-984.

[79] Thorens, HGJB. The extended GLUT-family of sugar/polyol transport facilitators: nomenclature, sequence characteristics, and potential function of its novel members. *Molecular Membrane Biology*, 2001 18(4), 247-256.

[80] Nualart, F; Garcia, MA; Medina, RA; Owen, GI. Glucose transporters in sex steroid hormone related cancer. *Current Vascular Pharmacology* 2009 7(4), 534-48.

[81] Mueckler, M. Facilitative glucose transporters. *European Journal of Biochemistry*, 1994 219(3), 713-725.

[82] Godoy, A; Ulloa, V; Rodríguez, F; Reinicke, K; Yañez, AJ; García, MA; Medina, RA; Carrasco, M; Barberis, S; Castro, T. Differential subcellular distribution of glucose transporters GLUT1-6 and GLUT9 in human cancer: ultrastructural localization of GLUT1 and GLUT5 in breast tumor tissues. *Journal of Cellular Physiology*, 2006 207(3), 614-627.

[83] Li, Q; Manolescu, A; Ritzel, M; Yao, S; Slugoski, M; Young, JD; Chen, XZ; Cheeseman, CI. Cloning and functional characterization of the human GLUT7 isoform SLC2A7 from the small intestine. *American Journal of Physiology-Gastrointestinal and Liver Physiology*, 2004 287(1), G236-G242.

[84] McVie-Wylie, AJ; Lamson, DR; Chen, Y. Molecular cloning of a novel member of the GLUT family of transporters, SLC2a10 (GLUT10), localized on chromosome 20q13. 1: a candidate gene for NIDDM susceptibility. *Genomics*, 2001 72(1), 113-117.

[85] Haberkorn, U; Bellemann, ME; Altmann, A; Gerlach, L; Morr, I; Oberdorfer, F; Brix, G; Doll, J; Blatter, J; van Kaick, G. PET 2-fluoro-2-deoxyglucose uptake in rat prostate adenocarcinoma during chemotherapy with gemcitabine. *Journal of Nuclear Medicine: official publication, Society of Nuclear Medicine*, 1997 38(8), 1215.

[86] Effert, P; Beniers, A; Tamimi, Y; Handt, S; Jakse, G. Expression of glucose transporter 1 (Glut-1) in cell lines and clinical specimens from human prostate adenocarcinoma. *Anticancer Research*, 2004 24(5A), 3057-3064.

[87] Stewart, GD; Gray, K; Pennington, CJ; Edwards, DR; Riddick, A; Ross, JA; Habib, FK. Analysis of hypoxia-associated gene expression in prostate cancer: lysyl oxidase and glucose transporter-1 expression correlate with Gleason score. *Oncology Reports*, 2008 20(6), 1561.

[88] Reinicke, K; Sotomayor, P; Cisterna, P; Delgado, C; Nualart, F; Godoy, A. Cellular distribution of Glut-1 and Glut-5 in benign and malignant human prostate tissue. *Journal of Cellular Biochemistry*, 2012 113(2), 553-562.
[89] Chandler, JD; Williams, ED; Slavin, JL; Best, JD; Rogers, S. Expression and localization of GLUT1 and GLUT12 in prostate carcinoma. *Cancer*, 2003 97(8), 2035-2042.
[90] Mabjeesh, NJ; Willard, MT; Frederickson, CE; Zhong, H; Simons, JW. Androgens Stimulate Hypoxia-inducible Factor 1 Activation via Autocrine Loop of Tyrosine Kinase Receptor/Phosphatidylinositol 3′-Kinase/Protein Kinase B in Prostate Cancer Cells. *Clinical Cancer Research*, 2003 9(7), 2416-2425.
[91] Horii, K; Suzuki, Y; Kondo, Y; Akimoto, M; Nishimura, T; Yamabe, Y; Sakaue, M; Sano, T; Kitagawa, T; Himeno, S. Androgen-dependent gene expression of prostate-specific antigen is enhanced synergistically by hypoxia in human prostate cancer cells. *Molecular Cancer Research*, 2007 5(4), 383-391.
[92] Wilson, JE. Isozymes of mammalian hexokinase: structure, subcellular localization and metabolic function. *Journal of Experimental Biology*, 2003 206(12), 2049-2057.
[93] Cornish-Bowden, A. Hexokinase and ′glucokinase' in liver metabolism. *Trends in Biochemical Sciences*, 1991 16, 281-282.
[94] Chehtane, M; Khaled, AR. Interleukin-7 mediates glucose utilization in lymphocytes through transcriptional regulation of the hexokinase II gene. *The American Journal of Physiology. Cell physiology*, 2010 298(6), C1560-C1571.
[95] Moon, JS; Jin, WJ; Kwak, JH; Kim, HJ; Yun, MJ; Kim, JW; Park, SW; Kim, KS. Androgen stimulates glycolysis for de novo lipid synthesis by increasing the activities of hexokinase 2 and 6-phosphofructo-2-kinase/fructose-2,6-bisphosphatase 2 in prostate cancer cells. *Biochemical Journal*, 2011 433(1), 225-33.
[96] Teahan, O; Bevan, CL; Waxman, J; Keun, HC. Metabolic signatures of malignant progression in prostate epithelial cells. *The International Journal of Biochemistry and Cell Biology* 2011 43(7), 1002-1009.
[97] Schwickert, G; Walenta, S; Sundfor, K; Rofstad, EK; Mueller-Klieser, W. Correlation of high lactate levels in human cervical cancer with incidence of metastasis. *Cancer Research*, 1995 55(21), 4757-9.
[98] Everse, J; Kaplan, NO. Lactate dehydrogenases: structure and function. *Advances in enzymology and related areas of molecular biology*, 1973 37, 61-133. USA: John Wiley and Sons, Inc.
[99] Halestrap, AP; Meredith, D. The SLC16 gene family-from monocarboxylate transporters (MCTs) to aromatic amino acid transporters and beyond. *European Journal of Physiology*, 2004 447(5), 619-628.
[100] Hao, J; Chen, H; Madigan, M; Cozzi, P; Beretov, J; Xiao, W; Delprado, W; Russell, P; Li, Y. Co-expression of CD147 (EMMPRIN), CD44v3-10, MDR1 and monocarboxylate transporters is associated with prostate cancer drug resistance and progression. *British Journal of Cancer*, 2010 103(7), 1008-1018.
[101] Pertega-Gomes, N; Vizcaino, JR; Miranda-Goncalves, V; Pinheiro, C; Silva, J; Pereira, H; Monteiro, P; Henrique, RM; Reis, RM; Lopes, C; Baltazar, F. Monocarboxylate transporter 4 (MCT4) and CD147 overexpression is associated with poor prognosis in prostate cancer. *BMC Cancer*, 2011 11(1), 312.

[102] Levin, YS; Albers, MJ; Butler, TN; Spielman, D; Peehl, DM; Kurhanewicz, J. Methods for metabolic evaluation of prostate cancer cells using proton and 13C HR-MAS spectroscopy and [3-13C] pyruvate as a metabolic substrate. *Magnetic Resonance in Medicine*, 2009 62(5), 1091-1098.

[103] Williamson, DH; Lund, P; Krebs, HA. The redox state of free nicotinamide-adenine dinucleotide in the cytoplasm and mitochondria of rat liver. *Biochemical Journal*, 1967 103(2), 514-27.

[104] Chaiswing, L; Zhong, W; Oberley, TD. Distinct Redox Profiles of Selected Human Prostate Carcinoma Cell Lines: Implications for Rational Design of Redox Therapy. *Cancers*, 2011 3(3), 3557-3584.

[105] Zhang, XD; Deslandes, E; Villedieu, M; Poulain, L; Duval, M; Gauduchon, P; Schwartz, L; Icard, P. Effect of 2-deoxy-D-glucose on various malignant cell lines in vitro. *Anticancer Research*, 2006 26(5A), 3561-3566.

[106] Sahra, IB; Tanti, JF; Bost, F. The combination of metformin and 2 deoxyglucose inhibits autophagy and induces AMPK-dependent apoptosis in prostate cancer cells. *Autophagy*, 2010 6(5), 670-671.

[107] Stein, M; Lin, H; Jeyamohan, C; Dvorzhinski, D; Gounder, M; Bray, K; Eddy, S; Goodin, S; White, E; DiPaola, RS. Targeting tumor metabolism with 2-deoxyglucose in patients with castrate- resistant prostate cancer and advanced malignancies. *The Prostate*, 2010 70(13), 1388-1394.

[108] Sahra, IB; Laurent, K; Giuliano, S; Larbret, F; Ponzio, G; Gounon, P; Le Marchand-Brustel, Y; Giorgetti-Peraldi, S; Cormont, M; Bertolotto, C. Targeting cancer cell metabolism: the combination of metformin and 2-deoxyglucose induces p53-dependent apoptosis in prostate cancer cells. *Cancer Research*, 2010 70(6), 2465-2475.

[109] Wei, S; Kulp, SK; Chen, CS. Energy restriction as an antitumor target of thiazolidinediones. *Journal of Biological Chemistry*, 2010 285(13), 9780-9791.

[110] Chen, CH; Huang, PH; Chu, PC; Chen, MC; Chou, CC; Wang, D; Kulp, SK; Teng, CM; Wang, Q; Chen, CS. Energy restriction-mimetic agents induce apoptosis in prostate cancer cells in part through epigenetic activation of KLF6 tumor suppressor gene expression. *Journal of Biological Chemistry*, 2011 286(12), 9968-9976.

[111] Lin, HY; Kuo, YC; Weng, YI; Lai, IL; Huang, THM; Lin, SP; Niu, DM; Chen, CS. Activation of silenced tumor suppressor genes in prostate cancer cells by a novel energy restriction-mimetic agent. *The Prostate*, 2012 72(16), 1767-1778.

[112] Macheda, ML; Rogers, S; Best, JD. Molecular and cellular regulation of glucose transporter (GLUT) proteins in cancer. *Journal of Cellular Physiology*, 2005 202(3), 654-662.

[113] Flaig, TW; Gustafson, DL; Su, LJ; Zirrolli, JA; Crighton, F; Harrison, GS; Pierson, AS; Agarwal, R; Glodé, LM. A phase I and pharmacokinetic study of silybin-phytosome in prostate cancer patients. *Investigational New Drugs*, 2007 25(2), 139-146.

[114] Fang, JS; Gillies, RD; Gatenby, RA. Adaptation to hypoxia and acidosis in carcinogenesis and tumor progression. *Seminars in Cancer Biology*, 2008 18(5), 330-7.

[115] Marino, M; Fais, S; Djavaheri-Mergny, M; Villa, A; Meschini, S; Lozupone, F; Venturi, G; Della Mina, P; Pattingre, S; Rivoltini, L. Proton pump inhibition induces autophagy as a survival mechanism following oxidative stress in human melanoma cells. *Cell death and Disease*, 2010 1(10), e87.

[116] Porporato, PE; Dhup, S; Dadhich, RK; Copetti, T; Sonveaux, P. Anticancer targets in the glycolytic metabolism of tumors: a comprehensive review. *Frontiers in Pharmacology*, 2011 2, 49.

[117] Fantin, VR; St-Pierre, J; Leder, P. Attenuation of LDH-A expression uncovers a link between glycolysis, mitochondrial physiology, and tumor maintenance. *Cancer Cell*, 2006 9(6), 425-434.

[118] Le, A; Cooper, CR; Gouw, AM; Dinavahi, R; Maitra, A; Deck, LM; Royer, RE; Vander Jagt, DL; Semenza, GL; Dang, CV. Inhibition of lactate dehydrogenase A induces oxidative stress and inhibits tumor progression. *Proceedings of the National Academy of Sciences*, 2010 107(5), 2037-2042.

[119] Sonveaux, P; Végran, F; Schroeder, T; Wergin, MC; Verrax, J; Rabbani, ZN; De Saedeleer, CJ; Kennedy, KM; Diepart, C; Jordan, BF. Targeting lactate-fueled respiration selectively kills hypoxic tumor cells in mice. *The Journal of Clinical Investigation*, 2008 118(12), 3930.

[120] Gallagher, SM; Castorino, JJ; Philp, NJ. Interaction of monocarboxylate transporter 4 with beta1-integrin and its role in cell migration. *American Journal of Physiology. Cell Physiology*, 2009 296(3), C414-21.

[121] Sarsour, EH; Kumar, MG; Chaudhuri, L; Kalen, AL; Goswami, PC. Redox control of the cell cycle in health and disease. *Antioxidants and Redox Signaling*, 2009 11(12), 2985-3011.

[122] Emonds, K; Swinnen, J; Mortelmans, L; Mottaghy, F. Molecular imaging of prostate cancer. *Methods*, 2009 48(2), 193-199.

[123] Liu, Y; Zuckier, LS; Ghesani, NV. Dominant uptake of fatty acid over glucose by prostate cells: a potential new diagnostic and therapeutic approach. *Anticancer Research*, 2010 30(2), 369-374.

[124] Gatenby, RA; Gillies, RJ. Glycolysis in cancer: a potential target for therapy. *The International Journal of Biochemistry and Cell Biology*, 2007 39(7), 1358-1366.

[125] El Mjiyad, N; Caro-Maldonado, A; Ramírez-Peinado, S; Munoz-Pinedo, C. Sugar-free approaches to cancer cell killing. *Oncogene*, 2011 30(3), 253-264.

In: Glucose Uptake
Editors: Carter C. Johnson and Davis B. Williams
ISBN: 978-1-62618-670-5

Chapter 8

Regulation of Intestinal Glucose Uptake by Leptin

Jaione Barrenetxe[1], Carmen Fanjul[1], Robert Ducroc[2] and Maria Pilar Lostao[1]
[1]Department of Nutrition, Food Science and Physiology, University of Navarra, Pamplona, Spain
[2]Faculty of Medicine X. Bichat, Paris, France

Abstract

The endocrine and paracrine regulation of the intestinal absorption of nutrients is being subject of intense research leading to the identification of new actors. In this chapter, we present the findings regarding the regulation of intestinal glucose uptake by leptin, a hormone discovered in 1994 (Zhang et al. 1994) that is implicated in the control of food intake and energy balance (Tartaglia et al. 1995; Pelleymounter et al. 1995).

Shortly after leptin identification (Zhang et al. 1994), we demonstrated that leptin inhibits sugar absorption in rat intestine *in vitro* by short-term regulation of the Na^+/glucose cotransporter SGLT-1 (Lostao et al. 1998). In the same year, Bado et al. (1998) reported that the stomach chief cells secrete leptin into the gastric lumen in response to a meal or endocrine and neurocrine secretagogues. In line with those results, we described that leptin receptors are expressed in both the apical and basolateral membrane of human and murine enterocytes (Barrenetxe et al. 2002).

This distribution of leptin receptors led us to characterize the effects of leptin on SGLT-1 from luminal and serosal side of the intestinal mucosa, using the polarized Ussing chamber system. We demonstrated a rapid leptin receptor-mediated action of leptin from the luminal side and a slower effect from the serosal side that required leptin-induced CCK release (Ducroc et al. 2005).

This study was followed by the finding that the gastric chief cells also secrete the leptin soluble receptor, which binds to leptin in the secretory granules before its release into the gastric lumen (Cammisotto et al. 2006), and explains how leptin can remain stable in the gastric juice protected from this adverse proteolytic environment (Guilmeau et al. 2004). Further studies also showed the control by luminal leptin of glucose transporters from the GLUT family (Sakar et al. 2009; Sakar et al. 2010).

All these data supported our initial discovery about the role of luminal leptin as regulator of the intestinal sugar absorption and lead to the demonstration of the implication of PKC, PKA and AMPK on this leptin action (Barrenetxe et al. 2004; Ducroc et al. 2005; Sakar et al. 2009), and leptin modulation of sugar transport in health (Barrenetxe et al. 2004; Iñigo et al. 2007; Ducroc et al. 2005; Sakar et al. 2009; Fanjul et al. 2012) and disease (Iñigo et al. 2004; Osswald et al. 2005; Aït-Omar et al. 2011) that will be presented and discussed in this chapter.

1. Intestinal Glucose Absorption

One of the main functions of the small intestine is the absorption of nutrients arising from the diet. Among these nutrients, glucose is an essential substrate for the metabolism of most cells.

Glucose, galactose and fructose are delivered to the small intestine after carbohydrates, lactose and sucrose digestion in the intestinal lumen, and absorbed towards the blood circulation through specific transporters located in the apical and basolateral plasma membrane of the enterocytes.

1.1. Transporters Implicated on Intestinal Sugar Absorption

Glucose transport into the enterocytes is an active process that involves co-transport of one sugar molecule with two sodium ions through the SGLT-1 transporter. Thus, this transporter is electrogenic and able to concentrate glucose inside the cells using the sodium electrochemical gradient provided by the Na^{+}-K^{+}ATPase (Wrigth et al. 2007; Wright et al. 2011). SGLT-1 is constitutively expressed in the brush-border membrane of the enterocytes, but it can be rapidly mobilized from the apical membrane to the cytoplasm storages and vice versa after activation of different intracellular signals (Wright et al. 1997). SGLT-1 is a scavenger with high affinity/low capacity for glucose that becomes saturated around a concentration of 30 mM (Wright et al. 2011).

Galactose is also transported into the enterocytes by SGLT-1 (Wright et al. 2011) whereas fructose uses the facilitative transporter GLUT-5 (Barone et al. 2009). Glucose and fructose uptake can also occur through GLUT-7, another member of the GLUT facilitative family of transporters (Cheeseman 2008).

During post-prandial state, when glucose concentration reaches values over ~30 mM, the low affinity/high capacity glucose transporter GLUT-2, constitutively expressed in the basolateral membrane, can be recruited to the brush border membrane from intracellular compartments, providing the small intestine with an absorptive capacity to match dietary glucose uptake during meals (Kellet and Helliwell 2000; Kellet and Brot-Laroche 2005). This apical GLUT-2 is also able to transport fructose helping the main and fructose specific GLUT-5 transporter (Barone et al. 2009). Exit of the three monosaccharides from the enterocytes toward circulation takes place through GLUT-2 (Kellet et al. 2008).

1.2. Regulation of Glucose Intestinal Absorption

Intestinal glucose absorption can be regulated by several unspecific and specific mechanisms (Ferraris and Carey 2000). Unspecific mechanisms include changes in the absorptive surface by increasing the number (hyperplasia) or the size (hypertrophy) of the enterocytes. Thus, it has been demonstrated that a galactose enriched diet can increase the absorptive surface by enhancing the length of the intestinal microvilli (Smith et al. 1991), and in rats fed a hypercaloric diet, sugar absorption is increased by modifications on the intestinal mass (Planas et al. 1992). Diet can also induce changes in the lipidic composition of the plasma membrane, altering the lipid environment of the transporters and, in turn, their structure and activity. In this regard, an altered saturated/unsaturated fat ratio can lead to modifications of SGLT-1 function (Alessandri et al. 1990). Another unspecific mechanism would be modification of the Na^+/K^+ ATPase activity, which would change the sodium electrochemical gradient with consequences in the membrane potential and, therefore, on SGLT-1 activity (Ferraris and Diamond 1997).

Specific regulation of glucose absorption comprises transcriptional and post-transcriptional mechanisms which, for example, can modify the number of transporters present in the plasma membrane of the enterocytes (Ferraris and Diamond 1993) through modifications of their synthesis or degradation (Tsang and Cheeseman 1994). Post-transcriptional regulation of SGLT-1 also involves activation of protein kinases-dependent pathways which control insertion and recruitment of SGLT-1 in and from the plasma membrane respectively (Wright et al. 1997; Ishikawa et al. 1997; Vayro and Silverman, 1999). In this regard, several hormones such as GLP-2 (Au et al. 2002; Cheeseman et al. 1997), Glucagon-37 (Stümpel et al. 1997), EGF (Chung et al. 1999) and CCK (Hirsh and Cheeseman, 1998) have been found to regulate SGLT-1 activity in a short-term manner by regulating its expression in the brush border membrane.

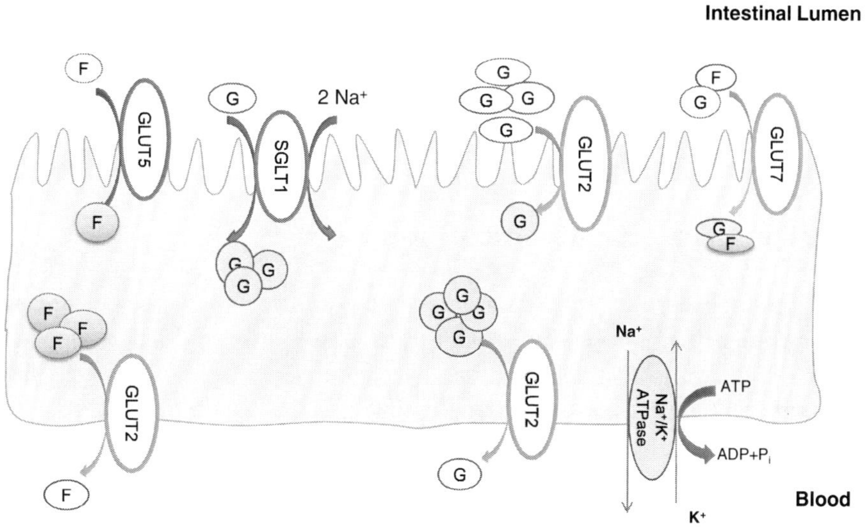

Figure 1.Transporters implicated on intestinal sugar absorption (see text for explanation). F: fructose; G: glucose/galactose.

2. Leptin as a Multifunctional Hormone

Between 1950 and 1965, the Jackson laboratories (Bar Harbor, Maine, USA) identified two mice strains, the *ob/ob* (obese) and the *db/db* (diabetic) mice, that presented identical phenotypes: morbid obesity, insulin resistance, infertility and lethargy (Kennedy et al. 1953; Hervey 1959). Two decades later, Coleman's laboratory characterized the differences between these two mice strains by joining both circulatory systems with normal mice or between them. These studies showed that the *ob/ob* strain was deficient in a blood-borne factor, while the *db/db* mice were deficient on the receptor for this factor (Coleman 1973, 1978 and 2010). Both strains thus displayed major food intake alterations with related complications leading to morbid obesity.

In 1994, the *ob* gene was identified and cloned from rodent adipose tissue. Its product was named leptin after the greek "leptos" meaning thin (Zhang et al. 1994). Initially, leptin receptors were only localized in areas of the hypothalamus, suggesting that leptin action was restricted to this tissue and that the hormone was only involved in the regulation of food intake and energy expenditure (Tartaglia et al. 1995).

Leptin was then considered as an adipostatic signal acting from the adipose tissue to the brain, as a part of a negative feedback loop, to regulate the size of the energy stores and energy balance. This theory was supported by the fact that when the leptin deficient *ob/ob* mice were treated with leptin, their weight returned to control values displaying a normal food intake and an increased energy expenditure (Campfiel et al. 1995; Pelleymounter et al. 1995).

In addition, the complications related to the immune system and fertility associated with leptin deficiency in the *ob/ob* mice disappeared, which suggested that the hormone could have different physiological roles in the organism.

Leptin plasma concentration is related to the amount of adipose tissue, being therefore higher in women than in men (Kennedy et al. 1997) and also higher in obese individuals (~ 2-8 nM) than in lean (normal weight) subjects (0.2-0.8 nM) (Considine et al. 1996). Nevertheless, nowadays it is known that leptin is also secreted by other tissues as placenta (Masuzaki et al. 1997), fetal tissues (Shimon et al. 1998), muscle (Wang et al. 1998), ovary (Ryan et al. 2002) and kidney (Martinez-Ansó et al. 1999) among others.

2.1. Leptin and Its Physiological Role

Since leptin discovery in 1994, a substantial body of information has been amassed about the functions mediated by the hormone and its receptors in central and peripheral tissues. Regarding the central effects, it is known that once leptin is released into the circulation, it is able to reach the central nervous system by crossing the blood-brain barrier through receptor-mediated endocytosis (Baumann et al. 1996; Banks et al. 1996; Golden et al. 1997). In the brain, leptin regulates the synthesis of a broad range of neuropeptides involved in the regulation of food intake and energy balance, such as neuropeptide Y (Erickson et al. 1996) and corticotropine releasing hormone (Schwartz et al. 1997), among others (Meister et al. 2000). Leptin is also implicated in the regulation of fertility and reproduction acting on the hypothalamic-pituitary-axis (Gonzalez et al. 2000b). Leptin treatment restores fertility of *ob/ob* female mice by increasing the levels of plasmatic luteinizing hormone, as well as the uterus and ovaric weight, and in *ob/ob* male mice by increasing the levels of folliculestimulating hormone and the weight of testis and seminal vesicles (Barash et al. 1996).

At the peripheral level, leptin is able to regulate several processes such as angiogenesis (Park et al. 2001), hematopoiesis (Fantuzzi and Fangionni 2000), immune function (Matarese 2000), testicular and ovaric function (Caprio et al. 2001), embryonic development (Gonzalez et al. 2000a), bone formation (Fleet et al. 2000), wound healing (Frank et al. 2000) and lipid and glucidic metabolism (Hynes and Jones 2001; Ceddia et al. 2001).

2.2. Leptin Receptors and Signaling

The leptin receptors, Ob-R, are members of the class I cytokine receptor family, and to date, there are six alternatively spliced isoforms (Ob-Ra-f) (Tartaglia 1997; Frühbeck 2006). All these receptor isoforms come from the same gene (*Ob* gene) and have the same extracellular domain and the same affinity for leptin. They differ in the amino acid sequence and the length of their transmembrane and intracellular domains (Chen et al. 1996; Lee et al. 1996). Ob-Ra, c, d and f possess a relatively short cytoplasmic domain whereas Ob-Rb, the so-called "long" isoform, has an extended C-terminal region and it is the only variant capable of complete intracellular signal transduction (Baumann et al. 1996). Nevertheless, a degree of intracellular signaling has also been demonstrated for the short isoforms (Bjorbaek et al. 1997; Yamashita et al. 1998; Han et al. 2001; Barrenetxe et al. 2003).Ob-Re, also known as the soluble receptor, is a truncated form of the leptin receptor which is secreted into the circulation and it is thought to play a role as a binding protein, regulating leptin bioavailability.

Thus, Ob-Re receptor increases leptin half-life, regulates its serum concentration and acts as a carrier protein delivering the hormone to its membrane receptors (Yang et al. 2004). In addition, some studies have implicated OB-Re in the regulation of leptin transport across the blood-brain barrier (Tu et al. 2007).

Binding of leptin to its receptor results in auto-phosphorylation of JAK-2 and subsequent activation of the JAK/STAT-3 pathway (Banks et al. 2000). Parallel signaling pathways are also activated including PI3K and MAPK/ERK pathways. SOC-3 is highly induced in response to leptin (Bjorbaek et al. 1998), and acts as a negative regulator of leptin signaling pathway. AMP-activated protein kinase is known as a key sensor of energy status within cells. Activated by an increase in the AMP/ATP ratio, AMPK acts to limit cellular energy depletion by down-regulating selective ATP-dependent processes including active transport. In hypothalamus, leptin suppresses AMPK activity and restricts food intake. In intestinal mucosa, leptin increases the phosphorylation of α2 regulatory subunit of AMPK (Walker et al. 2004; Sakar et al. 2009).

2.3. Leptin Actions on the Gastrointestinal Tract

At present, several leptin actions on the gastrointestinal tract have been described in the literature and leptin is regarded as a new effective gut peptide.

2.3.1. Leptin and the Stomach

Although adipose tissue is the major source of leptin, the hormone is also produced by different peripheral tissues such as stomach (Bado et al. 1998), placenta (Masuzaki et al. 1997) and kidney (Martinez-Ansó et al. 1999) where leptin receptors are also expressed (De Matteis et al. 1998; Groschl et al. 2001; Margetic et al. 2002; Gertler et al. 2009). It has been reported that the stomach produces and secretes leptin into the gastric lumen by pepsinogen-containing secretory granules of the chief cells (Bado et al. 1998).

These granules also contain the leptin soluble receptor, Ob-Re, that is released into the gastric lumen together with leptin (Cammisotto et al. 2006). Leptin remains stable in the gastric juice because the binding to its soluble receptor protects it from the acidic pH and proteolytic activity of the gastric lumen (Guilmeau et al. 2003 and 2004).

Endocrine cells of the gastric mucosa also secrete leptin to the circulation, in particular, after meal consumption, contributing to the plasma leptin levels (Cammisotto et al. 2005; Bado et al. 1998; Cinti et al. 2000).

Exocrine secretion of leptin by the gastric mucosa is regulated by the nutritional status of the body (Coleman and Hermann 1999).

Food intake is also a strong stimulus for gastric leptin secretion together with acetylcholine release by the vagus nerve (Sobhani et al. 2002), secretin, CCK (Sobhani et al. 2000), insulin, glucocorticoids and trans-retinoic acids (Goïot et al. 2005). Interestingly, luminal fructose can also stimulate leptin production by the gastric mucosa of mice without modifying plasma leptin levels (Sakar et al. 2009).

2.3.2. Leptin and the Intestine

Leptin receptors are present in both the apical and the basolateral membrane of the enterocytes (Barrenetxe et al. 2002), suggesting that the hormone can act from both sides of the intestinal epithelium: the luminal side and the blood. Indeed, once secreted into the gastric lumen, leptin can reach the intestine where it can bind to its brush border receptors (Barrenetxe et al. 2002; Aparicio et al. 2004) to regulate different processes of the intestinal wall such as nutrient absorption, mucus secretion and cell proliferation (Cammisotto et al. 2010b). Circulating leptin can also act as an endocrine hormone on the enterocytes by binding to the receptors, present on the basolateral membrane, and modify transporters activity (Stan et al. 2001, Fanjul et al. 2013).

Plasma leptin rises significantly and rapidly within 15 min after the intake of food or upon CCK stimulation (Bado et al. 1998).

This rapid increase in leptin plasma levels is not likely due to an increase in leptin secretion from the adipose tissue, since it takes over 60 minutes for the adipocytes to respond to any secretory stimuli (Cammissotto et al. 2010a).

At the ultrastructural level, leptin has been found to be associated with enterocytes microvilli in endocytic vesicles and Golgi apparatus and at the level of the basolateral membrane (Cammisotto et al. 2010a). Based on these data, it has been suggested that this quick increase in plasma leptin levels could be due to the gastric leptin that reaches blood circulation after crossing the intestinal mucosa by transcytosis (Cammisotto et al. 2007 and 2010a).

2.3.3. Effects of Leptin on the Gastrointestinal Tract

2.3.3.1. Leptin and Nutrients Absorption

Many studies published during the past 15 years have shown that gut leptin is a major regulator of intestinal functions and, in particular, of nutrients absorption. In rodents *in vivo* and in the model of human intestinal epithelial cells Caco-2, it has been shown that luminal leptin, in 15 min, enhances peptides absorption (Buyse et al. 2001a). This effect is due to an increase on the density of the proton-coupled peptide transporter PEPT1 in the apical plasma membrane. Similarly, after 12 or 24 h incubation of Caco-2 cells with leptin, an increase on butyrate uptake was observed due to an augment of the intracellular pools and translocation to the brush-border membrane of the monocarboxylate transporter MCT-1 (Buyse et al. 2002). In contrast, leptin acting from the basal membrane does not have any effect on peptides uptake (Buyse et al. 2001a). Nevertheless, it has been reported also in Caco-2 cells, that leptin acting from the basolateral membrane affects lipid handling by decreasing the export of triglycerides to the basolateral medium without affecting monoglyceride, diglyceride and cholesterolester classes. It also decreases the release of *de novo* synthesized apolipoproteins B-100 and B-48, as well as the newly formed chylomicrons and low density lipoproteins (Stan et al. 2001). Interestingly, *in vivo* intravenous injection of leptin attenuates apolipoprotein A-IV increase elicited by intra-duodenal infusion of lipids (Morton et al. 1998; Doi et al. 2001).

We have also demonstrated by *in vitro* techniques that in rat, luminal leptin can inhibit glutamine and phenylalanine uptake after 5 and 30 min incubation. This effect is associated with a decreased expression in the brush-border membrane of the transporters ASCT2 and B^0AT1 and a reduction of their mRNA levels after 1 h treatment of the intestine with the hormone (Ducroc et al. 2010). These data have been confirmed in Caco-2 cells where also inhibition of proline and β-alanine uptake has been found (Fanjul et al. 2012).

2.3.3.2. Leptin and Intestinal Motility

Afferent and efferent vagus nerve endings contain leptin receptors (Buyse et al. 2001b). In the small intestine, leptin can cause excitatory and inhibitory effects on mechanoreceptors thus leading to complex effects on intestinal motility. It has been shown that leptin deficiency increases the rate of gastric empting (Asakawa et al. 1999) and the transit activity in the jejunum, reducing total transit time in the small intestine (Kiely et al. 2005). CCK itself increases the release of leptin from gastric glands, suggesting that leptin and CCK are involved in a positive feedback loop. Local intra-arterial administration of leptin stimulates motility in small intestine pretreated with CCK, whereas leptin alone does not induce any activity (Guilmeau et al. 2003; Gaigé et al. 2003). These data suggest that modulation of vagal fibers by CCK may play a role on the leptin effects on intestinal motility.

2.3.3.3. Leptin and Intestinal Growth

A few studies have shown that leptin acts as a trophic factor in the gastrointestinal tract and can stimulate intestinal epithelial cell proliferation when administered exogenously (Hardwick et al. 2001; Fitzgerald et al. 2005). Moreover, leptin might have a role in the regulation of the anti-tumoral immune response in colonic cells (Abolhassani et al. 2008; Aloulou et al. 2008).

2.3.3.4. Leptin and Secretion of Intestinal Mucus

Mucus is a constitutive component of the intestinal barrier and contributes to its defense. In particular, goblet cells from the epithelium of the small and large intestine produce mucin MUC2. Mucin production is regulated by a process controlled by nerve activation being carbamylcholine the most potent known secretagogue of intestinal mucus. We observed that luminal leptin resulted in massive secretion of mucus in rat colon. Leptin dose-dependently stimulates mucin secretion with an EC50 of 0.8 nM and maximal response observed with 10 nM leptin (Plaisancié et al. 2006). Further studies in the human model of mucus secreting cell line HT29 MTX confirmed this stimulatory effect of leptin on mucus secretion (El Homsi et al. 2007). Furthermore, regulation of gene expression of mucin MUC5AC by leptin in this human cell line is dependent on MAPK, PKC and PI3-kinase pathways.

2.3.3.5. Leptin and Intestinal Inflammation and Immune Function

Leptin has important effects on the intestinal immune system by modulating the inflammatory and anti-inflammatory responses of the gastrointestinal tract (Gabay et al. 2001). Leptin might also prevent gastric ulcer formation by increasing the activities of the cyclooxygenase and nitric oxide pathways and enhancing mucus secretion (Adeyemi et al. 2005). Other authors have reported that serum concentration of leptin is low in patients with Crohn´s disease due to the increased levels of TNF-α in these patients (Franchimont et al. 2005; Karmiris et al. 2006). A role of leptin on intestinal injury and healing has also been described (Tasdelen et al. 2004).

3. Leptin Regulation of Intestinal Glucose Uptake

Among the physiological actions of leptin on the gastrointestinal tract, the effect of leptin to regulate intestinal absorption of glucose/galactose appears as a major function. Here, we present the data obtained by our groups in the past 15 years.

3.1. Leptin Effect on Intestinal Glucose Uptake

Shortly after leptin identification, we demonstrated in rat intestine, using intestinal everted rings, that leptin inhibited by ~40 % galactose uptake after 5 and 30 min incubation of the rings with the hormone (0.2-78 nM).This effect was accompanied by a decrease on the apparent V_{max} (20 *vs.* 4.8 μmol/g wet weight) and $K_{0,5}$ (15.8 *vs.* 5.3 mM) of the active sugar transport, inhibited by phlorizin, without any change in the phlorizin-resistant component. These results indicated a short-term regulation of the Na^+/glucose co-transporter SGLT-1 by leptin (Lostao et al. 1998). Similar results were obtained in mice jejunum using the same technique (Barrenetxe et al. 2001).

As mentioned before, in the same year, Bado and colleagues (1998) demonstrated that the stomach chief cells secrete leptin into the gastric lumen in response to a meal or endocrine and neurocrine secretagogues. In line with those results, a few years later we demonstrated by immunohistochemistry that the long leptin receptor isoform (Ob-Rb) is localized in both apical and basolateral membranes of human and murine enterocytes (Barrenetxe et al. 2002).

This distribution of leptin receptors led us to characterize the effects of leptin on SGLT-1 from luminal and serosal side of the intestinal mucosa using the polarized Ussing chamber system, and measuring short-circuit currents (Isc).

Ussing chamber system is a classical physiological device appropriate to study topical contribution of hormones and peptides in polarized tissues. Isolated intestinal epithelium can be mounted in the chamber and synthetic leptin can be added to either mucosal or serosal reservoir (Ducroc et al. 2005). Thus, using isolated rat jejunum, 10 mM glucose present in the luminal compartment induced short-circuit current. Addition of 10 nM leptin to the same compartment dramatically decreased this Isc by 90 % in 5 min. The inhibitory effect of leptin was dose-dependent with an IC50 of 0.13 nM. This effect of leptin was not observed in receptor-deficient *fa/fa* rats, indicating the requirement of functional leptin receptors for the control of SGLT-1 activity. In addition, this rapid inhibition was associated with a parallel decrease in the abundance of SGLT-1 in the brush-border membrane of the enterocytes, as demonstrated with Western blots. On the other hand, serosal addition of leptin only inhibited glucose Isc after 10 min exposure and was blocked by a CCK receptor antagonist, which indicated that the action of leptin from the serosal side of the mucosa was indirect and mediated by the endogenous release of CCK (Ducroc et al. 2005).

Once we demonstrated that leptin inhibited galactose and glucose transport in rat intestine using *in vitro* techniques, we studied whether luminal leptin could also regulate sugar absorption *in vivo* both at low (basal) and high (post-prandial) sugar concentration. *In vivo* intestinal sugar absorption in rat was measured with the recirculating and single pass perfusion systems and absorption determined by its disappearance from the perfusion medium. Luminal leptin (25 nM) inhibited by 25 % sugar absorption at low galactose concentration (1 mM), indicating that the hormone regulated SGLT-1 activity (Iñigo et al. 2007). Inhibition was reverted within 15 min after the removal of the hormone from the perfusion medium. At high glucose concentration (75 mM) leptin also inhibited (33 %) the phloretin-insensitive component of sugar absorption, mediated by SGLT-1, but had no effect on the apical passive component of the absorption, mediated by GLUT-2. This study confirmed our previous findings and supported the view that leptin exerts a regulatory role on intestinal sugar absorption in the post-prandial state by regulating SGLT-1 (Iñigo et al. 2007).

Further studies also showed the control by luminal leptin of facilitative glucose transporters of the GLUT family (Sakar et al. 2009; Sakar et al. 2010). *In vitro* luminal leptin (5 nM) enhanced 100 mM galactose and 30 mM fructose uptake by increasing GLUT-2 activity and GLUT-5 insertion into the plasma membrane respectively (Sakar et al. 2009). *In vivo* oral fructose administration induced a rapid and potent release of leptin into the gastric juice without any change on leptin plasma levels. Oral administration of leptin stimulated fructose transport by GLUT-5 and fructose-induced increase in blood glucose and mRNA levels of several gluconeogenesis key enzymes (Sakar et al. 2009). These data identified for the first time a positive regulatory control loop between gut leptin and fructose, in which fructose triggers the release of gastric leptin which, in turn, up-regulates GLUT5 and concurrently modulates metabolic functions in the liver.

Similarly to what happens in rodents, in the human model of intestinal cells Caco-2, apical leptin (0.2 and 8 nM) inhibited by ~30 % the uptake of α-methyl-glucoside (α-MG), a specific substrate of SGLT-1, after 5 or 30 min treatment. This effect was accompanied by a decrease on SGLT-1 protein abundance in the apical membrane of the cells (Fanjul et al. 2012).

Altogether these data support the vision of the gastric leptin as a regulator of intestinal nutrients absorption and in particular of glucose absorption.

3.2. Intracellular Mechanisms Implicated in Leptin Regulation of Intestinal Glucose Uptake

Since protein kinase C (PKC) and protein kinase A (PKA) are involved in the regulation of SGLT-1 (Wright et al. 1997; Ishikawa et al. 1997; Vayro and Silverman, 1999; Ducroc et al. 2005) and it is known that leptin is able to activate these kinases (Sweeney, 2002; Yamagishi et al. 2001; Takekoshi et al. 2001; Maingrette et al. 2003), we investigated whether PKC and PKA could be implicated in the inhibition of sugar absorption by leptin in rat small intestine. Inhibition of 1 mM galactose by 0.2 nM leptin was blocked by chelerythrine, a specific inhibitor of PKC. However H-89, a PKA inhibitor, did not block leptin effect. Biochemical assays showed that the inhibitory effect of leptin was related with a 2-fold increase in PKC and PKA activity (Barrenetxe et al. 2004).

Table 1. Summary of the main findings about the regulation by leptin of the intestinal sugar uptake. BB: Brush border; α -MG: α -methyl-D-glucoside

Animal/Tissue	Method	Sugar Conc.	Leptin (nM) Exposure time (min)	Effect	Reference
RAT/MOUSE Intestine	Everted jejunal rings	1 mM galactose	Luminal (0.2-78 nM) 5/30 min	↓ SGLT1-mediated sugar uptake	*Lostao et al. 1998* *Barrenetxe et al. 2001*
RAT Intestine	Ussing chambers	10 mM glucose	Luminal (10 nM) 5 min	↓ SGLT1-mediated Isc ↓ SGLT1 BB protein expression level	*Ducroc et al. 2005*
RAT Intestine	Ussing chambers	10 mM glucose	Serosal (10 nM) 10 min	↓ SGLT1-mediated sugar uptake through CCK	*Ducroc et al. 2005*
RAT Intestine	*In vivo*	1 mM galactose 75 mM glucose	Luminal (25 and 100 nM) 25 min	↓ SGLT1-mediated sugar uptake (Reversible)	*Iñigo et al. 2007*
RAT Intestine	Jejunal loops	100 mM galactose	Luminal (5 nM) 8 min	↑ GLUT-2-mediated sugar uptake	*Sakar et al. 2009*
RAT Intestine	Jejunal loops	30 mM fructose	Luminal (0.1-10 nM) 8 min	↑ fructose uptake	*Sakar et al. 2009*
Human Caco-2 cells	Apical uptake	0.1 mM α-MG	Luminal (0.2 and 8 nM) 5/30 min	↓ SGLT1-mediated sugar uptake ↓ SGLT1 BB protein expression level	*Fanjul et al. 2012*
Rat intestine	Western blot	30 mM Fructose	Luminal (1 nM) 8 min	↑ GLUT5 BB protein expression level	*Sakar et al. 2009*

PKC is a family of isozymes that are grouped into three subclasses i.e. conventional, atypical and novel. We demonstrated that inhibition by luminal leptin of glucose-induced Isc in Ussing chambers can be blunted by an inhibitor of the conventional PKC isoforms but not by an inhibitor of the novel isoforms (Ducroc et al. 2005). These findings suggested that activation of PKC is more relevant than PKA activation in the inhibition of galactose absorption by leptin in rat. In contrast, PKA did not block the inhibitory effect of leptin on α-MG uptake in Caco-2 cells, suggesting that PKA is not implicated on the leptin effect in human intestine (Fanjul et al. 2012).

Leptin stimulation of fructose uptake in rat was mediated by increase on the phosphorylation/activation of PKC-βII and 5′AMP-activated protein kinase (AMPK), that in turn, enhanced GLUT-2 and GLUT-5 insertion in the brush border membrane of the enterocytes and reduced the insertion of SGLT-1 (Sakar et al. 2009).

Similar intracellular pathways are implicated in the regulation of amino acid absorption by leptin, e.g. inhibition of glutamine and β-alanine uptake by leptin is mediated by a PKA-dependent intracellular pathway (Fanjul et al. 2012). From all these studies, it is becoming clear that leptin signalling in the small intestine involves PKA, AMPK and overall PKC activation that, in turn, leads to modulation of the expression levels of SGLT1, GLUT-2 and GLUT-5 in the plasma membrane of the enterocytes.

4. Leptin Regulation of Glucose Uptake in Obesity

Obesity is a multifactorial syndrome resulting from the alteration of food intake regulation and energy expenditure by environmental and neuroendocrine factors, in subjects with a genetic predisposition (Bray and York, 1997). In this context, leptin is clearly implicated in the development of obesity, since its major role at the central level is the regulation of appetite and energy expenditure.

The role of leptin in obesity seems to be related to its plasma circulating levels. Since leptin regulates body weight, differences in leptin sensibility or synthesis could lead to differences in body weight. Several studies have shown that obese animals, except the *ob/ob* mice, show higher leptin plasma levels than normal weight animals (Maffei *et al.* 1995). Therefore, it has been suggested that in obese subjects high leptin plasma levels are associated with leptin resistance, being this fact the cause of developing obesity. However, little is known about the mechanisms that lead to leptin resistance. Besides, there is a 10% of the obese population that show normal leptin plasma levels and, even in some cases, obesity can be due to a diminished leptin production by the adipose tissue (Ioffe et al. 1998).

As plasma leptin levels can differ in obesity, our group decided to analyze whether leptin regulation of sugars absorption could be altered in this pathology. We studied the effect of exogenous leptin on intestinal galactose absorption in the genetically obese *ob/ob* (leptin deficient) and *db/db* (leptin-resistant) mice, by the intestinal everted rings technique. Basal galactose uptake was similar in wild-type mice and the two obese groups. Contrarily to what happens in wild-type mice, 0.2 and 0.4 nM leptin increased galactose uptake in *db/db* mice. In the *ob/ob* strain, leptin at a concentration of 0.2 nM also increased galactose uptake while at 0.4 nM did not show any effect (Iñigo et al. 2004). These data indicates that leptin regulation of sugar absorption differ between lean and these genetically obese animals. Since the *db/db* strain is leptin resistant due to the absence of the functional long leptin receptor isoform, the stimulatory effect of leptin on galactose absorption should be mediated by the short receptor isoforms signaling. In contrast, in normal mice, the activation of the long and short receptor isoforms and the resulting cross-talk between their intracellular signaling in which several kinases are involved would produce a decrease on galactose uptake (Iñigo et al. 2004).

Interestingly, 0.2 nM leptin also increases galactose uptake in *ob/ob* mice. This strain does not secrete leptin suggesting that *ob/ob* mice tissues may have different regulation of leptin receptors expression and, therefore, different sensitivity and response to the hormone (Iñigo et al. 2004).

In summary, these data suggest a different leptin regulatory mechanism on glucose intestinal absorption in wild-type and obese animals. In this sense, other study has shown that morbid obese subjects exhibit a different expression pattern of the GLUT-2 transporter when compared with lean subjects. GLUT-2 was present in the apical membrane of the enterocytes in obese subjects even during fasting state, while in lean subjects the presence of GLUT-2 in the brush border membrane is usually limited to post-prandial state. This fact is related to the clinical parameters of the obese subjects such as insulin resistance and diabetes (Ait-Omar et al. 2011). In addition, Osswald and colleagues (2005) demonstrated that genetically modified mice lacking RS1, a peptide implicated in the regulation of SGLT-1, developed obesity with high levels of the transporter expression, which was associated with increased sugar absorption.

These findings indicate that the apical expression of glucose transporters differs between lean and obese subjects and that an impaired regulation of nutrients absorption, among other factors, could also be implicated in the onset of obesity.

Table 2. Differences in leptin regulation of intestinal sugar uptake between obese and lean mice strains

Mice Strain	ob/ob *(lacks Ob gene)*	lean	db/db *(lacks Ob-Rb)*
Basal conditions	No differences in basal galactose uptake		
Leptin addition	↑ Galactose uptake	↓ Galactose uptake	↑ Galactose uptake

Conclusion

Leptin acting from the apical or the basal membrane of the enterocytes modulates in a short-term manner and in a reversible way, *in vivo* and *in vitro* glucose uptake in rat intestine and human Caco-2 cells by inhibiting SGLT-1. The mechanisms involved include direct effects on the transporter membrane expression through activation of PKC and, in the case of basal leptin, indirect processes through other gastrointestinal hormones such as CCK. Taken together, the findings presented in this chapter contribute to the vision of leptin as an important hormonal signal for the regulation of intestinal absorption of glucose which impairment can help to the development of obesity.

Acknowledgments

The works performed by our group here cited were supported by Inserm (Institut National de la Santé et de la Recherche Médicale); grants PM98-0034 and BFU2007 60420/BFI from the Spanish Government; grants from the "Departamento de Educación y Cultura" of the Navarra Government, PIUNA (University of Navarra) and "Fundación Marcelino Botín".

The Spanish group is a member of the Network for Cooperative Research on Membrane Transport Proteins (REIT), cofunded by the "Ministerio de Educación y Ciencia" and the European Regional Development Fund (ERDF) (Grant BFU2007-30688-E/BFI).

References

[1] Abolhassani M, Aloulou N, Chaumette MT, Aparicio T, Martin-Garcia N, et al. (2008) Leptin receptor-related immune response in colorectal tumors: the role of colonocytes and interleukin-8. *Cancer Res.,* 68: 9423-9432.

[2] Adeyemi EO, Bastaki SA, Chandranath IS, Hasan MY, Fahim M, et al. (2005) Mechanisms of action of leptin in preventing gastric ulcer. *World J. Gastroenterol.,* 11: 4154-4160.

[3] Aït-Omar A, Monteiro-Sepulveda M, Poitou C, Le Gall M, Cotillard A, et al. (2011) GLUT2 accumulation in enterocyte apical and intracellular membranes: a study in morbidly obese human subjects and ob/ob and high fat-fed mice. *Diabetes,* 60: 2598-2607.

[4] Alessandri JM, Arfi TS, Thevenoux J, Leger CL (1990) Diet-induced alterations of lipids during cell differentiation in the small intestine of growing rats: effect of an essential fatty acid deficiency. *Journal of Pediatric Gastroenterology and Nutrition,* 10: 504-515.

[5] Aloulou N, Bastuji-Garin S, Le Gouvello S, Abolhassani M, Chaumette MT, et al. (2008) Involvement of the leptin receptor in the immune response in intestinal cancer. *Cancer Res.,* 68: 9413-9422.

[6] Aparicio T, Guilmeau S, Goiot H, Tsocas A, Laigneau JP, et al. (2004) Leptin reduces the development of the initial precancerous lesions induced by azoxymethane in the rat colonic mucosa. *Gastroenterology,* 126: 499-510.

[7] Asakawa A, Inui A, Ueno N, Makino S, Fujino MA, et al.(1999) Urocortin reduces food intake and gastric emptying in lean and ob/ob obese mice. *Gastroenterology,* 116: 1287-1292.

[8] Au A, Gupta A, Schembri P, Cheeseman CI (2002) Rapid insertion of GLUT2 into the rat jejunal brush-border membrane promoted by glucagon-like peptide 2. *Biochem. J.,* 367: 247-254.

[9] Bado A, Levasseur S, Attoub S, Kermorgant S, Laigneau JP, et al. (1998) The stomach is a source of leptin. *Nature,* 394: 790-793.

[10] Banks AS, Davis SM, Bates SH, Myers MG, Jr. (2000) Activation of downstream signals by the long form of the leptin receptor. *J. Biol. Chem.,* 275: 14563-14572.

[11] Banks WA, Kastin AJ, Huang W, Jaspan JB, Maness LM (1996) Leptin enters the brain by a saturable system independent of insulin. *Peptides,* 17: 305-311.

[12] Barash IA, Cheung CC, Weigle DS, Ren H, Kabigting EB, et al. (1996) Leptin is a metabolic signal to the reproductive system. *Endocrinology,* 137: 3144-3147.

[13] Barone S, Fussell SL, Singh AK, Lucas F, Xu J, et al. (2009) Slc2a5 (Glut5) is essential for the absorption of fructose in the intestine and generation of fructose-induced hypertension. *J. Biol. Chem.,* 284: 5056-5066.

[14] Barrenetxe J, Barber A, Lostao MP (2001) Leptin effect on galactose absorption in mice jejunum. *J. Physiol. Biochem.,* 57: 345-346.

[15] Barrenetxe J, Villaro AC, Guembe L, Pascual I, Muñoz-Navas M, et al. (2002) Distribution of the long leptin receptor isoform in brush border, basolateral membrane, and cytoplasm of enterocytes. *Gut,* 50: 797-802.

[16] Barrenetxe J, Palacios R, Barber A, Lostao MP (2003) Functional expression of the short isoform of the murine leptin receptor Ob-Rc (muB1.219) in Xenopus laevis oocytes. *J. Physiol. Biochem.,* 59: 119-126.

[17] Barrenetxe J, Sainz N, Barber A, Lostao MP (2004) Involvement of PKC and PKA in the inhibitory effect of leptin on intestinal galactose absorption. *Biochemical and Biophysical Research Communications,* 317: 717-721.

[18] Baumann H, Morella KK, White DW, Dembski M, Bailon PS, et al. (1996) The full-length leptin receptor has signaling capabilities of interleukin 6-type cytokine receptors. *Proc. Nat. Acad. of Sciences USA,* 93: 8374-8378.

[19] Bjorbaek C, Uotani S, da Silva B, Flier JS (1997) Divergent signaling capacities of the long and short isoforms of the leptin receptor. *J. Biol. Chem.*, 272: 32686-32695.

[20] Bjorbaek C, Elmquist JK, Frantz JD, Shoelson SE, Flier JS (1998) Identification of SOCS-3 as a potential mediator of central leptin resistance. *Mol. Cell.*,1: 619-625.

[21] Bray GA, York DA (1997) Clinical review 90: Leptin and clinical medecine: a new piece in the puzzle of obesity. *J. Clin. Endocrinol. Metab.,* 82: 2771-2776.

[22] Buyse M, Berlioz F, Guilmeau S, Tsocas A, Voisin T, et al. (2001a) PepT1-mediated epithelial transport of dipeptides and cephalexin is enhanced by luminal leptin in the small intestine. *J. Clin. Invest.,* 108: 1483-1494.

[23] Buyse M, Ovesjo ML, Goiot H, Guilmeau S, Peranzi G, et al. (2001b) Expression and regulation of leptin receptor proteins in afferent and efferent neurons of the vagus nerve. *Eur. J. Neurosci.,* 14: 64-72.

[24] Buyse M, Sitaraman SV, Liu X, Bado A, Merlin D (2002) Luminal leptin enhances CD147/MCT-1-mediated uptake of butyrate in the human intestinal cell line Caco2-BBE. *J. Biol. Chem.,* 277: 28182-28190.

[25] Cammisotto PG, Renaud C, Gingras D, Delvin E, Levy E, et al. (2005) Endocrine and exocrine secretion of leptin by the gastric mucosa. *J. Histochem. Cytochem.,* 53: 851-860.

[26] Cammisotto PG, Gingras D, Renaud C, Levy E, Bendayan M (2006) Secretion of soluble leptin receptors by exocrine and endocrine cells of the gastric mucosa. *AmJ of Physiol Gastrointest and Liver Physiol.,* 290: G242-249.

[27] Cammisotto PG, Gingras D, Bendayan M (2007) Transcytosis of gastric leptin through the rat duodenal mucosa. *Am. J. Physiol. Gastrointest. Liver. Physiol.,* 293: G773-779.

[28] Cammisotto PG, Bendayan M, Sane A, Dominguez M, Garofalo C, et al. (2010a) Receptor-Mediated Transcytosis of Leptin through Human Intestinal Cells In Vitro. *Int. J. Cell Biol.,* 2010: 928169.

[29] Cammisotto PG, Levy E, Bukowiecki LJ, Bendayan M (2010b) Cross-talk between adipose and gastric leptins for the control of food intake and energy metabolism. *Prog. Histochem. Cytochem.,* 45: 143-200.

[30] Campfield LA, Smith FJ, Mackie G, Tenenbaum R, Sassano ML, et al. (1995) Insulin normalization as an approach to the pharmacological treatment of obesity. *Obesity research,* 3 Suppl 4: 591S-603S.

[31] Caprio M, Fabbrini E, Isidori AM, Aversa A, Fabbri A (2001) Leptin in reproduction. *Trends Endocrinol. Metab.,* 12: 65-72.

[32] Ceddia RB, William WN, Jr., Curi R (2001) The response of skeletal muscle to leptin. *Front. Biosci.,* 6: D90-97.

[33] Cheeseman C (2008) GLUT7: a new intestinal facilitated hexose transporter. *Am. J. Physiol. Endocrinol. Metab.,* 295: E238-241.

[34] Cheeseman CI (1997) Upregulation of SGLT-1 transport activity in rat jejunum induced by GLP-2 infusion in vivo. *Am. J. Physiol.,* 273: R1965-1971.

[35] Chen H, Charlat O, Tartaglia LA, Woolf EA, Weng X, et al. (1996) Evidence that the diabetes gene encodes the leptin receptor: identification of a mutation in the leptin receptor gene in db/db mice. *Cell,* 84: 491-495.

[36] Chung BM, Wong JK, Hardin JA, Gall DG (1999) Role of actin in EGF-induced alterations in enterocyte SGLT1 expression. *Am. J. Physiol.,* 276: G463-469.

[37] Cinti S, Matteis RD, Pico C, Ceresi E, Obrador A, et al. (2000) Secretory granules of endocrine and chief cells of human stomach mucosa contain leptin. *Int. J. Obes. Relat. Metab. Disord.,* 24: 789-793.

[38] Coleman DL (1973) Effects of parabiosis of obese with diabetes and normal mice. *Diabetologia,* 9: 294-298.

[39] Coleman DL (1978) Obese and diabetes: two mutant genes causing diabetes-obesity syndromes in mice. *Diabetologia,* 14: 141-148.

[40] Coleman DL (2010) A historical perspective on leptin. *Nature medicine,* 16: 1097-1099.

[41] Coleman RA, Herrmann TS (1999) Nutritional regulation of leptin in humans. *Diabetologia,* 42: 639-646.

[42] Considine RV, Sinha MK, Heiman ML, Kriauciunas A, Stephens TW, et al. (1996) Serum immunoreactive-leptin concentrations in normal-weight and obese humans. *N. Engl. J. Med.,* 334: 292-295.

[43] De Matteis R, Cinti S (1998) Ultrastructural immunolocalization of leptin receptor in mouse brain. *Neuroendocrinology,* 68: 412-419.

[44] Doi T, Liu M, Seeley RJ, Woods SC, Tso P (2001) Effect of leptin on intestinal apolipoprotein AIV in response to lipid feeding. *Am. J. Physiol. Regul. Integr. Comp. Physiol.,* 281: R753-759.

[45] Ducroc R, Guilmeau S, Akasbi K, Devaud H, Buyse M, et al. (2005) Luminal leptin induces rapid inhibition of active intestinal absorption of glucose mediated by sodium-glucose cotransporter 1. *Diabetes,* 54: 348-354.

[46] Ducroc R, Sakar Y, Fanjul C, Barber A, Bado A, et al. (2010) Luminal leptin inhibits L-glutamine transport in rat small intestine: involvement of ASCT2 and B0AT1. *Am. J. Physiol. Gastrointest. Liver Physiol.,* 299: G179-185.

[47] El Homsi M, Ducroc R, Claustre J, Jourdan G, Gertler A, et al. (2007) Leptin modulates the expression of secreted and membrane-associated mucins in colonic epithelial cells by targeting PKC, PI3K, and MAPK pathways. *Am. J. Physiol. Gastrointest. Liver Physiol.*, 293: G365-373.

[48] Erickson JC, Clegg KE, Palmiter RD (1996) Sensitivity to leptin and susceptibility to seizures of mice lacking neuropeptide Y. *Nature,* 381: 415-421.

[49] Fanjul C, Barrenetxe J, Inigo C, Sakar Y, Ducroc R, et al. (2012) Leptin regulates sugar and amino acids transport in the human intestinal cell line Caco-2. *Acta Physiologica,* (Oxford, England) 205: 82-91.

[50] Fanjul C, Barrenetxe J, and Lostao MP (2013) Basal leptin regulates amino acid uptake in polarized *Caco-2 cells J. Physiol. Biochem., (in press).*

[51] Fantuzzi G, Faggioni R (2000) Leptin in the regulation of immunity, inflammation, and hematopoiesis. *J. Leukoc. Biol.,* 68: 437-446.

[52] Ferraris RP, Carey HV (2000) Intestinal transport during fasting and malnutrition. *Annu. Rev. Nutr.,* 20: 195-219.

[53] Ferraris RP, Diamond J (1997) Regulation of intestinal sugar transport. *Physiol. Rev.,* 77: 257-302.

[54] Ferraris RP, Diamond JM (1993) Crypt/villus site of substrate-dependent regulation of mouse intestinal glucose transporters. *Proc. Natl. Acad. Sci. USA,* 90: 5868-5872.

[55] FitzGerald AJ, Mandir N, Goodlad RA (2005) Leptin, cell proliferation and crypt fission in the gastrointestinal tract of intravenously fed rats. *Cell Prolif.,* 38: 25-33.

[56] Fleet JC (2000) Leptin and bone: does the brain control bone biology? *Nutr. Rev.,* 58: 209-211.

[57] Franchimont D, Roland S, Gustot T, Quertinmont E, Toubouti Y, et al. (2005) Impact of infliximab on serum leptin levels in patients with Crohn's disease. *J. Clin. Endocrinol. Metab.,* 90: 3510-3516.

[58] Frank S, Stallmeyer B, Kampfer H, Kolb N, Pfeilschifter J (2000) Leptin enhances wound re-epithelialization and constitutes a direct function of leptin in skin repair. *J. Clin. Invest.,* 106: 501-509.

[59] Fruhbeck G (2006) Intracellular signalling pathways activated by leptin. *Biochem. J.,* 393: 7-20.

[60] Gabay C, Dreyer M, Pellegrinelli N, Chicheportiche R, Meier CA (2001) Leptin directly induces the secretion of interleukin 1 receptor antagonist in human monocytes. *J. Clin. Endocrinol. Metab.,* 86: 783-791.

[61] Gaige S, Abysique A, Bouvier M (2003) Effects of leptin on cat intestinal motility. *J. Physiol.,* 546: 267-277.

[62] Goiot H, Laigneau JP, Devaud H, Sobhani I, Bado A (2005) Similarities and differences in the transcriptional regulation of the leptin gene promoter in gastric and adipose cells. *FEBS Lett.,* 579: 1911-1916.

[63] Golden PL, Maccagnan TJ, Pardridge WM (1997) Human blood-brain barrier leptin receptor.Binding and endocytosis in isolated human brain microvessels. *J. Clin. Invest.,* 99: 14-18.

[64] Gonzalez RR, Caballero-Campo P, Jasper M, Mercader A, Devoto L, et al. (2000) Leptin and leptin receptor are expressed in the human endometrium and endometrial leptin secretion is regulated by the human blastocyst. *J. Clin. Endocrinol. Metab.,* 85: 4883-4888.

[65] Gonzalez RR, Simon C, Caballero-Campo P, Norman R, Chardonnens D, et al. (2000) Leptin and reproduction. *Human Reproduction Update* 6: 290-300.

[66] Groschl M, Rauh M, Wagner R, Neuhuber W, Metzler M, et al. (2001) Identification of leptin in human saliva. *J. Clin. Endocrinol. Metab.,* 86: 5234-5239.

[67] Guilmeau S, Buyse M, Bado A (2004) Gastric leptin: a new manager of gastrointestinal function. *Current opinion in pharmacology,* 4: 561-566.

[68] Guilmeau S, Buyse M, Tsocas A, Laigneau JP, Bado A (2003) Duodenal leptin stimulates cholecystokinin secretion: evidence of a positive leptin-cholecystokinin feedback loop. *Diabetes,* 52: 1664-1672.

[69] Han DC, Isono M, Chen S, Casaretto A, Hong SW, et al. (2001) Leptin stimulates type I collagen production in db/db mesangial cells: glucose uptake and TGF-beta type II receptor expression. *Kidney Int.,* 59: 1315-1323.

[70] Hardwick JC, Van Den Brink GR, Offerhaus GJ, Van Deventer SJ, Peppelenbosch MP (2001) Leptin is a growth factor for colonic epithelial cells. *Gastroenterology,* 121: 79-90.

[71] Hervey GR (1959) The effects of lesions in the hypothalamus in parabiotic rats. *J. Physiol.,* 145: 336-352.

[72] Hirsh AJ, Cheeseman CI (1998) Cholecystokinin decreases intestinal hexose absorption by a parallel reduction in SGLT1 abundance in the brush-border membrane. *J. Biol. Chemistry,* 273: 14545-14549.

[73] Hynes GR, Jones PJ (2001) Leptin and its role in lipid metabolism. *Curr. Opin. Lipidol.,* 12: 321-327.

[74] Iñigo C, Barber A, Lostao MP (2004) Leptin effect on intestinal galactose absorption in ob/ob and db/db mice. *J. Physiol. Biochem.,* 60: 93-97.

[75] Iñigo C, Patel N, Kellett GL, Barber A, Lostao MP (2007) Luminal leptin inhibits intestinal sugar absorption in vivo. *Acta Physiologica,* (Oxford, England) 190: 303-310.

[76] Ioffe E, Moon B, Connolly E, Friedman JM (1998) Abnormal regulation of the leptin gene in the pathogenesis of obesity. *Proc. Natl. Acad. Sci. USA,* 95: 11852-11857.

[77] Ishikawa Y, Eguchi T, Ishida H (1997) Mechanism of beta-adrenergic agonist-induced transmural transport of glucose in rat small intestine. Regulation of phosphorylation of SGLT1 controls the function. *Biochim. et Biophys. Acta.,* 1357: 306-318.

[78] Karmiris K, Koutroubakis IE, Xidakis C, Polychronaki M, Voudouri T, et al. (2006) Circulating levels of leptin, adiponectin, resistin, and ghrelin in inflammatory bowel disease. *Inflamm. Bowel. Dis.,* 12: 100-105.

[79] Kellett GL, Brot-Laroche E (2005) Apical GLUT2: a major pathway of intestinal sugar absorption. *Diabetes,* 54: 3056-3062.

[80] Kellett GL, Brot-Laroche E, Mace OJ, Leturque A (2008) Sugar absorption in the intestine: the role of GLUT2. *Annual Review of Nutrition,* 28: 35-54.

[81] Kellett GL, Helliwell PA (2000) The diffusive component of intestinal glucose absorption is mediated by the glucose-induced recruitment of GLUT2 to the brush-border membrane. *Biochem. J.,* 350 Pt 1: 155-162.

[82] Kennedy A, Gettys TW, Watson P, Wallace P, Ganaway E, et al. (1997) The metabolic significance of leptin in humans: gender-based differences in relationship to adiposity, insulin sensitivity, and energy expenditure. *J. Clin. Endocrinol. Metab.,* 82: 1293-1300.
[83] Kennedy GC (1953) The role of depot fat in the hypothalamic control of food intake in the rat. *Proceedings of the Royal Society of London Series B, Containing papers of a Biological Character Royal Society,* (Great Britain) 140: 578-596.
[84] Kiely JM, Noh JH, Graewin SJ, Pitt HA, Swartz-Basile DA (2005) Altered intestinal motility in leptin-deficient obese mice. *J. Surg. Res.,* 124: 98-103.
[85] Lee GH, Proenca R, Montez JM, Carroll KM, Darvishzadeh JG, et al. (1996) Abnormal splicing of the leptin receptor in diabetic mice. *Nature,* 379: 632-635.
[86] Lostao MP, Urdaneta E, Martinez-Ansó E, Barber A, Martinez JA (1998) Presence of leptin receptors in rat small intestine and leptin effect on sugar absorption. *FEBS letters,* 423: 302-306.
[87] Maffei M, Halaas J, Ravussin E, Pratley RE, Lee GH, et al. (1995) Leptin levels in human and rodent: measurement of plasma leptin and ob RNA in obese and weight-reduced subjects. *Nat. Med.,* 1: 1155-1161.
[88] Maingrette F, Renier G (2003) Leptin increases lipoprotein lipase secretion by macrophages: involvement of oxidative stress and protein kinase C. *Diabetes,* 52: 2121-2128.
[89] Margetic S, Gazzola C, Pegg GG, Hill RA (2002) Leptin: a review of its peripheral actions and interactions. *Int. J. Obes. Relat. Metab. Disord.,* 26: 1407-1433.
[90] Martínez-Ansó E, Lostao MP, Martínez JA (1999) Immunohistochemical localization of leptin in rat kidney. *Kidney Int.,* 55: 1129-1130.
[91] Masuzaki H, Ogawa Y, Sagawa N, Hosoda K, Matsumoto T, et al.(1997) Nonadipose tissue production of leptin: leptin as a novel placenta-derived hormone in humans. *Nat. Med.,* 3: 1029-1033.
[92] Matarese G (2000) Leptin and the immune system: how nutritional status influences the immune response. *Eur. Cytokine. Netw.,* 11: 7-14.
[93] Meister B (2000) Control of food intake via leptin receptors in the hypothalamus. *Vitamins and Hormones,* 59: 265-304.
[94] Morton NM, Emilsson V, Liu YL, Cawthorne MA (1998) Leptin action in intestinal cells. *J. Biol. Chem.,* 273: 26194-26201.
[95] Osswald C, Baumgarten K, Stumpel F, Gorboulev V, Akimjanova M, et al. (2005) Mice without the regulator gene Rsc1A1 exhibit increased Na+-D-glucose cotransport in small intestine and develop obesity. *Mol. and Cell. Biol.,* 25: 78-87.
[96] Park HY, Kwon HM, Lim HJ, Hong BK, Lee JY, et al. (2001) Potential role of leptin in angiogenesis: leptin induces endothelial cell proliferation and expression of matrix metalloproteinases *in vivo* and *in vitro*. *Exp. Mol. Med.,* 33: 95-102.
[97] Pelleymounter MA, Cullen MJ, Baker MB, Hecht R, Winters D, et al. (1995) Effects of the obese gene product on body weight regulation in ob/ob mice. *Science,* (New York, NY) 269: 540-543.
[98] Plaisancie P, Ducroc R, El Homsi M, Tsocas A, Guilmeau S, et al. (2006) Luminal leptin activates mucin-secreting goblet cells in the large bowel. *Am. J. of Physiol. Gastrointest. Liver physiol.,* 290: G805-812.

[99] Planas B, Pons S, Nicolau MC, Lopez-Garcia JA, Rial R (1992) Morphofunctional changes in gastrointestinal tract of rats due to cafeteria diet. *Revista Española de Fisiología,* 48: 37-43.

[100] Ryan NK, Woodhouse CM, Van der Hoek KH, Gilchrist RB, Armstrong DT, et al. (2002) Expression of leptin and its receptor in the murine ovary: possible role in the regulation of oocyte maturation. *Biol. Reprod.,* 66: 1548-1554.

[101] Sakar Y, Nazaret C, Letteron P, Ait-Omar A, Avenati M, et al. (2009) Positive regulatory control loop between gut leptin and intestinal GLUT2/GLUT5 transporters links to hepatic metabolic functions in rodents. *PloS One,* 4: e7935.

[102] Sakar Y., Nazaret C., Bado A.andDucroc R. (2010) Leptin and insulin control the activity and expression of intestinal hexose transporter GLUT7. Gene and Nutrition, 5(1), S71 European Intestinal Transport Group, 23th meeting, Salerno, Italy.

[103] Schwartz MW, Seeley RJ, Woods SC, Weigle DS, Campfield LA, et al. (1997) Leptin increases hypothalamic pro-opiomelanocortin mRNA expression in the rostral arcuate nucleus. *Diabetes,* 46: 2119-2123.

[104] Shimon I, Yan X, Magoffin DA, Friedman TC, Melmed S (1998) Intact leptin receptor is selectively expressed in human fetal pituitary and pituitary adenomas and signals human fetal pituitary growth hormone secretion. *J. Clin. Endocrinol. Metab.,* 83: 4059-4064.

[105] Smith MW, James PS, Peacock MA (1991) Galactose effects on enterocyte differentiation in the mouse jejunum. *Biochim. et Biophys. Acta.,* 1093: 144-146.

[106] Sobhani I, Bado A, Vissuzaine C, Buyse M, Kermorgant S, et al. (2000) Leptin secretion and leptin receptor in the human stomach. *Gut,* 47: 178-183.

[107] Sobhani I, Buyse M, Goiot H, Weber N, Laigneau JP, et al. (2002) Vagal stimulation rapidly increases leptin secretion in human stomach. *Gastroenterology,* 122: 259-263.

[108] Stan S, Levy E, Bendayan M, Zoltowska M, Lambert M, et al. (2001) Effect of human recombinant leptin on lipid handling by fully differentiated Caco-2 cells. *FEBS Lett.,* 508: 80-84.

[109] Stumpel F, Scholtka B, Jungermann K (1997) A new role for enteric glucagon-37: acute stimulation of glucose absorption in rat small intestine. *FEBS Lett.,* 410: 515-519.

[110] Sweeney G (2002) Leptin signalling. *Cell Signal.,* 14: 655-663.

[111] Takekoshi K, Ishii K, Nanmoku T, Shibuya S, Kawakami Y, et al. (2001) Leptin stimulates catecholamine synthesis in a PKC-dependent manner in cultured porcine adrenal medullary chromaffin cells. *Endocrinology,* 142: 4861-4871.

[112] Tartaglia LA (1997) The leptin receptor. *J. Biol. Chem.*, 272: 6093-6096.

[113] Tartaglia LA, Dembski M, Weng X, Deng N, Culpepper J, et al. (1995) Identification and expression cloning of a leptin receptor, OB-R. *Cell,* 83: 1263-1271.

[114] Tasdelen A, Algin C, Ates E, Kiper H, Inal M, et al. (2004) Effect of leptin on healing of colonic anastomoses in rats. *Hepatogastroenterology,* 51: 994-997.

[115] Tsang R, Ao Z, Cheeseman C (1994) Influence of vascular and luminal hexoses on rat intestinal basolateral glucose transport. *Can. J. Physiol.and Pharmacol.,* 72: 317-326.

[116] Tu H, Pan W, Feucht L, Kastin AJ (2007) Convergent trafficking pattern of leptin after endocytosis mediated by ObRa-ObRd. *J. Cell Physiol.,* 212: 215-222.

[117] Vayro S, Silverman M (1999) PKC regulates turnover rate of rabbit intestinal Na^{+}-glucose transporter expressed in COS-7 cells. *Am. J. Physiol.*, 276: C1053-1060.

[118] Walker CD, Salzmann C, Long H, Otis M, Roberge C, et al. (2004) Direct inhibitory effects of leptin on the neonatal adrenal and potential consequences for brain glucocorticoid feedback. *Endocr. Res.,* 30: 837-844.
[119] Wang J, Liu R, Hawkins M, Barzilai N, Rossetti L (1998) A nutrient-sensing pathway regulates leptin gene expression in muscle and fat. *Nature,* 393: 684-688.
[120] Wright EM, Hirsch JR, Loo DD, Zampighi GA (1997) Regulation of Na+/glucose cotransporters. *J. Exp. Biol.,* 200: 287-293.
[121] Wright EM, Loo DD, Hirayama BA (2011) Biology of human sodium glucose transporters. *Physiol. Rev.,* 91: 733-794.
[122] Yamagishi SI, Edelstein D, Du XL, Kaneda Y, Guzman M, et al. (2001) Leptin induces mitochondrial superoxide production and monocyte chemoattractant protein-1 expression in aortic endothelial cells by increasing fatty acid oxidation via protein kinase A. *J. Biol. Chem.,* 276: 25096-25100.
[123] Yamashita T, Murakami T, Otani S, Kuwajima M, Shima K (1998) Leptin receptor signal transduction: OBRa and OBRb of fa type. *Biochem. Biophys. Res. Com.,* 246: 752-759.
[124] Yang G, Ge H, Boucher A, Yu X, Li C (2004) Modulation of direct leptin signaling by soluble leptin receptor. *Mol. Endocrinol.,* 18: 1354-1362.
[125] Zhang Y, Proenca R, Maffei M, Barone M, Leopold L, et al. (1994) Positional cloning of the mouse obese gene and its human homologue. *Nature,* 372: 425-432.

In: Glucose Uptake
Editors: Carter C. Johnson and Davis B. Williams
ISBN: 978-1-62618-670-5

Chapter 9

The Ras Superfamily of Small GTP-Binding Proteins in Glucose Transporter Type 4-Mediated Glucose Uptake in Insulin-Responsive Tissues

Takaya Satoh*
Laboratory of Cell Biology, Department of
Biological Science, Graduate School of Science,
Osaka Prefecture University, Osaka, Japan

Abstract

Insulin stimulates the uptake of glucose in several target tissues, including skeletal muscle and adipose tissue, thereby contributing to the decline in the blood glucose level. Insulin signaling is mediated by a highly integrated network composed of a variety of signal transducing proteins. Induction of glucose uptake in skeletal muscle and adipose tissue primarily depends on the redistribution of the glucose transporter type 4 (GLUT4) from intracellular storage sites to the plasma membrane. This process consists of several discernible steps, such as vesicle transport, docking, and fusion, each of which is regulated rigorously by the signal transduction machinery. Despite intensive research for decades, mechanisms underlying the regulation of GLUT4-mediated glucose uptake remain largely elusive. The Ras superfamily of small GTP-binding proteins serves as a molecular switch in many facets of signal transduction processes in various types of mammalian cells. Emerging evidence supports the notion that insulin also employs small GTP-binding proteins in the intracellular signaling network to exert its actions in skeletal muscle and adipose cells. Two types of proteins, guanine nucleotide exchange proteins (GEFs) and GTPase-activating proteins (GAPs), regulate on/off switching of small GTP-binding proteins, respectively. In fact, diverse GEFs and GAPs have been characterized as key components in insulin signaling.

---

* Corresponding author: Takaya Satoh. Mailing address: 1-1 Gakuen-cho, Naka-ku, Sakai, Osaka, 599-8531, Japan. E-mail: tkysato@b.s.osakafu-u.ac.jp.

Regulation of the Ras Superfamily of Small GTP-Binding Proteins

The Ras superfamily is comprised of over 150 signal transducing small GTP-binding proteins, which are classified into five subfamilies, i.e. Ras, Rho, Rab, Arf, and Ran families (Takai et al., 2001; Wennerberg et al., 2005). The molecular switch function of Ras superfamily small GTP-binding proteins depends on their bound guanine nucleotide species (Figure 1).

Virtually all small GTP-binding proteins exist in complex with GDP or GTP, and these two conformations interact with a different set of proteins. If the affinities of the protein for GDP and GTP are similar (and this is the case for many small GTP-binding proteins), the protein preferentially binds to GTP because the intercellular concentration of GTP is higher than that of GDP. Small GTP-binding proteins themselves catalyze the hydrolysis of the bound GTP (and thus are also called "GTPases"), generating the GDP-bound conformation and orthophosphate.

In general, only the GTP-bound conformation interacts with downstream effectors or target proteins, which are responsible for conveying the signal further downstream.

Therefore, guanine nucleotide exchange from GDP to GTP is considered to be a "switch on" process, whereas the hydrolysis of GTP is considered to be a "switch off" process. These two processes are regulated respectively by diverse regulatory proteins.

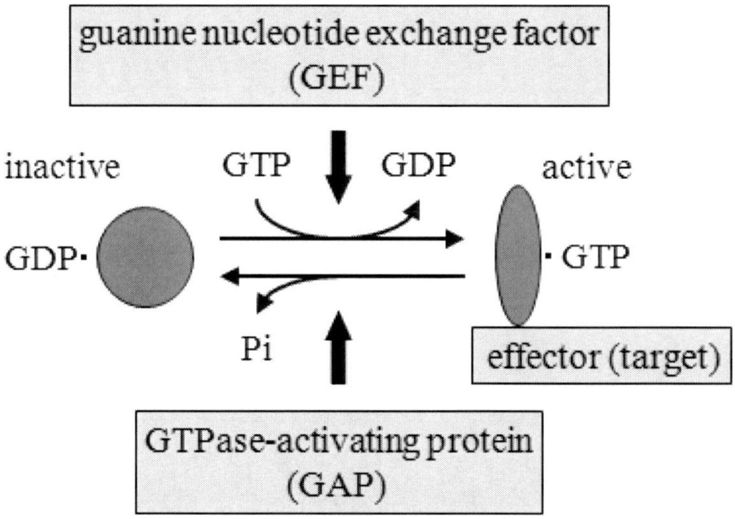

Figure 1. The GDP/GTP cycle of the small GTP-binding protein and its regulators. Small GTP-binding proteins exist in either the GDP-bound inactive or GTP-bound active conformation. GEFs stimulate the release of non-covalently bound GDP from the protein, leading to the accumulation of the GTP-bound conformation. The GTP-bound conformation specifically interacts with and activates downstream effector (target) proteins. GAPs enhance the GTPase activity of the GTP-binding protein. Thus, GEFs and GAPs act as positive and negative regulators for signal transduction, respectively. GAP, GTPase-activating protein; GEF, guanine nucleotide exchange factor.

The protein that enhances guanine nucleotide exchange from GDP to GTP is termed "guanine nucleotide exchange factor (GEF)", whereas the protein that enhances GTP hydrolysis is termed "GTPase-activating protein (GAP)".In many signal transduction pathways, the upstream signal is transmitted to a GTP-binding protein through a GEF, but in some cases, a GAP acts as an upstream regulator.

Various mutations naturally occur in genes that encode small GTP-binding proteins, causing cellular dysfunction. The most famous example of this is a GTPase-deficient mutant of Ras found in a variety of cancer cells.

In addition to such pathologically characterized mutants, many types of mutants have been generated by recombinant DNA techniques, being employed for the analysis of signal transduction mechanisms. These mutants of small GTP-binding proteins can be classified into two categories; one shows GTPase deficiency and another shows high affinity to GEFs. GTPase-deficient mutants exist in a GTP-bound conformation, continuously conveying the signal to downstream targets. Thus, this type of mutants is "constitutively active" in terms of signal transduction.

However, it should be noted that this is not always the case because, in some cases, both GTP binding and hydrolysis (i.e. continual GDP/GTP cycling) is required for exerting the signaling function. Mutants with high affinity to GEFs usually act in a "dominant-negative" manner by sequestering their GEFs.

GLUT4 in Insulin-Dependent Glucose Uptake Signaling

The glucose transporter type 4 (GLUT4) mediates insulin-dependent glucose uptake in skeletal muscle and adipose tissue (Huang and Czech, 2007; Saltiel and Kahn, 2001). GLUT4 is a 12-transmembrane protein that catalyzes glucose transport across the cell membrane by an ATP-independent facilitative diffusion mechanism (Huang and Czech, 2007). A pivotal role for GLUT4 in whole-body glucose homeostasis is strongly suggested by various genetically engineered mouse models (Huang and Czech, 2007).

For instance, mice with conditional deletion of GLUT4 in skeletal muscle or adipose tissue display impaired insulin responsiveness and diabetic phenotypes (Abel et al., 2001; Zisman et al., 2000). Thus, it is important to understand the signaling mechanisms whereby insulin stimulates glucose uptake through the modulation of GLUT4 activity.

GLUT4 is distributed in various membrane structures and dynamically recycles between intracellular stores and the plasma membrane (Figure 2). In unstimulated cells, GLUT4 is sequestered away from the endosomal recycling compartment (ERC) into a specialized intracellular compartment called GLUT4 storage vesicles (GSV) due to slow exocytosis (Karylowski et al., 2004).

Insulin accelerates exocytosis of GSV to the plasma membrane, leading to a net accumulation of GLUT4 on the cell surface. Multiple steps of exocytosis, including trafficking, tethering, docking, and fusion, are subjected to insulin regulation. Recently, it has been demonstrated that insulin also regulates GLUT4 vesicle formation, which is the earliest step of GLUT4 recycling (Li et al., 2012). Endocytosis of GLUT4, on the other hand, is attenuated upon insulin stimulation, also contributing to plasma membrane distribution of GLUT4.

Following insulin stimulation, the activated insulin receptor triggers downstream signaling by phosphorylating a variety of target proteins. The adaptor protein IRS1 mediates the formation of the signaling complex.

A protein kinase cascade consisting of phosphoinositide 3-kinase (PI3K), PDK1, and Akt2 plays a pivotal role downstream of IRS1 for the induction of glucose uptake (Huang and Czech, 2007; Saltiel and Kahn, 2001; Figure 3). PI3K yields 3-phosphoinositides such as phosphatidylinositol 3,4,5-trisphosphate, which induce translocation of PDK1 and Akt2 to the plasma membrane.

Subsequently, PDK1 and the mammalian target of rapamycin complex 2 phosphorylate and activate Akt2. Akt2 has been implicated in several GLUT4 exocytic processes by phosphorylating specific targets (Franke, 2008; Vanhaesebroeck et al., 2012).

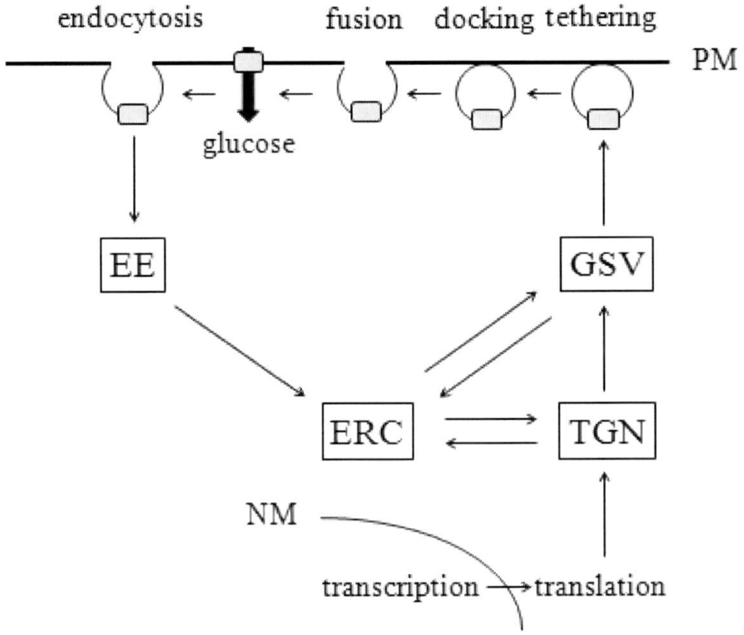

Figure 2. GLUT4 trafficking pathways. GLUT4 is sequestered in the specialized intracellular compartment GSV in unstimulated cells. Upon insulin stimulation, GLUT4 is redistributed, accumulating in the plasma membrane. Various endocytic and exocytic processes are subjected to insulin regulation. EE, early endosome; ERC, endosomal recycling compartment; GSV, GLUT4 storage vesicle; NM, nuclear membrane; PM, plasma membrane; TGN, trans-Golgi network. Gray squares represent GLUT4.

Rab Proteins

The Rab family of small GTP-binding proteins consists of over 60 members found in humans. Rab proteins have been implicated in the regulation of consecutive steps in intracellular membrane trafficking, including vesicle budding, delivery, tethering, and fusion with the target (acceptor) membrane. The specificity of membrane tethering and fusion is particularly important for the proper flow of cargo within the cell (Zerial and McBride, 2001; Grosshans et al., 2006).

Like other small GTP-binding proteins, Rab proteins act as molecular switches, cycling between GDP-bound and GTP-bound states (Figure 4). In addition, Rab proteins undergo a cycle of membrane insertion and extraction, which is coupled with the above GDP/GTP cycle. The post-translational modification of C-terminal cysteines with geranylgeranyl moieties is prerequisite for membrane insertion.

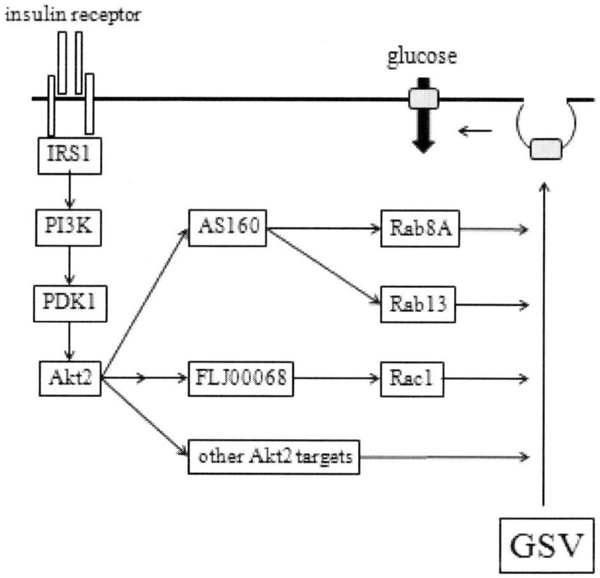

Figure 3. Signaling pathways downstream of the insulin receptor that regulate GLUT4 translocation to the plasma membrane in muscle cells. The ligand-bound insulin receptor stimulates a kinase cascade consisting of PI3K, PDK1, and Akt2. Akt2 is a master switch that regulates GLUT4 translocation in muscle cells. GSV, GLUT4 storage vesicle; PI3K, phosphoinositide 3-kinase.

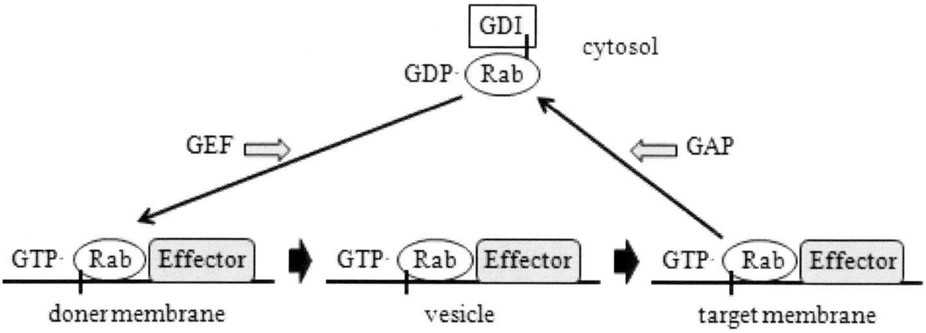

Figure 4. Membrane trafficking and the Rab protein cycle. Rab proteins cycle between membrane and cytosolic compartments. In membrane compartments, these proteins regulate vesicle transport through the action of various effectors. GDI recognizes the lipid moiety of Rab proteins, preserving these proteins in the cytosolic compartment. GAP, GTPase-activating protein; GDI, GDP dissociation inhibitor; GEF, guanine nucleotide exchange factor.

Extraction from the membrane and preservation in the cytosol of lipid-modified Rab proteins require a protein termed GDP dissociation inhibitor (GDI), which masks the lipid moiety of the GDP-bound form of the Rab protein.

Once associated with membrane Rab proteins are subjected to the action of GEF, which catalyzes GTP binding to Rab proteins. The active, membrane-bound Rab protein interacts with specific effectors, exerting its functions. The switch is then turned off via hydrolysis of bound GTP, which is accelerated by specific GAPs, and the GDP-bound inactive Rab protein is recycled back to the cytosol (Zerial and McBride, 2001; Grosshans et al., 2006).

In adipocytes and skeletal muscle, Rab4 is localized not only to early endosomes, but also GLUT4-containing vesicles, suggesting a role for Rab4 in GLUT4 translocation (Cormont et al., 1996; Sherman et al., 1996; Vollenweider et al., 1997).

In fact, decrease in insulin-dependent GLUT4 translocation was observed upon inhibition of Rab4, and insulin stimulated GTP loading to Rab4 in PI3K-dependent manner (Cormont et al., 1996; Shibata et al., 1997). Activated Rab4 interacts with the kinesin motor KIF3, regulating GLUT4 vesicle transport along microtubules (Imamura et al., 2003). Therefore, Rab4 may participate in the subcellular redistribution of GLUT4 in insulin-stimulated conditions. However, molecular mechanisms whereby insulin regulates Rab4 remain obscure.

Rab5 regulates the early endocytic transport from the plasma membrane to early endosomes mediated by clathrin-coated vesicles (Zerial and McBride, 2001).

Inhibition of Rab5 increased the basal level of surface GLUT4 and reduced GLUT4 internalization after insulin removal (Huang et al., 2001). Thus, Rab5 may be involved in endocytic removal of cell surface GLUT4.

Another Rab family member Rab11 is required for GLUT4 trafficking from recycling endosomes to the specialized compartment and the insulin-induced translocation of GLUT4 to the plasma membrane (Uhlig et al., 2005; Zeigerer et al., 2002).

Rab31 is a member of the Rab5 subfamily, which is implicated in the regulation of membrane trafficking from the trans-Golgi network to endosomes. Dominant-negative Rab31 increased the basal level of GLUT4 accumulation in the plasma membrane, whereas overexpression of wild-type or constitutively active Rab31 blocked insulin-stimulated GLUT4 translocation (Lodhi et al., 2007).

Furthermore, knockdown of Rab31 significantly potentiated GLUT4 translocation stimulated by insulin. The formation of the GTP-bound form of Rab31 was inhibited by insulin. Taken together, insulin may permit GLUT4 vesicles to escape the futile intracellular cycle between endosomal fractions and to be translocated to the plasma membrane by inactivating Rab31 (Lodhi et al., 2007). A current model for the mechanism by which insulin inhibits Rab31 will be described below.

Rab10 is also present in GLUT4 vesicles, and is reported to have a pivotal role in insulin-stimulated translocation of GLUT4 to the adipocyte plasma membrane (Sano et al., 2007). GTPase-deficient Rab10 increased the cell surface level of GLUT4, whereas insulin-dependent GLUT4 translocation was attenuated when Rab10 was knocked down.

Importantly, Rab10 is a substrate of AS160, a Rab-GAP regulated through phosphorylation by Akt2, and the increase in the basal level of GLUT4 in the plasma membrane due to AS160 knockdown was restored by knocking down Rab10. Collectively, Rac10 is a insulin-responsive regulator of GLUT4 translocation, acting downstream of AS160 (Sano et al., 2007; see below). Rab8A, Rab10, and Rab13 belong to the same subfamily of

Rab GTP-binding proteins, and are substrates of AS160. Thus, it is likely that Rab8A and Rab13 are also implicated in the regulation of insulin-dependent GLUT4 trafficking.

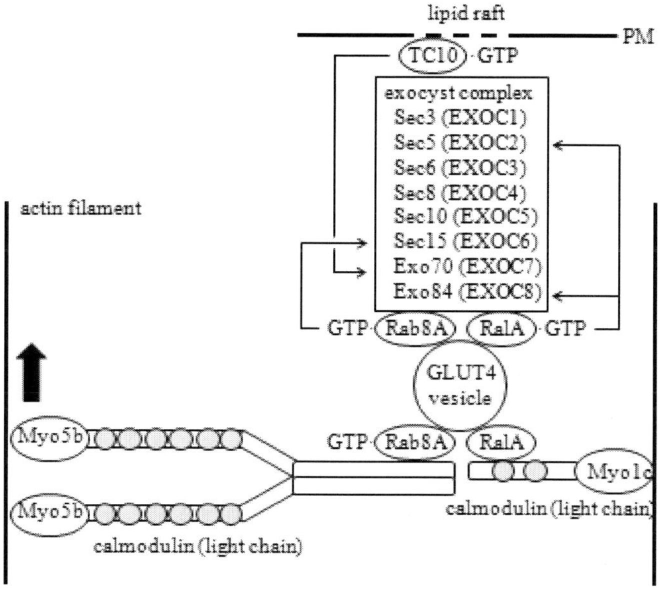

Figure 5. GULT4 vesicle transport driven by myosin motor proteins and exocyst complex-mediated tethering of the GLUT4 vesicle to the plasma membrane. Plasma membrane-localized TC10 anchors the exocyst complex beneath the plasma membrane. Rab8A and RalA contribute to the association of the GLUT4 vesicle to the exocyst complex. Rab8A and RalA also link the GLUT4 vesicle to Myo5b and Myo1c motors, respectively.

In fact, Rab8A and Rab13 are expressed in skeletal muscle cells, and become activated upon insulin stimulation (Sun et al., 2010). Furthermore, Rab8A and Rab13 can suppress the inhibitory effect of a dominant-negative AS160 mutant (Sun et al., 2010).

Taken together, Rab10 in adipocytes and Rab8A and Rab13 in muscle cells, respectively, are likely to be responsible for insulin-induced GLUT4 translocation downstream of AS160.

Rab8A can interact with the exocyst component Sec15 (Wu et al., 2005; Bryant et al., 2010), and thus it may contribute to GLUT4 vesicle tethering to the plasma membrane in skeletal muscle (Figure 5). However, it remains unclear whether Rab10 is also responsible for the interaction between the GLUT4 vesicle and the exocyst complex in adipocytes. Another effector for Rab8A and Rab10 is the motor protein Myo5b (Roland et al., 2007; Roland et al., 2011). In fact, engagement of Rab8A and Myo5b is important for insulin-induced GLUT4 translocation in muscle cells (Ishikura and Klip, 2008; Figure 5).

Rab GEFs

Although various Rab proteins have been implicated in GLUT4 vesicle trafficking, underlying regulatory mechanisms remain largely unknown. Given that many small GTP-binding proteins are activated in response to upstream signals through the action of their specific GEFs, Rab GEFs are possible candidates for direct activators. Indeed, a GEF for

Rab10 called Dennd4C is localized in GLUT4 vesicles, and its knockdown markedly diminished insulin-induced GLUT4 translocation (Sano et al., 2011). Thus, this GEF may be responsible for insulin-dependent activation of Rab10 in adipocytes. To my knowledge, GEFs that act on Rab8A and Rab13 in insulin-stimulated muscle cells remain unidentified. Another example of the involvement of a GEF in Rab regulation is insulin-dependent sequestration of the GEF Gapex-5 from its substrate Rab31 in adipocyte-specific TC10-mediated signaling (Lodhi et al., 2007; see below).

Rab GAPs

Through screening Akt substrates, a TBC domain-containing Rab-GAP termed AS160 (TBC1D4) was identified (Kane et al., 2002). A subfamily of Rab proteins consisting of Rab8A, Rab10, and Rab13 is known to be a target of AS160, being involved in insulin signaling. AS160 is associated with GLUT4 vesicles in unstimulated cells, and becomes dissociated following insulin stimulation (Larance et al., 2005). Five amino acid residues of AS160 are phosphorylated by Akt in insulin-stimulated adipocytes (Sano et al., 2003). Phosphorylation of AS160 is also detected in skeletal muscle in response to insulin and contraction (Bruss et al., 2005; Kramer et al., 2006a). An AS160 mutant in which four of the phosphorylation sites are altered to alanine, but the Rab-GAP catalytic domain remains intact, markedly inhibited insulin-induced GLUT4 translocation in a dominant-negative manner in both adipocytes and skeletal muscle (Kramer et al., 2006b; Sano et al., 2003). Furthermore, depletion of AS160 by RNA interference increased the fraction of GLUT4 on the surface of unstimulated cells (Eguez et al., 2005; Larance et al., 2005). Collectively, insulin stimulates phosphorylation of AS160, leading to its dissociation from GLUT4 vesicles and the attenuation of the Rab-GAP activity, both of which may increase the active GTP-bound form of Rab proteins in GLUT4 vesicles (Figure 3). 14-3-3 proteins directly bind to AS160 in a phosphorylation-dependent manner (Ramm et al., 2006). Rab-GAP activity of AS160 may be negatively affected by the association with 14-3-3 proteins. TBC1D1, a close relative of AS160, is also a substrate of Akt2, and is characterized as a Rab-GAP that regulates GLUT4 traffic in response to insulin stimulation (Roach et al., 2007).

The TC10 Protein

In adipocytes, two independent signaling pathways are triggered by engagement of the insulin receptor (Kanzaki and Pessin, 2003). One is a PI3K-dependent pathway that leads to the activation of the downstream effector kinase Akt2, and the other is an adipocyte-specific pathway involving the Rho family small GTP-binding protein TC10 (Chiang et al., 2001; Kanzaki and Pessin, 2003; Figure 6). These two distinct signaling pathways act coordinately to elicit plasma membrane translocation of GLUT4 mediated by vesicle trafficking. The activation of TC10 takes place through the formation of a complex consisting of Cbl, CAP, CrkII, and C3G depending on tyrosine phosphorylation of Cbl in response to insulin stimulation (Chiang et al., 2001). This complex is localized specifically in caveolin-enriched lipid raft microdomains, in which TC10 is also localized. The specific localization in lipid rafts may account for the specific function of the TC10 pathway in fully differentiated adipocytes. The activated TC10 in turn recruits the exocyst complex to the plasma membrane through the interaction between activated TC10 and the exocyst component Exo70 (Inoue et al., 2003; Figures 5 and 6). The exocyst complex is then required for the GLUT4 vesicle to be tethered to the appropriate sites of the plasma membrane before the final fusion step. Another mechanism by which TC10 enhances GLUT4 translocation involves Rab31 (Lodhi et al., 2007; Figure 6). A GEF for Rab31 called Gapex-5 is a binding partner of CIP4, which interacts with the GTP-bound active form of TC10. Insulin sequesters the CIP4/Gapex-5 complex away from its substrate Rab31 via the activation of TC10. The recruitment of the CIP4/Gapex-5 complex to the plasma membrane results in the failure of Rab31 activation, and thereby causes the escape of GLUT4 from the retention in the endosomal compartments (Lodhi et al., 2007). In addition, TC10 is involved in the regulation of actin cytoskeletal rearrangements like other Rho family GTP-binding proteins. Studies using constitutively activated and dominant-interfering mutants revealed that TC10 differentially regulates cortical and perinuclear actin polymerization in adipocytes (Kanzaki et al., 2002).

The Rac1 Protein

From an array of studies by the use of cultured myocytes and skeletal muscle, a pivotal role of reorganization of cytoskeletal structures in insulin-stimulated glucose uptake was clarified (Brozinick et al., 2004; Chiu et al., 2011; Kanzaki and Pessin, 2001; Tsakiridis et al., 1994).

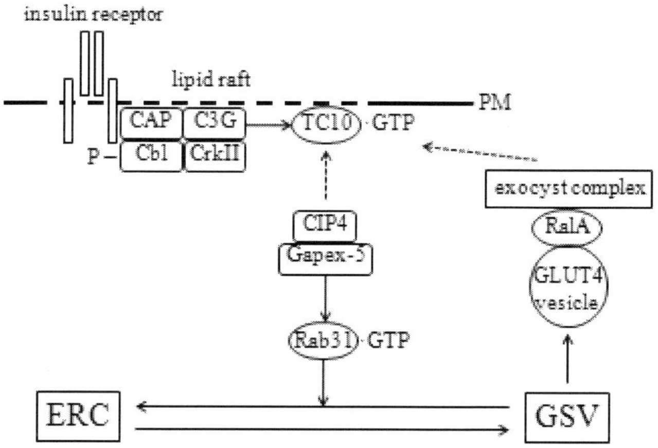

Figure 6. Role of TC10 downstream of the insulin receptor in adipocytes. Insulin triggers an adipocyte-specific TC10-mediated pathway in addition to the PI3K-dependent pathway. The GEF C3G forms a complex with the insulin receptor and other signaling proteins in response to insulin stimulation. TC10, when activated by C3G, recruits the GEF for Rab31 Gapex-5 to the plasma membrane through the interaction with its effector CIP4, thereby inhibiting the activation of Rab31. Inactivation of Rab31 permits GLUT4 trafficking from ERC to GSV. Moreover, TC10 tethers the GLUT4 vesicle near the plasma membrane through the interaction with the exocyst complex. ERC, endosomal recycling compartment; GSV, GLUT4 storage vesicle; PM, plasma membrane.

Reorganized microtubules and actin polymers serve as filaments to which GLUT4 vesicles are connected by kinesin (KIF3 and KIF5b) and myosin (Myo1c, Myo5a, and Myo5b) motor proteins, respectively (Bose et al., 2002; Imamura et al., 2003; Ishikura and Klip, 2008; Semiz et al., 2003; Yip et al., 2008; Yoshizaki et al., 2007). Considering that diverse Rho family GTP-binding proteins act as a regulator of cytoskeletal rearrangements in many types of cells, these proteins are likely to be involved in the regulation of GLUT4 translocation also in muscle cells.

Whereas TC10 has been implicated in insulin-stimulated GLUT4 translocation in adipocytes, TC10 may not play an important role in muscle cells (JeBailey et al., 2004). Instead, a role of another Rho family GTPase Rac1 in muscle cells has been proposed (Chiu et al., 2011; JeBailey et al., 2004; JeBailey et al., 2007; Khayat et al., 2000; Ueda et al., 2008). Actually, Rac1 activation was observed following insulin stimulation, and insulin-dependent GLUT4 translocation was almost completely abrogated by knockdown of Rac1.

Moreover, ectopic expression of an activated Rac1 mutant by itself induced GLUT4 translocation. By using muscle-specific *rac1* knockout mice, a pivotal role of Rac1 in inulin signaling is demonstrated in skeletal muscle as well (Ueda et al., 2010). Redistribution of GLUT4 in skeletal muscle was evaluated by immunostaining of an exofacial epitope-tagged GLUT4 reporter and immunogold electron microscopic analysis of endogenous GLUT4. Both assay clearly demonstrated that insulin-induced GLUT4 translocation to the skeletal muscle cell membrane was markedly reduced by muscle-specific *rac1* knockout.

Furthermore, ectopic expression of a constitutively activated Rac1 in mouse skeletal muscle in fact induced GLUT4 translocation as shown in the L6 cell line. Transgenic mice carrying a similar exofacial epitope-tagged GLUT4 reporter are also instrumental in visualizing GLUT4 translocation in response to contraction (Fazakerley et al., 2009; Schertzer et al., 2009). Recently, Sylow et al. (2013a) reported that insulin-stimulated glucose uptake was indeed lowered in *rac1* knockout mouse skeletal muscle. Importantly, *rac1* knockout mice showed decreased insulin and glucose tolerance, and glucose-stimulated plasma insulin concentrations in *rac1* knockout mice trended higher. Moreover, Rac1-dependent signaling was in fact impaired in insulin resistant human skeletal muscle. These findings strongly support the idea that Rac1 is a key regulator of insulin-dependent glucose uptake in skeletal muscle.

To my knowledge, FLJ00068 is a unique GEF that has been implicated in insulin-dependent Rac1 activation in muscle cells to date (Ueda et al., 2008). Overexpression of this GEF, but not other GEFs for Rac1 tested, enhanced insulin-stimulated Rac1 activation and GLUT4 translocation. A constitutively activated FLJ00068 mutant, when ectopically expressed, induced Rac1 activation and GLUT4 translocation. Moreover, knockdown of FLJ00068 largely inhibited insulin-dependent GLUT4 translocation. We recently observed that GLUT4 translocation to sarcolemma was indeed induced by constitutively activated FLJ00068 also in mouse skeletal muscle (unpublished results). Thus, FLJ00068 is considered to be a crucial Rac1 regulator in signaling that controls glucose uptake in skeletal muscle.

Insulin-induced Rac1 activation is mediated by PI3K (JeBailey et al., 2004; Ueda et al., 2008). However, precise mechanisms whereby PI3K induces the activation of Rac1 remain to be elucidated. Constitutively activated Rac1, when ectopically expressed, does not induce phosphorylation of Akt2, and insulin-stimulated Akt2 phosphorylation is not affected by muscle-specific *rac1* knockout (Ueda et al., 2008; Ueda et al., 2010). Therefore, Rac1 acts downstream or independently of Akt2. Our recent observations support the notion that Akt2 may be involved in PI3K-mediated Rac1 activation in muscle cells (Nozaki et al., 2013). Detailed mechanisms for Rac1 activation will be revealed in future studies.

Recently, a pivotal role of Rac1 in contraction-stimulated glucose uptake in skeletal muscle has been revealed (Sylow et al., 2013b). AMP-activated protein kinase-independent Rac1 activation was observed after exercise in both mice and humans. Furthermore, contraction-stimulated glucose uptake in mouse muscle was reduced by *rac1* knockout. Thus, Rac1 may play a central role in the regulation of glucose uptake in response to various physiological stimuli in skeletal muscle. Mechanisms for Rac1 activation, however, may differ depending on upstream signals. A GEF involved in contraction-stimulated Rac1 activation, for instance, remains obscure, and may be distinct from FLJ00068.

An important role of Rac1 in GLUT4 translocation in muscle cells is the regulation of actin cytoskeleton (Chiu et al., 2011). Cortical actin filaments serve as a track for GLUT4 vesicle transport (Loubéry and Coudrier, 2008). Myo5a and Myo5b act as a motor for GLUT4 vesicle trafficking along actin filaments in adipocytes and muscle cells, respectively (Ishikura and Klip, 2008; Yoshizaki et al., 2007). Additionally, cortical actin remodeling is required for tethering of GLUT4 vesicles near the plasma membrane and subsequent fusion events. In fact, Rac1 has been reported to participate in GLUT4 vesicle retention beneath the plasma membrane via actin cytoskeletal remodeling (Randhawa et al., 2008). The motor protein Myo1c may have a role as a link between the GLUT4 vesicle and actin filaments in the tethering step (Bose et al., 2002; Chen et al., 2007; Loubéry and Coudrier, 2008).

Ectopic expression of a constitutively activated Rac1 mutant induces GLUT4 translocation in muscle cultured cells and mouse skeletal muscle (Ueda et al., 2008; Ueda et al., 2010). Hence, it is likely that Rac1 participates not only in the regulation of cytoskeletal rearrangements, but also in other processes necessary for GLUT4 translocation although the mechanism underlying constitutively activated Rac1-dependent GLUT4 translocation remains to be clarified. Recently, it has been shown that the Ras family small GTP-binding protein RalA plays an important role in the control of GLUT4 translocation downstream of Rac1 in myoblast cells (Nozaki et al., 2012). Actually, constitutively activated Rac1 stimulated GTP loading on RalA and induced the redistribution of activated RalA to the membrane ruffling area, whereas knockdown of endogenous RalA abrogated constitutively activated Rac1-induced GLUT4 translocation. RalA may be responsible for the tethering step of Rac1-induced GLUT4 translocation in muscle cells because RalA regulates insulin-stimulated GLUT4 vesicle tethering to the plasma membrane through the interaction with exocyst and Myo1c complexes in adipocytes (Chen et al., 2007, see below). Detailed mechanisms whereby Rac1 regulates RalA activity and function, however, remain obscure, and will be clarified in future research.

In adipocytes, constitutively activated Rac1 does not induce GLUT4 translocation, and dominant-negative Rac1 does not affect insulin-induced GLUT4 translocation. Thus, in contrast to skeletal muscle, Rac1 may not participate in GLUT4 vesicle trafficking in adipocytes (Hou et al., 2006). However, a recent study highlights the role of Rac1 also in adipocytes. A Rac1-specific GEF termed P-Rex1, which is known to act downstream of PI3K, enhanced insulin-induced GLUT4 translocation and membrane ruffle formation (Balamatsias et al., 2011). This effect was indeed dependent on Rac1. Furthermore, knockdown of P-Rex1 inhibited insulin-induced glucose uptake.

Considering these results, Rac1 and its regulator P-Rex1 may be involved in insulin-dependent actin cytoskeletal remodeling and GLUT4 translocation to the plasma membrane

(Balamatsias et al., 2011). Interestingly, the *P-Rex1* gene has been mapped to a type 2 diabetes susceptibility locus, further supporting a role in insulin signaling (Lewis et al., 2010).

Many effector proteins for Rac1 have been identified in various types of cells. However, the effector that is responsible for transducing the Rac1 signal leading to glucose uptake remains unknown. Interestingly, the serine/threonine kinase PAK1, originally characterized as an effector for Rac1 and Cdc42, has been implicated in insulin-stimulated GLUT4 translocation in skeletal muscle (Wang et al., 2011). Actually, *PAK1* knockout mice exhibited peripheral insulin resistance coupled to this defect (Wang et al., 2011). Thus, it is possible that PAK1 plays a role in Rac1-mediated signaling in skeletal muscle, although Cdc42, another activator of PAK1, is not involved in insulin signaling (Ueda et al., 2008).

The RalA Protein

RalA resides in intracellular vesicles of exocytic compartments, participating in the regulation of polarized membrane transport and secretion, such as the basolateral delivery of membrane components in epithelial cells and synaptic vesicle trafficking in neuronal cells (Feig, 2003). RalA is also involved in the process of receptor-mediated endocytosis. The possibility that RalA participates in GLUT4 vesicle transport in insulin-responsive cells arose when the RalA function in vesicle trafficking was proposed. Actually, in adipocytes, RalA has been shown to regulate GLUT4 targeting to the plasma membrane in response to insulin (Chen et al., 2007). RalA is present in GLUT4 vesicles, and is activated following insulin stimulation. Activated RalA directly binds to Sec5 and Exo84 subunits of the exocyst complex, which mediates tethering of post-Golgi exocytic vesicles to the plasma membrane prior to membrane fusion (Chen et al., 2007; Moskalenko et al., 2002; Sugihara et al., 2002). The Myo1c motor protein, which connects the GLUT4 vesicle to the actin filament (Bose et al., 2002), also interacts with RalA. The binding of RalA to Myo1c is independent of the GDP/GTP status of RalA. Thereby, RalA directs the cooperation between Myo1c and the exocyst complex for GLUT4 vesicle transport along actin filaments and tethering to the plasma membrane (Chen et al., 2007).

A GTPase-deficient mutant of RalA, when ectopically expressed, increased glucose uptake in adipocytes and myocytes (Chen et al., 2011a; Nozaki et al., 2012). These results suggest that inactivation of RalA (hydrolysis of bound GTP to GDP) is not required for the disengagement of RalA from the exocyst complex before the fusion event. Alternatively, protein kinase C-mediated phosphorylation of the Ral-binding domain of Sec5 is reported to contribute to the dissociation of RalA (Chen et al., 2011a). Thus, not only the GDP/GTP cycle of RalA, but also the phosphorylation/ dephosphorylation cycle of its effector, are involved in the regulation of GLUT4 vesicle transport and tethering to the plasma membrane prior to membrane fusion. Mechanisms underlying insulin-dependent activation of RalA may be different between cell types. In adipocytes, inactivation of the Ral GAP complex through Akt2-catalyzed phosphorylation of the RGC2 subunit accounts for insulin-induced activation of RalA (Chen et al., 2011b). Although the mechanism for insulin-dependent activation of RalA in muscle cells remains incompletely understood, the involvement of Rac1 in insulin-dependent regulation of RalA may be a notable difference between muscle and adipose tissue (Nozaki et al., 2012).

The Arf Protein

Recently, the Arf subfamily member Arf6 has been implicated in insulin-stimulated GLUT4 recycling (Li et al., 2012). SecinH3 is a small molecule antagonist of the cytohesin family of brefeldin-A-insensitive GEFs for Arf GTP-binding proteins (Hafner et al., 2006). This compound was found to potently inhibit insulin-dependent cell-surface translocation of GLUT4 in adipocytes, suggesting the involvement of a cytohesin family GEF and its target Arf protein (Li et al., 2012). Through a series of experiments, the cytohesin family member Grp1 (also called cytohesin-3) and Arf6 have been identified as key players in insulin-induced GLUT4 vesicle formation as well as later steps of the specialized GLUT4 recycling pathway. Furthermore, phosphorylation of Grp1 by Akt has been proposed to be a mechanism for insulin regulation of Grp1.

Conclusion

Glucose uptake in adipose tissue and skeletal muscle is mediated by GLUT4 translocation from intracellular storage sites to the plasma membrane, and its precise regulation is critical for whole-body glucose homeostasis. Small GTP-binding proteins and their regulators play a pivotal role in many aspects of signal transduction that controls GLUT4 translocation as described above. Our understanding of the regulatory mechanisms, however, is still incomplete. Efforts to further delineate the molecular basis of the signaling mechanisms may provide clues for new potential therapeutic targets for diabetes.

References

Abel, E. D., Peroni, O., Kim, J. K., Kim, Y. B., Boss, O., Hadro, E., Minnemann, T., Shulman, G. I., and Kahn, B. B. (2001). Adipose-selective targeting of the GLUT4 gene impairs insulin action in muscle and liver. *Nature*, 409, 729-733.

Balamatsias, D., Kong, A. M., Waters, J. E., Sriratana, A., Gurung, R., Bailey, C. G., Rasko, J. E., Tiganis, T., Macaulay, S. L., and Mitchell, C. A. (2011). Identification of P-Rex1 as a novel Rac1-guanine nucleotide exchange factor (GEF) that promotes actin remodeling and GLUT4 protein trafficking in adipocytes. *J. Biol. Chem.*, 286, 43229-43240.

Bose, A., Guilherme, A., Robida, S. I., Nicoloro, S. M., Zhou, Q. L., Jiang, Z. Y., Pomerleau, D. P., and Czech, M. P. (2002). Glucose transporter recycling in response to insulin is facilitated by myosin Myo1c. *Nature*, 420, 821-824.

Brozinick, J. T. Jr, Hawkins, E. D., Strawbridge, A. B., and Elmendorf, J. S. (2004). Disruption of cortical actin in skeletal muscle demonstrates an essential role of the cytoskeleton in glucose transporter 4 translocation in insulin-sensitive tissues. *J. Biol. Chem.* 279, 40699-40706.

Bruss, M. D., Arias, E. B., Lienhard, G. E., and Cartee, G. D. (2005). Increased phosphorylation of Akt substrate of 160 kDa (AS160) in rat skeletal muscle in response to insulin or contractile activity. *Diabetes*, 54, 41-50.

Bryant, D. M., Datta, A., Rodríguez-Fraticelli, A. E., Peränen, J., Martín-Belmonte, F., and Mostov, K. E. (2010). A molecular network for de novo generation of the apical surface and lumen. *Nat. Cell Biol.*, 12, 1035-1045.

Chen, X. W., Leto, D., Chiang, S. H., Wang, Q., and Saltiel, A. R. (2007). Activation of RalA is required for insulin-stimulated Glut4 trafficking to the plasma membrane via the exocyst and the motor protein Myo1c. *Dev. Cell,* 13, 391-404.

Chen, X. W., Leto, D., Xiao, J., Goss, J., Wang, Q., Shavit, J. A., Xiong, T., Yu, G., Ginsburg, D., Toomre, D., Xu, Z., and Saltiel, A. R. (2011a). Exocyst function is regulated by effector phosphorylation. *Nat. Cell Biol.*, 13, 580-588.

Chen, X. W., Leto, D., Xiong, T., Yu, G., Cheng, A., Decker, S., and Saltiel, A. R. (2011b). A Ral GAP complex links PI 3-kinase/Akt signaling to RalA activation in insulin action. *Mol. Biol. Cell*, 22, 141-152.

Chiang, S. H., Baumann, C. A., Kanzaki, M., Thurmond, D. C., Watson, R. T., Neudauer, C. L., Macara, I. G., Pessin, J. E., and Saltiel, A. R. (2001). Insulin-stimulated GLUT4 translocation requires the CAP-dependent activation of TC10. *Nature*, 410, 944-948.

Chiu, T. T., Jensen, T. E., Sylow, L., Richter, E. A., and Klip, A. (2011). Rac1 signalling towards GLUT4/glucose uptake in skeletal muscle. *Cell. Signal.*, 23, 1546-1554.

Cormont, M., Bortoluzzi, M. N., Gautier, N., Mari, M., van Obberghen, E., and Le Marchand-Brustel, Y. (1996). Potential role of Rab4 in the regulation of subcellular localization of Glut4 in adipocytes. *Mol. Cell. Biol.,* 16, 6879-6886.

Eguez, L., Lee, A., Chavez, J. A., Miinea, C. P., Kane, S., Lienhard, G. E., and McGraw, T. E. (2005). Full intracellular retention of GLUT4 requires AS160 Rab GTPase activating protein. *Cell Metab.*, 2, 263-272.

Fazakerley, D. J., Lawrence, S. P., Lizunov, V. A., Cushman, S. W., and Holman, G. D. (2009). A common trafficking route for GLUT4 in cardiomyocytes in response to insulin, contraction and energy-status signalling. *J. Cell Sci.*, 122, 727-734.

Feig, L. A. (2003). Ral-GTPases: approaching their 15 minutes of fame. *Trends Cell Biol.*, 13, 419-425.

Franke, T. F. (2008). PI3K/Akt: getting it right matters. *Oncogene*, 27, 6473-6488.

Grosshans, B. L., Ortiz, D. and Novick, P. (2006). Rabs and their effectors: achieving specificity in membrane traffic. *Proc. Natl. Acad. Sci. US*, 103, 11821-11827.

Hafner, M., Schmitz, A., Grüne, I., Srivatsan, S. G., Paul, B., Kolanus, W., Quast, T., Kremmer, E., Bauer, I., and Famulok, M. (2006). Inhibition of cytohesins by SecinH3 leads to hepatic insulin resistance. *Nature*, 444, 941-944.

Hou, J. C., Shigematsu, S., Crawford, H. C., Anastasiadis, P. Z., and Pessin, J. E. (2006). Dual regulation of Rho and Rac by p120 catenin controls adipocyte plasma membrane trafficking. *J. Biol. Chem.*, 281, 23307-23312.

Huang, J., Imamura, T. and Olefsky, J. M. (2001). Insulin can regulate GLUT4 internalization by signaling to Rab5 and the motor protein dynein. *Proc. Natl. Acad. Sci. US,* 98, 13084-13089.

Huang, S. and Czech, M. P. (2007). The GLUT4 glucose transporter. *Cell Metab.,* 5, 237-252.

Imamura, T., Huang, J., Usui, I., Satoh, H., Bever, J. and Olefsky, J. M. (2003). Insulin-induced GLUT4 translocation involves protein kinase C-lambda-mediated functional coupling between Rab4 and the motor protein kinesin. *Mol. Cell. Biol.*, 23, 4892-4900.

Inoue, M., Chang, L., Hwang, J., Chiang, S. H., and Saltiel, A. R. (2003). The exocyst complex is required for targeting of Glut4 to the plasma membrane by insulin. *Nature*, 422, 629-633.

Ishikura, S. and Klip, A. (2008). Muscle cells engage Rab8A and myosin Vb in insulin-dependent GLUT4 translocation. *Am. J. Physiol. Cell Physiol.*, 295, C1016-C1025.

JeBailey, L., Rudich, A., Huang, X., Di Ciano-Oliveira, C., Kapus, A., and Klip, A. (2004). Skeletal muscle cells and adipocytes differ in their reliance on TC10 and Rac for insulin-induced actin remodeling. *Mol. Endocrinol.*, 18, 359-372.

JeBailey, L., Wanono, O., Niu, W., Roessler, J., Rudich, A., and Klip, A. (2007). Ceramide- and oxidant-induced insulin resistance involve loss of insulin-dependent Rac-activation and actin remodeling in muscle cells. *Diabetes*, 56, 394-403.

Kane, S., Sano, H., Liu, S. C., Asara, J. M., Lane, W. S., Garner, C. C., and Lienhard, G. E. (2002). A method to identify serine kinase substrates. Akt phosphorylates a novel adipocyte protein with a Rab GTPase-activating protein (GAP) domain. *J. Biol. Chem.*, 277, 22115-22118.

Kanzaki, M. and Pessin, J. E. (2001). Insulin-stimulated GLUT4 translocation in adipocytes is dependent upon cortical actin remodeling. *J. Biol. Chem.*, 276, 42436-42444.

Kanzaki, M. and Pessin, J. E. (2003). Insulin signaling: GLUT4 vesicles exit via the exocyst. *Curr. Biol.*, 13, R574-R576.

Kanzaki, M., Watson, R. T., Hou, J. C., Stamnes, M., Saltiel, A. R. and Pessin, J. E. (2002). Small GTP-binding protein TC10 differentially regulates two distinct populations of filamentous actin in 3T3L1 adipocytes. *Mol. Biol. Cell,* 13, 2334-2346.

Karylowski, O., Zeigerer, A., Cohen, A., and McGraw, T. E. (2004). GLUT4 is retained by an intracellular cycle of vesicle formation and fusion with endosomes. *Mol. Biol. Cell,* 15, 870-882.

Khayat, Z. A., Tong, P., Yaworsky, K., Bloch, R. J., and Klip, A. (2000). Insulin-induced actin filament remodeling colocalizes actin with phosphatidylinositol 3-kinase and GLUT4 in L6 myotubes. *J. Cell Sci.*, 113, 279-290.

Kramer, H. F., Witczak, C. A., Fujii, N., Jessen, N., Taylor, E. B., Arnolds, D. E., Sakamoto, K., Hirshman, M. F., and Goodyear, L. J. (2006a). Distinct signals regulate AS160 phosphorylation in response to insulin, AICAR, and contraction in mouse skeletal muscle. *Diabetes*, 55, 2067-2076.

Kramer, H. F., Witczak, C. A., Taylor, E. B., Fujii, N., Hirshman, M. F., and Goodyear, L. J. (2006b). AS160 regulates insulin- and contraction-stimulated glucose uptake in mouse skeletal muscle. *J. Biol. Chem.,* 281, 31478-31485.

Larance, M., Ramm, G., Stöckli, J., van Dam, E. M., Winata, S., Wasinger, V., Simpson, F., Graham, M., Junutula, J. R., Guilhaus, M., and James, D. E. (2005). Characterization of the role of the Rab GTPase-activating protein AS160 in insulin-regulated GLUT4 trafficking. *J. Biol. Chem.*, 280, 37803-37813.

Lewis, J. P., Palmer, N. D., Ellington, J. B., Divers, J., Ng, M. C., Lu, L., Langefeld, C. D., Freedman, B. I., and Bowden, D. W. (2010). Analysis of candidate genes on chromosome 20q12-13.1 reveals evidence for BMI mediated association of PREX1 with type 2 diabetes in European Americans. *Genomics*, 96, 211-219.

Li, J., Malaby, A. W., Famulok, M., Sabe, H., Lambright, D. G., and Hsu, V. W. (2012). Grp1 plays a key role in linking insulin signaling to glut4 recycling. *Dev. Cell*, 22, 1286-1298.

Lodhi, I. J., Chiang, S. H., Chang, L., Vollenweider, D., Watson, R. T., Inoue, M., Pessin, J. E., and Saltiel, A. R. (2007). Gapex-5, a Rab31 guanine nucleotide exchange factor that regulates Glut4 trafficking in adipocytes. *Cell Metab.*, 5, 59-72.

Loubéry, S. and Coudrier, E. (2008). Myosins in the secretory pathway: tethers or transporters? *Cell. Mol. Life Sci.*, 65, 2790-2800.

Moskalenko, S., Henry, D. O., Rosse, C., Mirey, G., Camonis, J. H. and White, M. A. (2002). The exocyst is a Ral effector complex. *Nat. Cell Biol.*, 4, 66-72.

Nozaki, S., Ueda, S., Takenaka, N., Kataoka, T., and Satoh, T. (2012). Role of RalA downstream of Rac1 in insulin-dependent glucose uptake in muscle cells. *Cell. Signal.*, 24, 2111-2117.

Nozaki, S., Takeda, T., Kitaura, T., Takenaka, N., Kataoka, T. and Satoh, T. (2013). Akt2 regulates Rac1 activity in the insulin-dependent signaling pathway leading to GLUT4 translocation to the plasma membrane in skeletal muscle cells. *Cell. Signal.*, in press.

Ramm, G., Larance, M., Guilhaus, M., and James, D. E. (2006). A role for 14-3-3 in insulin-stimulated GLUT4 translocation through its interaction with the RabGAP AS160. *J. Biol. Chem.*, 281, 29174-29180.

Randhawa, V. K., Ishikura, S., Talior-Volodarsky, I., Cheng, A. W., Patel, N., Hartwig, J. H., and Klip, A. (2008). GLUT4 vesicle recruitment and fusion are differentially regulated by Rac, AS160, and Rab8A in muscle cells. *J. Biol. Chem.*, 283, 27208-27219.

Roach, W. G., Chavez, J. A., Mîinea, C. P., and Lienhard, G. E. (2007). Substrate specificity and effect on GLUT4 translocation of the Rab GTPase-activating protein Tbc1d1. *Biochem. J.*, 403, 353-358.

Roland, J. T., Bryant, D. M., Datta, A., Itzen, A., Mostov, K. E., and Goldenring, J. R. (2011). Rab GTPase-Myo5B complexes control membrane recycling and epithelial polarization. *Proc. Natl. Acad. Sci. US*, 108, 2789-2794.

Roland, J. T., Kenworthy, A. K., Peranen, J., Caplan, S., and Goldenring, J. R. (2007). Myosin Vb interacts with Rab8a on a tubular network containing EHD1 and EHD3. *Mol. Biol. Cell*, 18, 2828-2837.

Saltiel, A. R. and Kahn, C. R. (2001). Insulin signalling and the regulation of glucose and lipid metabolism. *Nature*, 414, 799-806.

Sano, H., Eguez, L., Teruel, M. N., Fukuda, M., Chuang, T. D., Chavez, J. A., Lienhard, G. E., and McGraw, T. E. (2007). Rab10, a target of the AS160 Rab GAP, is required for insulin-stimulated translocation of GLUT4 to the adipocyte plasma membrane. *Cell Metab.*, 5, 293-303.

Sano, H., Kane, S., Sano, E., Mîinea, C. P., Asara, J. M., Lane, W. S., Garner, C. W., and Lienhard, G. E. (2003). Insulin-stimulated phosphorylation of a Rab GTPase-activating protein regulates GLUT4 translocation. *J. Biol. Chem.*, 278, 14599-14602.

Sano, H., Peck, G. R., Kettenbach, A. N., Gerber, S. A., and Lienhard, G. E. (2011). Insulin-stimulated GLUT4 protein translocation in adipocytes requires the Rab10 guanine nucleotide exchange factor Dennd4C. *J. Biol. Chem.*, 286, 16541-16545.

Schertzer, J. D., Antonescu, C. N., Bilan, P. J., Jain, S., Huang, X., Liu, Z., Bonen, A., and Klip, A. (2009). A transgenic mouse model to study glucose transporter 4myc regulation in skeletal muscle. *Endocrinology*, 150, 1935-1940.

Semiz, S., Park, J. G., Nicoloro, S. M., Furcinitti, P., Zhang, C., Chawla, A,, Leszyk, J., and Czech, M. P. (2003). Conventional kinesin KIF5B mediates insulin-stimulated GLUT4 movements on microtubules. *EMBO J.*, 22, 2387-2399.

Sherman, L. A., Hirshman, M. F., Cormont, M., Le Marchand-Brustel, Y., and Goodyear, L. J. (1996). Differential effects of insulin and exercise on Rab4 distribution in rat skeletal muscle. *Endocrinology*, 137, 266-273.

Shibata, H., Omata, W. and Kojima, I. (1997). Insulin stimulates guanine nucleotide exchange on Rab4 via a wortmannin-sensitive signaling pathway in rat adipocytes. *J. Biol. Chem.*, 272, 14542-14546.

Sugihara, K., Asano, S., Tanaka, K., Iwamatsu, A., Okawa, K., and Ohta, Y. (2002). The exocyst complex binds the small GTPase RalA to mediate filopodia formation. *Nat. Cell Biol.*, 4, 73-78.

Sun, Y., Bilan, P. J., Liu, Z., and Klip, A. (2010). Rab8A and Rab13 are activated by insulin and regulate GLUT4 translocation in muscle cells. *Proc. Natl. Acad. Sci. US*, 107, 19909-19914.

Sylow, L., Jensen, T. E., Kleinert, M., Hojlund, K., Kiens, B., Wojtaszewski, J., Prats, C., Schjerling, P., and Richter, E. A. (2013a). Rac1 signaling is required for insulin-stimulated glucose uptake and is dysregulated in insulin resistant murine and human skeletal muscle. *Diabetes*, in press.

Sylow, L., Jensen, T. E., Kleinert, M., Mouatt, J. R., Maarbjerg, S. J., Jeppesen, J., Prats, C., Chiu, T. T., Boguslavsky, S., Klip, A., Schjerling, P., and Richter, E. A. (2013b). Rac1 Is a novel regulator of contraction-stimulated glucose uptake in skeletal muscle. *Diabetes*, in press.

Takai, Y., Sasaki, T. and Matozaki, T. (2001). Small GTP-binding proteins. *Physiol. Rev.*, 81, 153-208.

Tsakiridis, T., Vranic, M. and Klip, A. (1994). Disassembly of the actin network inhibits insulin-dependent stimulation of glucose transport and prevents recruitment of glucose transporters to the plasma membrane. *J. Biol. Chem.*, 269, 29934-29942.

Ueda, S., Kataoka, T. and Satoh, T. (2008). Activation of the small GTPase Rac1 by a specific guanine-nucleotide-exchange factor suffices to induce glucose uptake into skeletal-muscle cells. *Biol. Cell*, 100, 645-657.

Ueda, S., Kitazawa, S., Ishida, K., Nishikawa, Y., Matsui, M., Matsumoto, H., Aoki, T., Nozaki, S., Takeda, T., Tamori, Y., Aiba, A., Kahn, C. R., Kataoka, T., and Satoh, T. (2010). Crucial role of the small GTPase Rac1 in insulin-stimulated translocation of glucose transporter 4 to the mouse skeletal muscle sarcolemma. *FASEB J.*, 24, 2254-2261.

Uhlig, M., Passlack, W. and Eckel, J. (2005). Functional role of Rab11 in GLUT4 trafficking in cardiomyocytes. *Mol. Cell. Endocrinol.*, 235, 1-9.

Vanhaesebroeck, B., Stephens, L. and Hawkins, P. (2012). PI3K signalling: the path to discovery and understanding. *Nat. Rev. Mol. Cell. Biol.*, 13, 195-203.

Vollenweider, P., Martin, S. S., Haruta, T., Morris, A. J., Nelson, J. G., Cormont, M., Le Marchand-Brustel, Y., Rose, D. W., and Olefsky, J. M. (1997). The small guanosine triphosphate-binding protein Rab4 is involved in insulin-induced GLUT4 translocation and actin filament rearrangement in 3T3-L1 cells. *Endocrinology*, 138, 4941-4949.

Wang, Z., Oh, E., Clapp, D. W., Chernoff, J., and Thurmond, D. C. (2011). Inhibition or ablation of p21-activated kinase (PAK1) disrupts glucose homeostatic mechanisms in vivo. *J. Biol. Chem.*, 286, 41359-41367.

Wennerberg, K., Rossman, K. L. and Der, C. J. (2005). The Ras superfamily at a glance. *J. Cell Sci.*, 118, 843-846.

Wu, S., Mehta, S. Q., Pichaud, F., Bellen, H. J., and Quiocho, F. A. (2005). Sec15 interacts with Rab11 via a novel domain and affects Rab11 localization in vivo. *Nat. Struct. Mol. Biol.*, 12, 879-885.

Yip, M. F., Ramm, G., Larance, M., Hoehn, K. L., Wagner, M. C., Guilhaus, M., and James, D. E. (2008). CaMKII-mediated phosphorylation of the myosin motor Myo1c is required for insulin-stimulated GLUT4 translocation in adipocytes. *Cell Metab.*, 8, 384-398.

Yoshizaki, T., Imamura, T., Babendure, J. L., Lu, J. C., Sonoda, N., and Olefsky, J. M. (2007). Myosin 5a is an insulin-stimulated Akt2 (protein kinase Bbeta) substrate modulating GLUT4 vesicle translocation. *Mol. Cell. Biol.*, 27, 5172-5183.

Zeigerer, A., Lampson, M. A., Karylowski, O., Sabatini, D. D., Adesnik, M., Ren, M., and McGraw, T. E. (2002). GLUT4 retention in adipocytes requires two intracellular insulin-regulated transport steps. *Mol. Biol. Cell*, 13, 2421-2435.

Zerial, M. and McBride, H. (2001). Rab proteins as membrane organizers. *Nat. Rev. Mol. Cell. Biol.*, 2, 107-117.

Zisman, A., Peroni, O. D., Abel, E. D., Michael, M. D., Mauvais-Jarvis, F., Lowell, B. B., Wojtaszewski, J. F., Hirshman, M. F., Virkamaki, A., Goodyear, L. J., Kahn, C. R., and Kahn, B. B. (2000). Targeted disruption of the glucose transporter 4 selectively in muscle causes insulin resistance and glucose intolerance. *Nat. Med.*, 6, 924-928.

Index

A

B

C

D

G

H

I

J

K

L

M

N

O

P

Q

R

S

T

U

V

W

X

Y